T0137093

Wireless Networking Principles: From Terrestrial to Underwater Acoustic

Shengming Jiang

Wireless Networking Principles: From Terrestrial to Underwater Acoustic

 Springer

Shengming Jiang
Marine Internet Laboratory (MILAB),
    College of Information Engineering
Shanghai Maritime University
Shanghai
China

ISBN 978-981-13-3999-8   ISBN 978-981-10-7775-3   (eBook)
https://doi.org/10.1007/978-981-10-7775-3

© Springer Nature Singapore Pte Ltd. 2018
Softcover re-print of the Hardcover 1st edition 2018
This work is subject to copyright. All rights are reserved by the Publisher, whether the whole or part
of the material is concerned, specifically the rights of translation, reprinting, reuse of illustrations,
recitation, broadcasting, reproduction on microfilms or in any other physical way, and transmission
or information storage and retrieval, electronic adaptation, computer software, or by similar or dissimilar
methodology now known or hereafter developed.
The use of general descriptive names, registered names, trademarks, service marks, etc. in this
publication does not imply, even in the absence of a specific statement, that such names are exempt from
the relevant protective laws and regulations and therefore free for general use.
The publisher, the authors and the editors are safe to assume that the advice and information in this
book are believed to be true and accurate at the date of publication. Neither the publisher nor the
authors or the editors give a warranty, express or implied, with respect to the material contained herein or
for any errors or omissions that may have been made. The publisher remains neutral with regard to
jurisdictional claims in published maps and institutional affiliations.

Printed on acid-free paper

This Springer imprint is published by the registered company Springer Nature Singapore Pte Ltd.
part of Springer Nature
The registered company address is: 152 Beach Road, #21-01/04 Gateway East, Singapore 189721,
Singapore

To my wife, my son and my daughter

Preface

Today, wireless networking technologies deeply affect many social sectors and people's daily life. For example, mobile personal communication networks such as 4G and beyond as well as WiFi networks are the enabling technologies for mobile Internet used by almost everyone at anytime in terrestrial environments. These wireless networks provide the users with a great convenience for personal communication, social activities, shopping, and traveling as well as education, since many traditional activities and organizations now can be carried out simply through mobile smart terminals. The wireless networking technology as a whole is still developing. For example, mobile personal communication is under development toward 5G, aiming at providing much faster and reliable communications than the currently available one to eliminate the gap in service quality between wireless and wired networks. On the other hand, some wireless networks are still in early development stages. For example, a huge number of sensors and actuators as well as various types of vehicles have been deployed underwater, and this number is still growing. These underwater things usually equipped with communication facilities are able to construct an Internet of Underwater Thing (IoUT) [1]. How to link these nodes to cover large underwater areas and transfer the collected data to the surface processing center faces many challenges. Many issues are necessarily addressed for underwater wireless networks, which are still in enfant ages.

Since the radio wave cannot propagate long enough but only acoustic signal can in underwater environments to satisfy application requirements, currently most underwater wireless networks are based on acoustic waves. Therefore, in this book we call this kind of network underwater acoustic networks (UWANs), which include underwater acoustic sensor networks. UWANs have some special features not present in most terrestrial radio wireless networks (RWNs), such as much long propagation delays, very limited channel capacities, low channel reliability, and high dynamics of communication environments. These features greatly affect the design of underwater networking protocols and schemes, preventing those well-developed for RWNs from being used directly in UWANs.

Despite many differences in networking environments between RWNs and UWANs, they still have many similar issues to be addressed by both, and many approaches and design strategies developed for RWNs can be used as references in designing UWAN protocols, and some of them can be used in UWANs after proper modifications. Therefore, a good understanding of the well-developed RWN technologies can be helpful to understand network protocols and algorithms proposed for UWANs. There are several books systematically describing RWN technologies in the literature such as [2, 3, 4], while a couple of UWANs are also available such as [5, 6]. This book is unique in terms of discussing networking technologies for both RWNs and UWANs in one.

This book aims to highlight the key networking issues and explain the basic ideas of typical networking technologies, in which the principal challenging issues and research topics are addressed and discussed in detail. The book consists of thirteen chapters. The first chapter (Chap. 1) introduces the typical wireless networks and fundamental networking issues. The remaining chapters are divided into two parts. The first part will systematically describe the well-established RWN technologies that have been standardized or applied in practice, covering the following topics: error control (Chap. 2), medium access control (MAC) protocols (Chaps. 3 and 4), routing protocols (Chap. 5), end-to-end transmission control (Chap. 6), mobility (Chap. 7), and network security (Chap. 8). The majority of this part except TCP and vertical handoff in mobile networks are relatively mature and can provide a foundation to understand the counterparts of UWANs. The second part discusses the up-to-date networking technologies for UWANs, including underwater acoustic channels (Chap. 9), UWAN MAC protocols (Chap. 10), UWAN routing protocols (Chap. 11), UWAN transfer reliability control covering both error control and end-to-end transmission control (Chap. 12), and UWAN security (Chap. 13). This part is mainly based on the author's surveys on the related topics [7, 8, 9, 10].

I hope that the book can become a useful resource for new learners, researchers, and practitioners in RWNs and UWANs.

Shanghai, China   Shengming Jiang
January 2018

References

1. Domingo, M. C.: An overview of the internet of underwater things. J. Netw. Comput. Appl. **35**(1), 1879–1890 (2012)
2. Smith, C., Collins, D.: Wireless Networks: Design and Integration for LTE, EVDO, HSPA and WIMAX, 3rd edn. McGraw-Hill Education (2014). ISBN 978-0071819831
3. Beard, C., Stallings, W.: Wireless Communication Networks and Systems. Pearson Education (2015). ISBN 978-0133594171
4. Agha, K.A., Pujolle, G., Yahiha, T.A.: Mobile and Wireless Networks. Wiley-ISTE (2016). ISBN 978-1848217140

5. Xiao, Y. (ed.): Underwater Acoustic Sensor Networks. Auerbach Publications (2010). ISBN 978-1420067118
6. Cui, J.H., Gerla, M., Zhou, Z., Peng, Z.: Underwater Wireless Networks: Principles, Protocols and Implementations. Wiley (2016). ISBN 978-1118465264
7. Lu, Q., Liu, F., Zhang, Y., Jiang, S.M.: Routing protocols for underwater acoustic sensor networks: a survey from an application perspective. In: Zak, A. (ed.) Advances in Underwater Acoustics, chapter 2. INTECH (2017). ISBN 978-953-51-3609-5
8. Jiang, S.M.: State-of-The-Art Medium Access Control (MAC) protocols for underwater acoustic networks: a survey based on A MAC reference model. IEEE Commun. Surv. Tutorials **20**(1) (2018)
9. Jiang, S.M.: On reliable data transfer in underwater acoustic networks: a survey from networking perspective. IEEE Commun. Surv. Tutorials **PP**(99) (2018)
10. Jiang, S.M.: Securing underwater acoustic networks: a survey. IEEE Commun. Surv. Tutorials Submitted. (2017)

Acknowledgements

I would like to take this opportunity to sincerely thank my following students who help me to complete this book by downloading references and drafting figures: Fan Chao, Zhang Peng, Gan Xiaolong (graduated in 2015); Qian Yanzhen, Wang Xiyang, Chen Huihui, Jiang Shuchao, Yang Fang, Yang Kaijian, Wu Shidong, and Zhang Kai (graduated in 2016); Liu Jie, Wang Yiliang, Lu Qian, Bao Zhijie, Dai Yuxi, and Cao Jun (graduated in 2017); Luo Jiawei, Li Yongfeng, Liu Haiyang, Li Fuyong, Zheng Tao, Zhang Shaofeng, Wang Fei, Xia Jie, and Wu Bi (to graduate in 2018).

This work is supported by the National Natural Science Foundation of China (NSFC) for the project "Basic theory of open network architecture for the marine Internet" under Grant 61472237.

Contents

Acronyms

---

[1] It is the number of the page on which the corresponding acronym first appears.

---

[2]An acronym may refer to different definitions.

DCC	Dedicated control channel, 64	
DCCP	Datagram congestion control protocol, 137	
DCF	Distributed coordination function, 88	
DCH	Data channel, 84	
DCA	Dynamic channel allocation, 188	
DDoS	Distributed DoS, 204	
DIFS	DCF IFS, 89	
DoA	Direction of arrival, 356	
DoS	Denial of service, 204	
DP	Data period, 97	
DRP	Distributed reservation protocol, 97	
DSDV	Destination-sequenced distance vector, 115	
DSN	Destination sequence number, 124	
DSR	Dynamic source routing, 117	
DSSS	Direct-sequence SS, 60	
DTLS	Datagram transport layer security, 212	
DTN	Delay- and disruption-tolerant network, 11	
EAP	Extensible Authentication Protocol, 215	
EAPOL	EAP Over LANs, 216	
ECAES	Elliptic curve authenticated encryption scheme, 342	
ECC	Elliptic curve cryptography, 342	
ECDLP	Elliptic curve discrete logarithm problem, 342	
ECDSA	Elliptic curve digital signature algorithm, 342	
ECN	Explicit congestion notification, 156	
EIFS	Extended IFS, 90	
ELFN	Explicit link failure notification, 151	
ESP	Encapsulating security payload, 215	
ETSI	European Telecommunication Standards Institute, 2	
EU	European Union, 92	
FAMA	Floor acquisition multiple access, 54	
FCA	Fixed channel allocation, 188	
FCC	Federal Communications Commission, 96	
FCS	Frame check sequence, 38	
FDD	Frequency-division duplex, 319	
FDM	Frequency-division multiplexing, 60	
FDMA	Frequency-division multiple access, 63	
FEC	Forward error control, 21	
FFD	Full function device, 94	
FHSS	Frequency-hopping SS, 60	
FIFO	First in, first out, 247	
FS	Frame sensing, 62	
FTP	File transfer protocol, 137	
GC	Guard channel, 188	
GEO	Geostationary Earth orbit, 4	
GKH	Group key handshake, 226	

List of Figures

List of Tables

Chapter 1
Introduction and Overview

Abstract This chapter first introduces some typical wireless networks used today, and then discusses the fundamental networking approaches and issues.

1.1 Typical Wireless Networks

Many types of wireless networks have been deployed today, such as 2G ~ 5G mobile cellular networks, WiFi based on IEEE 802.11, Bluetooth based on IEEE 802.15, Iridium based on satellites etc. Accordingly, there are also many methods to classify them. One popular method classifies networks according to whether a network consists of specific networking units such as base stations (BS) and access points (AP). These units provide special networking functions, and the network cannot operate without them. If a wireless network consists of such specific units, it is called infrastructured wireless network, otherwise, called infrastructureless wireless network. This network is also called wireless ad hoc networks, in which terminals can communicate each other directly.

Another classification follows the space in which the networking units are deployed. In a satellite network or space network, the networking unit is satellite, which runs in an earth orbit. For a terrestrial wireless network, its major networking units are deployed on the earth such as BSs in mobile cellular networks. In an underwater network, most networking units such as hydrophones are deployed underwater, and some surface gateways may be used to connect these units with other networks.

Every wireless network relies on some wireless communication medium for communication between nodes. Typical wireless media include radio wave, optical wave and acoustic wave. Therefore, a wireless network is also often named after the used communication medium such as acoustic wireless networks.

These wireless networks are often used jointly to satisfy a particular networking requirement. With a hybrid of infrastructured and infrastructureless wireless networks, terminals can either communicate each other directly or through certain coordinators indirectly. Furthermore, wireless networks are widely used as access networks that provide the user with interfaces to wired backbone networks. Such combination can provide a large, flexible and sophisticated networks.

© Springer Nature Singapore Pte Ltd. 2018
S. Jiang, *Wireless Networking Principles: From Terrestrial to Underwater Acoustic*,
https://doi.org/10.1007/978-981-10-7775-3_1

1.1.1   Infrastructured Networks

Most wireless networks in use today require certain types of infrastructures such as base station and access point for reliable networking. Their major networking functions include (i) relaying the data transmitted from one terminal to another, (ii) coordinating terminals for network resource sharing and (iii) security control such as authentication. They can be further divided according to per-infrastructure coverage size.

1.1.1.1   Wireless Local Area Network (WLAN)

WLANs can often be found in coffee bars, canteens, hotels, libraries and airports. The typical one is IEEE 802.11 based WiFi networks, which provides wireless surf of the Internet through an access point (AP) as illustrated in Fig. 1.1. This network is designed for small environments to avoid wiring operations. The AP here mainly has two functions: (i) coordinating terminals to share the wireless medium, and (ii) bridge wireless terminals and other networks such as the Internet with some security control. Another similar network is the High Performance Local Area Network (HIPERLAN), which is standardized by the European Telecommunication Standards Institute (ETSI).

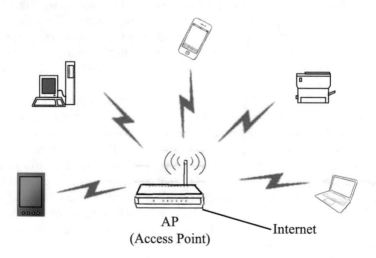

Fig. 1.1   A wireless local area network (WLAN)

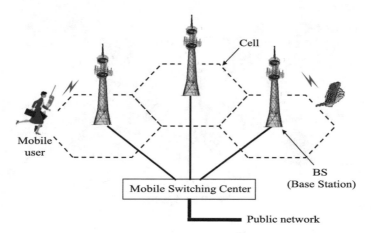

Fig. 1.2 A mobile cellular network

1.1.1.2   Mobile Cellular Network

Today, smart mobile phones are the most popular personal wireless devices with versatile functions. They have become an essential part of our daily life with an unprecedented large user population that any other types of mobile devices can have, such as Personal Digital Assistants (PDAs), palm computers and notebooks.

The network service of mobile phones is provided by mobile cellular networks, which actually is comprised by many small-sized cellular networks as illustrated in Fig. 1.2. Such network uses multiple small and low power transmitters to replace a single and high power transmitter to form flexible air interfaces for network access with mobility support. After about 45-year's development with more than 4 generations, this network is becoming one of the most important networks today. The major difference between generations is the transmission rate along with the enabling technologies as well as corresponding applications.

1.1.1.3   Wireless Metropolitan Area Network (WMAN)

For community areas such a residential area or a university campus, the WLAN is too small, and a more powerful WMAN is necessary with less implementation cost by avoiding digging and wiring operations. Several solutions are available such as Worldwide Interoperability for Microwave Access (WiMAX), which is standardized by IEEE 802.16. As illustrated in Fig. 1.3, with WiMAX, transmitting and receiving antennas are mounted in high buildings or towers so that they can cover large areas. Another solution is Long Term Evolution (LTE), which is motivated by 3GPP, and can provide larger coverage.

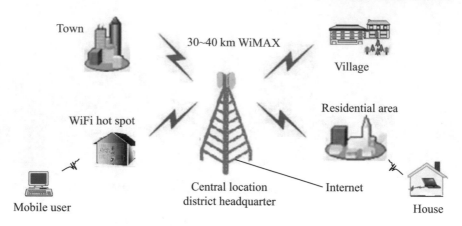

Fig. 1.3   A metropolitan area network based on WiMAX

1.1.1.4   Wireless Wide Area Network (WAN)

A WAN is used to cover a much larger area (e.g., a national wide area) than the above mentioned wireless networks can do for vast broadcast and communication especially in some special environments such as ocean and desert, where it is very difficult or expensive to install necessary infrastructures. Actually, the satellite network is the only option for such communication available today. This network is also widely used in navigation in ocean and space beside its traditional applications of long-distance communication such as across-continent or cross-ocean telephone. It has been also developed to provide personal mobile phone communication such as Iridium. As illustrated in Fig. 1.4, according to earth orbit attitudes, typical satellites include Low Earth Orbit (LEO), Medium Earth Orbit (MEO) and Geostationary Earth Orbit (GEO).

1.1.2   Infrastructureless Networks

Building network infrastructures in some environments for communication is neither feasible nor cost-effective. For example, in wild areas such as dessert and oceans, it is almost impossible to install base stations or access points for wireless communications. In remote rural areas, it is not economic to build infrastructure to cover a large area with only a small population of users. Some pre-built infrastructures may be destroyed by disasters such as earthquake, flood and hurricane as well as war so that they cannot work properly, while the time to rebuilt them is too long to meet communication demands. In these cases, infrastructureless wireless networks, or wireless ad hoc networks, can be used to provide networking services.

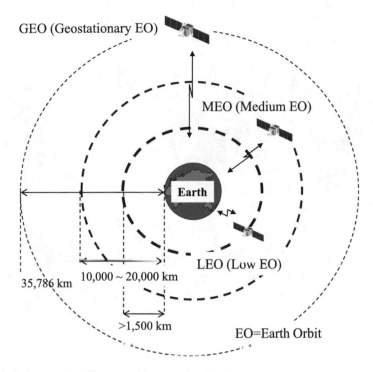

Fig. 1.4 Category of satellites according to earth orbit sizes

Such network is organized by user terminals themselves and can be re-organized anytime to maintain networking operation. Each node in such a network is functionally identical, which means that a node can be either a router, a data source or a data destination. The major advantages of such networks over infrastructured ones include: (i) easy and fast implementation, (ii) low implementation cost, and (iii) applicability in various environments. However, it is much more difficult to maintain transmission reliability and security than in infrastructured ones.

1.1.2.1 Mobile Ad Hoc Network (MANET)

A MANET is a collection of mobile terminals that dynamically form a temporary network without any aid from fixed infrastructure or centralized administration as illustrated in Fig. 1.5. This network is often used as a temporary network for special occasions such as tactical fields (which is the earliest application and termed as "packet radio network"), rescue operations and some occasions in which a group of people are gathering temporarily such as conference halls and classrooms. This kind of network supports terminal mobility without performance guarantee, and terminals are usually battery-operated. Actually, the popular WiFi based on IEEE 802.11 can

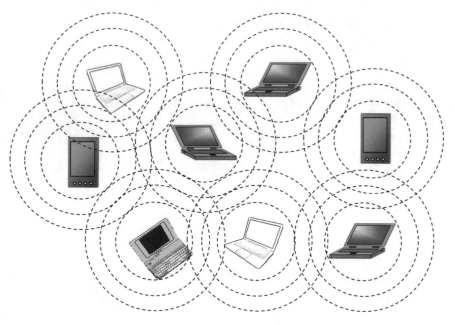

Fig. 1.5 A mobile ad hoc network (MANET)

either operate in the infrastructured mode with APs or the infrastructureless mode as a MANET.

An earlier example is Bluetooth [1], which is a wireless personal area network (WPAN) as illustrated in Fig. 1.6. Various digital devices can be linked together to form a home network, and a gateway links the home network and the public network. These devices typically include personal computer and various peripheral devices (e.g., printers and scanners), consumer electronics (e.g., TV, video player, telephone/fax, refrigerators), mobile devices and automotive networks. The Ultra-Wide Band (UWB) based WiMedia further promotes WPANs with much higher transmission rates up to 485 Mbps for more powerful indoor networking, such as wireless USB, which allows a USB interfaced device to be connected to a computer in a wireless mode.

Another typical example is the wireless body area network (WBAN) [2]. As illustrated in Fig. 1.7, a WBAN is used to connect various wearable devices and physiological sensors around a human body, and smart mobile phones or computers are used as data hub or gateway to connect a WBAN to the Internet. Wearable devices can be carried in clothe pockets, bags or by hand, and physiological sensors may be embedded inside or surface-mounted on the body. The WBAN is particularly useful for realtime monitoring patient health status, especially remote monitoring through the Internet. IEEE 802.15.6 [3] is a major standard for such kind of network.

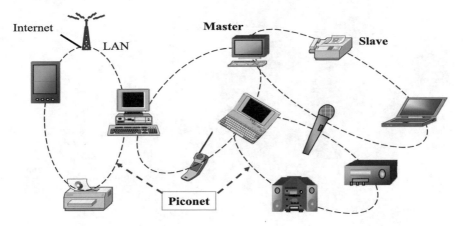

Fig. 1.6 An ad hoc wireless network: Bluetooth

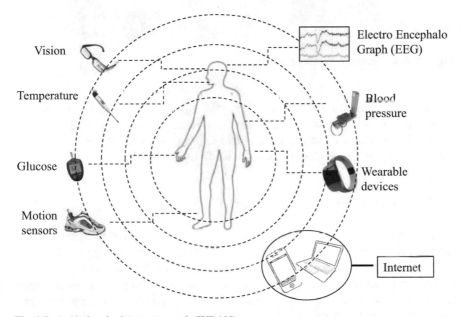

Fig. 1.7 A wireless body area network (WBAN)

1.1.2.2 Wireless Mesh Network (WMN)

In some environments such as remote rural areas, an almost-stationary wireless ad hoc network can be used as a backbone instead of a wired backbone as illustrated in Fig. 1.8 to reduce implementation cost. This kind of ad hoc network is called wireless mesh network (WMN). The WMN can provide much better performance than the MANET due to not only the low mobility but also an unlimit power supply. It is similar to the cellular network as a backbone with the following differences. (i) The

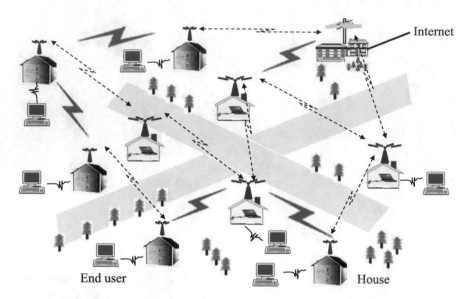

Fig. 1.8 A wireless mesh network (WMN)

link between nodes in a WMN is wireless while that between base stations in mobile cellular networks is wireline. (ii) The WMN can be self-organized and self-cured but not for the cellular network. (iii) The above differences lead to lower performance and small implementation complexity for WMNs.

1.1.2.3   Wireless Sensor Network (WSN)

A WSN is a special wireless ad hoc network for information collection rather than peer-to-peer communication between users, with the following major differences from other networks mentioned above. (i) The node in a WSN is usually much less capable in terms of processing and buffering, and much more energy-constrained. (ii) A WSN often has much larger node population and higher density than other networks so that the conventional addressing and routing schemes are not suitable for WSNs. Particularly from routing point of view, a WSN is data centric rather than address centric. That is, routing in a WSN aims to find a path to a particular data of interest rather than to a particular address. As illustrated in Fig. 1.9 for a fire alarm system in forest, the data that needs to be collected online is the area in which the sensed temperature goes to beyond the normal threshold.

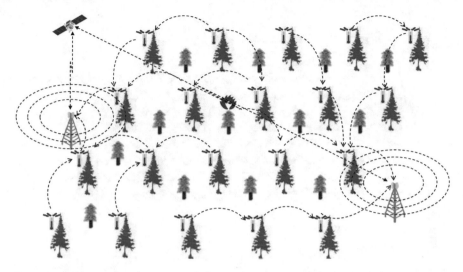

Fig. 1.9 A wireless sensor network for fire alarm system in forest

1.1.2.4 Opportunistic Network/Delay and Disruption Tolerant Network (DTN)

All the networks discussed so far have an assumption in mind, which assumes the existence of at least one path between any source-destination pair, and the major issue is how to find a best path. However, this assumption does not hold in an opportunistic network, in which network connectivity becomes partitioned because of periodic or intermittent disruptions caused by mobility or limit transmission ranges. Only a series of non-contemporaneous meetings between nodes may form a path to a destination opportunistically [4] as illustrated in Fig. 1.10. A typical example is Delay and Disruption Tolerant Networks (DTNs).

The NATO Interoperability Standards and Profiles (NISP) has identified DTNs as a way to provide network services when no end-to-end path exists, and enable standardized communications over long distances with the Bundle Protocol (BP), particularly for deep-space communications or connecting geographically remote clusters of nodes [4].

1.1.2.5 Vehicular Ad Hoc Network (VANET)

As illustrated in Fig. 1.11, a VANET is designed to support data exchange between vehicles and roadside equipment (i.e., sensors) for an Intelligent Transportation System (ITS) [5]. It is also used to provide inter-vehicle communication [6] through opportunistic networking. In this sense, the VANET can be regarded as a mixture of wireless sensor networks and opportunistic networks. Inter-vehicle communication

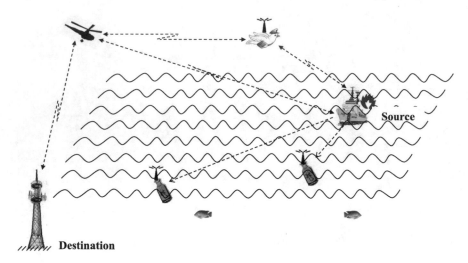

Fig. 1.10 An opportunistic network

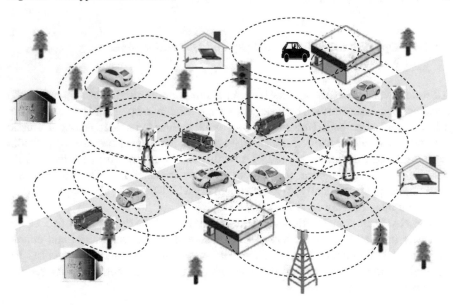

Fig. 1.11 A vehicular ad hoc network (VANET)

can also be realized through some infrastructure like road side units or mobile cellular networks.

Different from MANETs in which nodes move randomly in unpredictable manner, in VANETs, vehicles tend to move in a predictable track with limited randomness such as a highway. Usually energy consumption for wireless communication is not an issue here, while it is critical in other ad hoc networks.

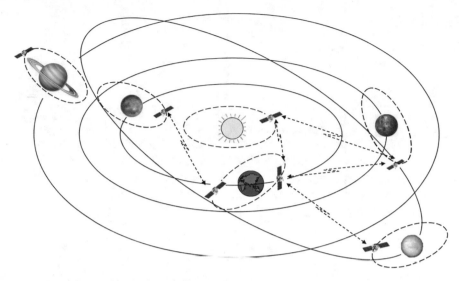

Fig. 1.12 A conceived interplanetary Internet (IPN)

1.1.3 Network Deployment Spaces

A wireless network can be deployed in space, terrestrial environments and underwater environments.

1.1.3.1 Space Wireless Network

It is deployed in space such as the satellite network mentioned earlier. A much larger one is the Interplanetary Internet (IPN) as illustrated in Fig. 1.12, which is a conceived communication network in space, consisting of a set of network nodes that can communicate with each other [7]. The major challenges for IPNs come from the vast interplanetary distance, much longer delays and more communication errors may be incurred than in the Earth-bound wired Internet, in which communication delay and errors can be negligible. Furthermore, in IPNs, the end-to-end connectivity may be not always available like in the Earth-bound Internet so that DTNs protocols have to be used.

1.1.3.2 Terrestrial Wireless Network

For terrestrial wireless networks, network nodes are deployed on the ground. The typical examples include IEEE 802.11 based WiFi and mobile cellular networks as

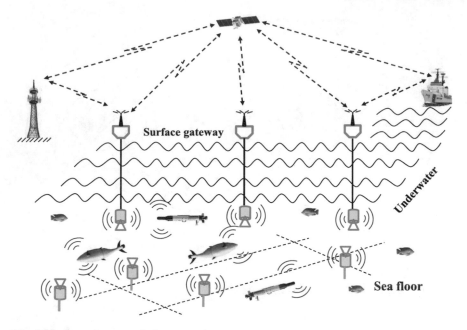

Fig. 1.13 An underwater wireless network

well as most of the infrastructureless networks mentioned above such as WMNs,
VANETs and WSNs.

1.1.3.3   Underwater Wireless Network

In such a network, network nodes are deployed underwater on the sea floor or in water
as illustrated in Fig. 1.13. Usually, some surface nodes are used to link underwater
nodes to satellites or terrestrial networks.

One of the most important underwater wireless networks is underwater wireless
sensor networks (UWSNs), which can be used for oceanographic data collection,
pollution monitoring, offshore exploration, disaster prevention, assisted navigation,
underwater surveillance of harbours, critical infrastructures and tactical surveillance
etc. Furthermore, multiple unmanned underwater vehicles (UUVs) or autonomous
underwater vehicles (AUVs) equipped with underwater sensors, can be used to
explore natural undersea resources and collect scientific data in collaborative monitoring missions [8].

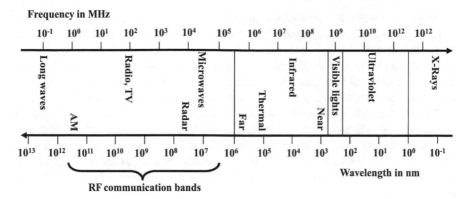

Fig. 1.14 Electromagnetic spectrum suitable for wireless communication

1.1.4 Communication Media

The major wireless communication media include radio wave, optical wave and acoustic wave.

1.1.4.1 Radio Wave

Radio media are the major communication media for most of wireless networks used today, such as satellite, WiFi and mobile cellular networks. As illustrated in Fig. 1.14, only a very small portion of the electromagnetic spectrum is suitable for wireless communication, and this part has been occupied by many applications, and become very crowded. Therefore, frequency band suitable for wireless communication becomes the most precious resource.

The major characteristics of radio media are summarized below. A radio medium is broadcast by nature, i.e., any node in a signal's coverage can receive the signal. Therefore, a radio medium is often shared by all nodes located in the signal's coverage. However, radio media are vulnerable to interferences from other electromagnetic sources. Due to much high attenuation in water, radio signal cannot propagate over a significant distance underwater, and is seldom used for underwater communication.

Radio media are further divided into radio-wave, microwave and millimetre-wave. The wavelength of radio wave ranges from $10^6 \sim 10^{10.5}$ mm, and its characteristics vary with frequencies. In general speaking, low-frequency wave has good penetration ability but fades with propagation distance in omni-directional propagation (i.e., spreading in all direction). High-frequency wave tends to travel in straight lines and is absorbed by rain, with poor penetration ability and vulnerability to interferences from electrical equipments. Usually licenses from the administerial authority are required to use radio spectrums except for Industrial, Scientific, Medical (ISM) bands.

The wave length for micro-wave ranges about $10^3 \sim 10^6$ mm. The signal travels in nearly straight line, and parabolic antennas are usually used to accurately align each other to achieve higher signal-to-noise ratio (SNR). It is often used in long-distance telephone transmission systems. However, multipath fading causes some signals refracted off low-lying atmospheric layers to cancel out the direct signal. It is also absorbed by water and waves with few centimeters long are absorbed by rain.

The wave length of millimeter wave is about $1 \sim 10^3$ mm. It can provide higher communication capacity with nearly line-of-sight (LOS) communication. It is relatively cheap and easily implemented but without penetration ability through solid objects. It is often used for short range communication such as remote control systems and communication between computers and peripherals without need of license for operation.

1.1.4.2   Optical Wave

Unlike radio signals, optical signals are not omnidirectional, and a sender-receiver pair needs to be aligned point-to-point or in LOS for optical communications. It does not suffer from electromagnetic interference and can provide large communication capacity without licensee for operation. However, it is susceptible to particles, sediment and marine fouling, and is affected by scattering so that it cannot propagate well in such environments.

Optical networks can be based on infrared, laser and visible light as illustrated in Figs. 1.15 and 1.16. Figure 1.15 depicts WLANs using infrared and visible light. With infrared, a sender-receiver pair needs to be aligned in a straight line while LOS with visible light. Figure 1.16 shows a laser-based backbone network using Light-Emitting Diode (LED). All these networks are practically implemented.

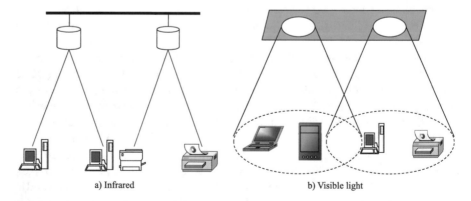

a) Infrared   b) Visible light

Fig. 1.15 Infrared and visible light based WLANs

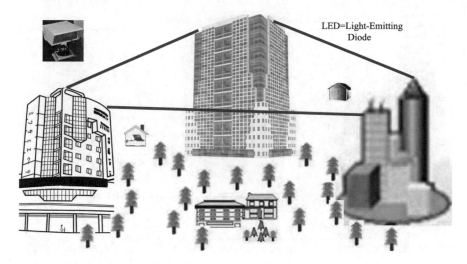

LED=Light-Emitting
Diode

Fig. 1.16 An LED-based optical network

Table 1.1 Comparison between radio, optical and acoustic media for underwater communication

Distance (km)	Transmission rate versus distance					Propagation
Medium	0.05	1	10	100[†]	200[‡]	Speed (km/s)
Radio (bps)[b]	300	3				29,000
Optical (Mbps)	10					29,000
Acoustic (kbps)	300	30	3	0.3 ~ 0.5		1.5

[†]For shallow water area, [‡]For deep water area.
[b]Roughly, Very-Low Frequency (VLF) (3 ~ 30 kHz): 70 bps over 20 m,
Extreme Low Frequency (ELF) (up to 3 kHz): several bps over thousand miles in depth of 400 m

1.1.4.3 Acoustic Wave

Since both radio and optical signals cannot propagate well in underwater environments, acoustic wave is the most popular underwater communication medium as listed in Table 1.1. However, it has some peculiar features such as long propagation delay, poor channel capacity, low channel reliability and dynamics of communication environment in comparison with RF and optical media as summarized in Fig. 9.3. More discussion on these issues can be found in Chap. 9.

1.2 Networking Fundamental Components

A communication network is used to transfer information between sources and destinations, and can be abstracted into a graph consisting of nodes and edges as illustrated in Fig. 1.17. An edge represents a physical link in either wireless medium or wireline

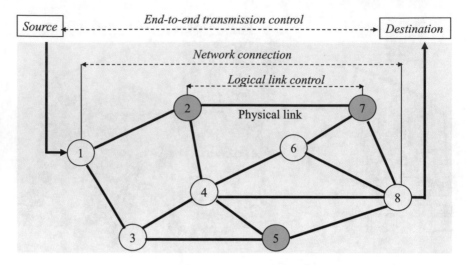

Fig. 1.17 Schematic illustration of a communication network [9]

between a pair of adjacent nodes, along which physical signals propagate from one node to the other. Each node is a networking unit, which may be a switch, a router or even a sub-network. To enable successful communication, data logic control is used to ensure error-free communication over physical links because the physical signal is vulnerable to various interferences. To extend communication beyond a link, a network path or connection composed of multiple physical links across the entire network is set up to transfer data packets from a source to a destination. Packet loss may happen during this journey, and end-to-end transmission control between a source-destination pair should be enforced to guarantee that the destination can receive what have been sent to it. These networking functions are transparent to the user.

A communication network is a complex system, and a layered structure is adopted to realize a modular implementation, and a formal definition of each layer is given below. This chapter briefly introduces a general network structure and networking issues, as well as the major differences between wireless networks and wired networks.

1.2.1 Networking Models

Typically, two networking models are available: one reference model defined by The International Organization for Standardization (ISO), and a practical model adopted by the Internet.

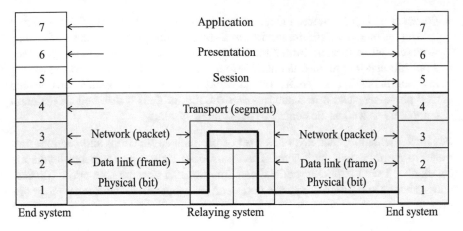

Fig. 1.18 The open system interconnection reference model of ISO [9]

1.2.1.1   ISO Reference Model

An Open System Interconnection (OSI) Reference Model is defined by ISO, consisting of the seven layers as illustrated in Fig. 1.18. The major functions of each layer is described below [10, 11].

- Physical layer handles physical signal communication via physical links. It provides the basic functions for digital communication such as signal modulation/demodulation, coding/decoding, frequency and time synchronization and error control. The protocol unit for digital communication on this layer is bit.
- Data link layer provides a frame communication over physical links. A frame is a formatted bit block. Since the physical layer cannot guarantee error-free communication, the primary function of this layer is to control error in order to provide error-free communication to the network layer. For a shared-medium network, in which multiple users share a common medium, a Medium Access Control (MAC) protocol is used to arbitrate medium sharing. The protocol unit of this layer is frame.
- Network layer provides packet communication. A packet is also a formatted bit block, and is carried by a frame in transmission. Addressing, routing, congestion control and network resource allocation are the major primitives of this layer. The protocol unit for this layer is packet.
- Transport layer is responsible for end-to-end transmission control, which mainly include flow control and transmission reliability control. This layer provides network-independent services to the layers above it. The protocol unit is segment for the Transmission Control Protocol (TCP).
- Session layer provides a control structure for communication between applications in participating terminals, exchanges terminal identifications, establishes and manages as well as terminates session between co-operating applications.

- Presentation layer provides independence to application processes from different data representations. It offers services to the application layer by transforming data structures into a standard format for the transmission among the agreeing partners.
- Application layer provides an interface to the user on an application process requiring communication service. It provides standard services for transmission between user processes, access to databases and running process distributed on different computers as well as the control of distributed systems.

For implementation, each layer has a set of pre-defined functions and services provided to its immediate high layer. The interfaces between two adjacent layers are standardized with explicitly defined call functions so that the networking devices and systems from different manufactures can inter-operate to work together in one system.

1.2.1.2   TCP/IP Model

In practice, the TCP/IP model is the most popular and successful implementation in the Internet. The mapping between this model and the OSI 7-layer model is depicted in Fig. 1.19. The key difference between them is that the ISO model adopts a connection-oriented approach for networking, and packet switching can be used to accelerate packet forwarding. The TCP/IP model adopts a connectionless approach using the Internet Protocol (IP), and routing is used to forward each packet [9]. A brief comparison between them can be found in [11].

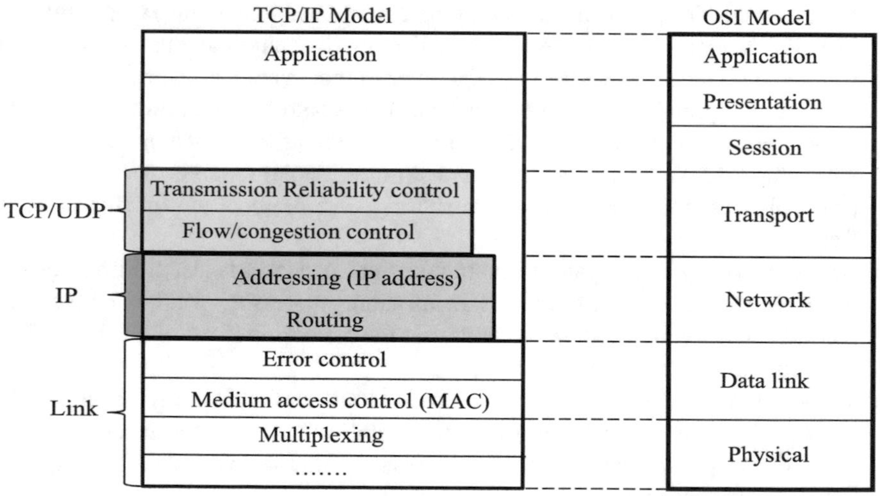

Fig. 1.19 Layered networking model: ISO versus TCP/IP [9]

1.2.2 Networking Approaches

For communication and networking, only layers 1 ∼ 4 in both networking models are involved. Particularly for communication, the major issues, such as signal processing, modulation and coding as well as RF technologies for wireless communication etc., are handled by the physical layer in other references [12]. This book mainly focuses on layers 2 ∼ 4 for networking issues, discussing fundamental networking approaches commonly present in various networks.

1.2.2.1   Shared-Medium Versus Switched-Network

In a shared-medium network, all nodes share a common medium, such as a bus, a ring or a band of electromagnetic spectrum, as illustrated in Fig. 1.20a. Typical shared-medium networks include classical Ethernet network using a bus and wireless networks using electromagnetic wave as a communication medium. In such network, a Medium Access Control (MAC) protocol is required to arbitrate medium sharing among all nodes. It is one of the key elements of the data link layer.

In a switched network, every node is linked to a central unit called switch or hub via separate line as illustrated in Fig. 1.20b. This unit centralizes medium sharing control into a switching fabric, which enables all linked nodes to have simultaneous medium access so that each linked node has a full wired speed. On the data link layer, no MAC protocol is used. Typical networks include the Asynchronous Transfer Mode (ATM) network and IP network.

1.2.2.2   Connectionless Versus Connection-Oriented

According to whether a network connection between a source-destination pair is required or not for data transmission, a network can be either connectionless (CL)

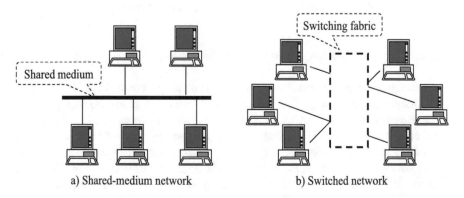

a) Shared-medium network   b) Switched network

Fig. 1.20 Shared-media networks versus switched networks [9]

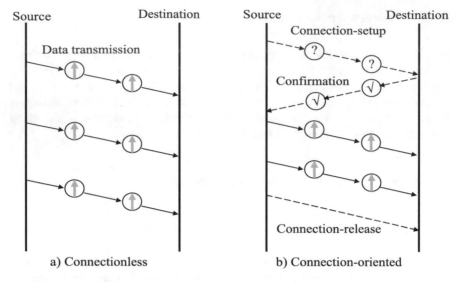

Fig. 1.21 Connectionless networks versus connection-oriented networks [9]

or connection-oriented (CO). In a CL network such as IP, data transmission can take place anytime whenever data is available. It does not require a pre-setup network connection, as illustrated in Fig. 1.21a, and routing is used for data forwarding. Such network is suitable for bursty data transfer. However, it is inefficient for Quality of Service (QoS) support since it cannot make resource reservation for packet transmission.

In a CO network, a network connection must be set up between a source-destination pair before data transmission can take place as illustrated in Fig. 1.21b. In such network, e.g., ATM, switching technology is used to speed up packet forwarding after a route has been established. Furthermore, during network connection setup, resource reservation for QoS provisioning can be carried out. The current locations of sources and destinations can be updated to each other for mobility support.

1.2.2.3 Circuit Switching Versus Packet Switching

According to the manner for resource allocation and usage, networks can be classified into circuit-switched and packet-switched. In a circuit-switched network, such as telephone networks and the Global System for Mobile Communication (GSM), the reserved network resource is dedicated to the corresponding user until the reserved resource is released. During this period, any unused part of the reserved resource cannot be used by other users. Such network is usually connection-oriented, and suitable for constant bit rate (CBR) traffic such as voice with QoS guarantee, but is not cost-effective for bursty traffic such as data application (Fig. 1.22a).

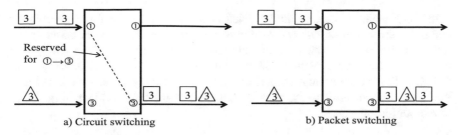

a) Circuit switching   b) Packet switching

Fig. 1.22 Circuit switching versus packet switching [9]

In a packet-switched network such as IP and ATM, the online resource allocation is performed at packet-level demand, and the unused resource reserved by a user can be shared by other users. Such network is better than circuit-switched networks to transfer bursty and non-delay sensitive traffic, but it is inefficient to support delay-sensitive traffic, such as voice (Fig. 1.22b).

1.2.3   Networking Issues

The fundamental networking issues to be discussed in this book include error control, Medium Access Control (MAC), routing/switching, flow/congestion control, end-to-end transfer reliability control and network security.

1.2.3.1   Error Control

Data transmitted in a error-prone channel may be corrupted due to erroneous physical signal as illustrated in Fig. 1.23. Error control schemes must be in place to detect and correct transmission error in order to provide error-free communication over wireless links between neighboring nodes. Error detection is usually conducted by a receiver through measuring the received signal and using schemes such as Cyclic Redundancy Check (CRC). Error correction can be carried out by a receiver and/or sender, using Forward Error Control (FEC) and Automatic Repeat reQuest (ARQ), respectively.

Sender   Receiver

Fig. 1.23 Transmission error of physical signals [9]

With FEC, which is usually implemented at the physical layer for more efficient and quick error control, error correction is performed by the receiver based on the redundant information transmitted by the sender along with data. With ARQ, error correction is carried out by the sender through retransmission and acknowledgement (ACK). ARQ is usually implemented above the physical layer, particularly the data link layer, while a similar scheme is also adopted by the transport layer (e.g., TCP) to recover lost packets.

1.2.3.2 Medium Access Control (MAC)

As mentioned earlier, in a shared-medium network, multiple nodes share a common medium. If more than one node uses the medium simultaneously, their signals may collide at the same receiver, resulting in a failed reception of all arriving signals. This phenomenon is analogous to the traffic situation on a crossroad as illustrated in Fig. 1.24, where only the traffic in parallels directions is allowed to pass the shared area simultaneously; otherwise, traffic collision may happen, and traffic lights have to be used. Particularly for wireless networks, when two or more nodes are transmitting simultaneously, probably none of their signals can be received clearly.

To avoid collision, a MAC protocol is needed to arbitrate the nodes to share the common medium in an efficient and fair manner. The design complexity of a MAC protocol mainly depends on network topology and application requirements. For example, with a star topology network, a centralized MAC protocol such as polling can be used, with which, a coordinator can fully control medium sharing by

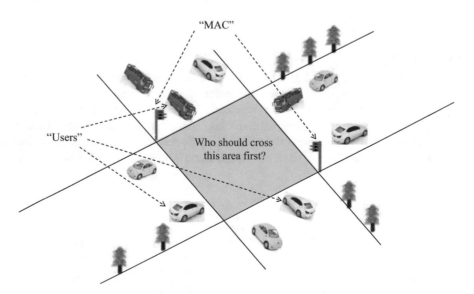

Fig. 1.24 Traffic situation on a crossroad versus MAC

regulating medium use according to a polling sequence. In a distributed topology, every node needs to join a negotiation or competition process in order to obtain a medium access opportunity. In a WSN, since energy is critical to prolong network lifetime, MAC protocols should also take into account energy efficiency besides protocol performance.

1.2.3.3 Routing and Switching

Both error control and MAC discussed above are related to the communication only between neighboring nodes. To deliver messages to any node in a network, a network path between any source-destination pair is needed. Network path searching and selection are carried out by routing. Once a network path is established, the information on this path can be either stored in a networking unit called router or carried by each packet using this route with source routing. With the former, which is also called table-driven routing, the basic operation to be performed by a router is to look up the pre-established routing table according to the destination address carried by packets. This address is used as the key index to look for the next hop corresponding to the destination of the packet in the routing table as illustrated in Fig. 1.25a. Table look-up processing delay depends on table sizes and searching algorithms, both of which are affected by the size of the network address space, which is equal to 2^{32} for IP version 4 (IPv4).

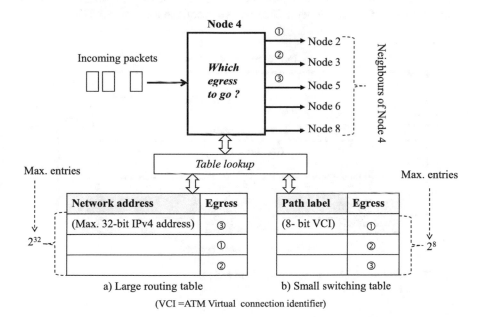

a) Large routing table   b) Small switching table

(VCI =ATM Virtual connection identifier)

Fig. 1.25 Routing versus switching

Consider a large number of packets to travel between the same source-destination pair along the same network path lasting for a certain time period. Upon a network path is established, the subsequent packets can be assigned a path label, and corresponding routing information is stored in the networking units (called switch) along this path. The label is carried by each packet and used as the index to look up the next hop from a much smaller switching table, as illustrated in Fig. 1.25b. This forwarding method is called switching. Usually the number of simultaneously active destination nodes is much smaller than the total number of network addresses. Thus, the label should be much shorter than the address length, which makes the switching table is much smaller than routing table. Therefore, switching is much faster than routing for packet forwarding.

1.2.3.4   Flow Control

It aims to control the traffic load of a flow between a source-destination pair to avoid congestion or comply with a traffic policy or service agreement. It can be simply performed end-to-end on the transport layer like TCP as illustrated in Fig. 1.26a. In this case, only the source and destination nodes are involved in flow control. Alternatively, flow control can also be conducted hop-by-hop on the network layer or even the data link layer if necessary as illustrated in Fig. 1.26b. For example, with the credit-based flow control [13], the downstream node of a link regulates the traffic of each flow from the upstream node. However, it is too complex and costly to implement such a control scheme because every node in the path must be involved.

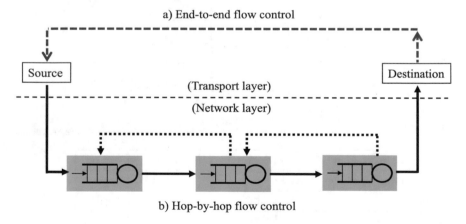

a) End-to-end flow control

(Transport layer)

(Network layer)

b) Hop-by-hop flow control

Fig. 1.26   Flow control: hop-by-hop versus end-to-end [9]

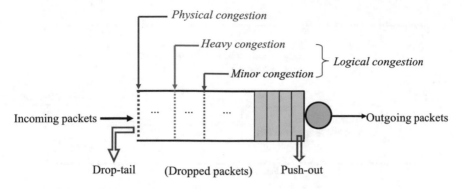

Fig. 1.27 Congestion control scenarios [9]

1.2.3.5 Congestion Control

It aims to relieve a congestion state in a congested networking unit as quickly as possible. A congestion state can be either physical or logical as illustrated in Fig. 1.27. By a physical congestion, the buffer is physically occupied and any newly arriving packet will be dropped. A logical congestion state is defined artificially to avoid physical congestions as much as possible, and can be further subdivided into different congestion levels if necessary.

Once a congestion occurs at a node, dropping packets is fastest to relieve the congestion state. In this case, which packets should be dropped is an issue to be addressed because dropping packets will affect related applications. There are three typical dropping policies. One is drop-tail, by which all newly arriving packets are dropped in the case of congestion. This policy privileges buffered packets over new arrivals, and causes unfair buffer allocation among traffic flows. To maintain fairness, selectively packet dropping is proposed, such as Random Early Discard (RED) [14, 15], which discards packets based on probabilities according to the following stages. (i) When the buffer is almost empty, dropping probability is zero. (ii) As the queue size grows, the probability for dropping an incoming packet grows too, proportional to the number of of packets that a flow has in the queue. (iii) When the buffer is full, the dropping probability is one. However, for delay-sensitive applications, if some packets are queued in a node too long to be meaningful even if they can reach the destination, dropping them on the head of line makes sense as illustrated in Fig. 1.27. This dropping policy is called push-out.

1.2.3.6 End-to-End Transfer Control

A packet may experience a congestion during its journey to the destination, and may be dropped so that the destination node cannot receive it. In this case, retransmitting the lost packets is necessary as illustrated in Fig. 1.28. Packet loss detection can be

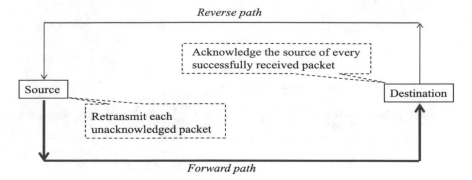

Fig. 1.28 End-to-end transmission control [9]

performed by the destination node using an acknowledgement scheme. With positive acknowledgement, if a destination node has successfully received a packet from a source node, it returns to the source an ACK. If the source node does not receive the expected ACK within a time period, it will retransmit the unacknowledged packet. With negative acknowledgement, a destination node should inform a source node of what it has not yet received so far. Since a packet loss may also happen to ACKs in the reverse path, a positive acknowledgement is much robust than a negative one.

1.2.3.7 Quality of Service (QoS)

Many applications have some specific requirements that a network should satisfy, such as end-to-end delay and packet loss probability that a packet may experience as it travels across the network. A set of parameters are defined to measure quantitatively QoS that a network can provide to the applications. Table 1.2 lists the settings of the primary QoS parameters for typical applications.

Traditionally, a communication network is designed to support one application with fixed QoS requirements. For example, the telephone network was originally designed to support only voice application by using virtual circuit switching to provide constant bit rate (CBR) service at 64-kbps to each voice session. The early IP

Table 1.2 QoS requirements of typical applications [16]

Traffic type	Mean traffic rate	QoS requirements		
		Delay (ms)	Jitter (ms)	Loss rate
Standard video	25 Mbps	250	10	10^{-3}
MPEG video	10 Mbps	250	10	10^{-9}
Voice audio	64 Kbps	250	10	10^{-1}
HiFi audio	2 Mbps	500	5	10^{-3}

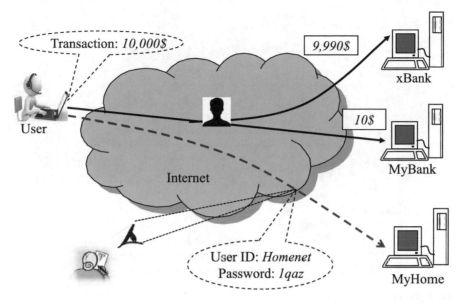

Fig. 1.29 Example of network security threats

network adopts packet switching to only support data applications (e.g., ftp and email) with best effort service. As developing the Integrated Service Data Network (ISDN), especially multimedia networks, supporting various applications with different QoS requirements over one network has been studied extensively and systematically, especially in ATM and new generations of IP networks.

1.2.3.8 Network Security

There are so many documented attacks to the Internet, which caused immense economical and social damage. As illustrated in Fig. 1.29, information may be modified, leaked out or even diverged during its journey toward its destination, and network security becomes upmost important to secure the information transferred across the network. Particularly for wireless networks, due to the exposure and broadcast nature of wireless media, such network is more vulnerable than wired networks to security threats. This weakness becomes the major hurdle of further deploying wireless networks into some important sectors, such as governmental departments and financial businesses.

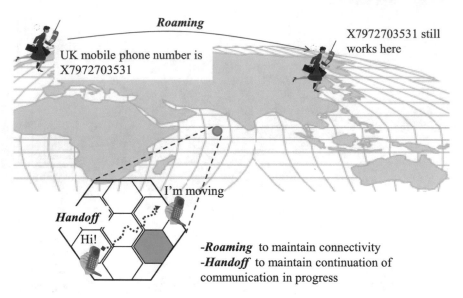

Fig. 1.30 Mobility support

1.2.3.9 Mobility

Since small but powerful mobile devices, such as smart mobile phones, Pads and notebook computers are, very popular, supporting communication anytime, anywhere and anyhow becomes an essential part of communication networks. Mobility support is a fundamental element for networking as the most important advantage of mobile network over other types of networks. The following schemes are devised to support mobility at different levels: maintaining user connectivity against location changes and maintaining the continuation of communication in progress against user's motion, as illustrated in Fig. 1.30. The first scheme is called location management, which allows mobile users to roam at large scale (e.g., traveling from China to Europe), without connectivity loss. The second scheme allows a mobile node to move during a communication in progress without suffering from communication interruption, and this process is called handoff support.

1.2.4 Wired Networks Versus Wireless Networks

Table 1.3 summarizes the major characteristics of wired optical networks and terrestrial RF wireless networks as well as underwater wireless acoustic networks (UWANs) in terms of communication media, user terminals and network structure. The primary difference between wired optical networks and wireless networks stems from their communication media. The optical fibre is extremely reliable with

Table 1.3 Wired networks versus wireless networks

Comparison items	Wired optical network	Wireless networks	
		Terrestrial RF	Underwater acoustic
Communication channel			
Reliability in BER	$\sim 10^{-9}$	$10^{-2} \sim 10^{-6}$	Higher
Channel capacity	$>Tbps$	Up to Gbps	Up to kbps (<10 km)
Channel stability	Stable	Unstable	
Channel security	Secure	Insecure	
Propagation speed	Almost 300,000 km/s		1.5 km/s (in seawater)
Broadcast/Unicast	Unicast	Broadcast	
Medium access	Switching	Medium access control (MAC)	
User terminal			
Mobility support	No	Yes	
Computing capability	Stronger	Weaker	
Buffer capacity	Larger	Smaller	
Power supply	Unlimited	Limited by battery	
Routing capability	No	Yes in ad hoc networks	

much higher transmission rates than wireless media, which are unreliable with small and unstable channel capacity. The exposure and broadcast nature of wireless media makes wireless communication channels insecure, while wired channels are well-protected and unicast. This broadcast and exposure feature of wireless media also makes the MAC protocol as an essential component of wireless networks. For UWANs, both transmission rate and propagation speeds are much slower than those in terrestrial RF wireless networks (RWNs).

The unique merit of wireless networks over wired networks is the capability that allows terminals to move at user's will, i.e., mobility support, which is one of the enabling technologies for pervasive computing that are popular today, such as personal mobile communication and mobile Internet. However, a mobile terminal has many limitations in terms of computing capability, buffer capacity and power supply. Particularly in Mobile Ad Hoc Networks (MANETs), each networking unit will also function as a router to relay packets.

The above-mentioned differences between wired networks and wireless networks are the major factors that prevent many networking protocols and schemes developed for wired networks form being used in wireless networks directly, and the same between terrestrial wireless networks and UWANs. The specific features of wireless network units (e.g., user terminals) also require protocols and algorithms designed for this kind of networks should be energy-efficient and secure as well as adaptive to dynamic networking environments.

1.3 Summary

Wireless networks play an important role in information and communication technology (ICT), and are the enabling technologies for pervasive computation available anywhere, anyhow and anytime. Some of them become part of our daily life, such as mobile cellular networks and WiFi, as access networks that bridge between user terminals and the Internet. Some others are still under research and development, especially underwater wireless networks. Actually, the development of wireless networks is not isolated from wired networks. Similarly for underwater wireless networks, their development also benefit from the achievement of terrestrial wireless networks because they face some similar networking issues, although they have different requirements in the corresponding networking environments. The major terrestrial wireless networks to be discussed in this book include mobile cellular networks, wireless local area networks (WLANs) and mobile ad hoc networks (MANETs). In comparison with these wireless networks, the link propagation delay in space networks and satellite networks are too large, whereas that in wireless body area networks (WBANs) is almost negligible. In both cases, some networking protocols, especially medium access control (MAC) and routing protocols should be designed differently.

   The above-mentioned networking issues and approaches except mobility have been extensively studied for wired networks. Wireless networks such as 3G and 4G as well as WiFi are widely used as access networks to the public network. In this case, a network connection often consists of both wireless and wired networks. Now, most wired networks as well as WLANs, MANETs and WMNs are packet-switched networks, while mobile cellular networks are developing toward full packet-switched networks. A wireless network can operate in either connectionless (CL) or connection-oriented (CO) mode. For example, for access networks, the mobile cellular network runs in CO mode to connect the CO telephone network, whereas WiFi is CL in matching with the CL IP network. Different from wired optical networks, a wireless network uses a shared-medium, which requires a MAC protocol on the data link layer, with an important unique networking feature of mobility, along with other networking issues also addressed in wired networks, including error control, routing, end-to-end transmission control and network security. All these issues will be discussed in the following chapters.

References

1. IEEE Std 802.15.4, *Wireless Medium Access Control (MAC) and Physical Layer (PHY) Specifications for Low-Rate Wireless Personal Area Networks (WPANs)* (2006)
2. Movassaghi, S., Abolhasan, M., Lipman, J., Smith, D., Jamalipour, A.: Wireless body area networks: a survey. IEEE Commun. Surv. Tutor. **16**(3), 1658–1686 (2014). Third Quarter
3. IEEE Std 802.15.6, *Standard for Wireless Body Area Networks* (2012). http://standards.ieee.org/findstds/standard/802.15.6-2012.html

4. Partan, J., Kurose, J., Levine, B.N.: A Survey of practical issues in underwater networks. In: Proceedings ACM International WS, Underwater Networks (WUWNet). Los Angeles, California, USA (2006)
5. Hartenstein, H., Laberteaux, K.P.: A tutorial survey on vehicular ad hoc networks. IEEE Commun. Mag. **46**(6), 164–171 (2008)
6. Satyajeet, D., Deshmukh, A.R., Dorle, S.S.: Heterogeneous approaches for cluster based routing protocol in vehicular ad hoc network (VANET). Int. J. Comput. Appl. **134**(2), 1–8 (2016)
7. Jackson, J.: The interplanetary internet. IEEE Spectrum (2005)
8. Berni, A., Merani, D., Potter, J., Been, R.: Heterogeneous system framework for underwater networking. Proc. IEEE Mil. Comm. Conf. (MILCOM) **5**, 2050–2056 (2011)
9. Jiang, S.M.: Future Wireless and Optical Networks: Networking Modes and Cross-Layer Design. Springer, London, UK (2012)
10. Peterson, L., Davie, B.: Computer Networks: A System Approach. Morgan Kaufmann Publishers, San Francisco (2000)
11. Tanenbaum, A.S.: Computer Networks, 4th edn. Pearson Education International (2003)
12. Sklar, B.: Digital Communications: Fundamentals and Applications, 2nd edn. Prentice-Hall, Upper Saddle River (2002). ISBN 7-5053-7870-8
13. Kung, N.T., Morris, R.: Credit-based flow control for ATM networks. IEEE Netw. Mag. **9**(2), 40–48 (1995)
14. Floyd, S: RED: Discussions of Setting Parameters (1997). http://www.aciri.org/floyd/REDparameters.txt
15. Rosolen, V., Bonaventure, O., Leduc, G.: A RED discard strategy for ATM networks and its performance evaluation with TCP/IP traffic. ACM SIGCOMM Comp. Commun, Rev. (CCR) (1999)
16. White, P.P.: ATM switching and IP routing integration: the next stage in internet evolution. IEEE Commun. Mag. 79–83 (1998)

Part I
Radio-Frequency Wireless Networks (RWNs)

Chapter 2
Error Control

Abstract In digital communication systems, all information to be transmitted is first digitalized into a series of binary bits of either 0 or 1. However, the wireless radio channel used to transmit these digital signals is error-prone (Fig. 1.23) because the channel is susceptible to impulse noises from many sources such as electromagnetic radiation and lightning strikes (Tanenbaum, Computer Networks, Pearson Education International, 2003, [1]). Therefore, error control schemes are necessarily used to detect errors and correct them if possible, which is essential to assure digital wireless communication quality.

An error control scheme usually consists of two parts: error detection and error correction. A detection scheme aims to find whether there are erroneous bits in a received bit block, which is usually carried out by the receiver. To this end, an amount of redundant information (simply redundancy) will be used along with the original information. A simplest method to detect a bit error from a bit block is parity check, and a more powerful error detection scheme that can detect multiple erroneous bits is Cyclic Redundancy Check (CRC).

Typical control schemes include Forward Error Control (FEC) and Automatic Repeat reQuest (ARQ). FEC is usually installed on the physical layer. It allows the receiver to detect and correct errors if any, using the redundancy transmitted together by the sender with the original signal. ARQ is often implemented above the physical layer, such as the data link layer, and the sender conducts error correction through simply retransmitting the original signal until the sender is acknowledged of the successful reception. Note that an ARQ-like scheme is also adopted by the TCP on the transport layer to recover the last packets transmitted on the network layer. Actually, FEC and ARQ are often used jointly. For example, FEC is implemented at the physical layer to reduce residual bit errors, while ARQ on the data link layer aims to eliminate such errors.

This chapter will introduce in detail the above mentioned error control schemes along with some analysis of their performance available in the literature such as [2].

© Springer Nature Singapore Pte Ltd. 2018

S. Jiang, *Wireless Networking Principles: From Terrestrial to Underwater Acoustic*,

https://doi.org/10.1007/978-981-10-7775-3_2

2.1 Error Detection

This section introduces parity check and Cyclic Redundancy Check (CRC).

2.1.1 Parity Check

It is the simplest method to detect errors of an information sample represented by a binary word. After examining a binary word, parity check adds an extra digital bit, called parity bit, to the binary word. The value of the parity bit depends upon the type of parity check in use, namely, even parity check and old parity check. With even parity check, the number of 1-bits in a codeword consisting of the original signal and one parity bit is even, while this number is odd with odd parity check. Therefore, with even parity check, if the number of 1-bits in the original signal is odd, the parity bit is set to 1 so that the total number of 1-bits in the codeword is even. Similarly, with odd parity check, the parity bit is set to 1 if the number of 1-bits in the original signal is even.

Assume a 7-bit word (i.e., $D_1 \sim D_7$) for original signal and one bit (C) for parity check. If $D_1 \sim D_7 = 0111101$, the total number of 1-bits is 5. With even parity check, C is set to 1 by the transmitter, resulting in a codeword $D_1 \sim D_7 C = 01111011$. At the receiver, if the number of 1-bits in a received codeword is odd, it means that at least one bit error occurs in the course of transmission. Similarly, with old parity check, C has to be set to 0, resulting in a codeword $D_1 \sim D_7 C = 01111010$. If the number of 1-bits in a received codeword is even, it means that at least one bit error occurs during transmission.

Parity check is a simple and fundamental method to detect errors. However, it cannot detect errors caused by an even number of erroneous bits since every two bit errors will cancel out each other. Furthermore, it cannot locate erroneous bits for error correction. A joint use of parity check can improve error detection and error correction capability such as Hamming code, which will be discussed later.

2.1.2 Cyclic Redundancy Check (CRC)

It can detect more erroneous bits with less redundancy by treating a bit string as a polynomial that has only two coefficients: 0 (for bit 0) and 1 (for bit 1). A k-bit frame is regarded as a coefficient list for a k-term polynomial, ranging from x^{k-1} on the leftmost to $x^0 = 1$ on the rightmost. For example, for a 7-bit block of 1011011, its polynomial is $x^6 + x^4 + x^3 + x + 1$. The remaining part of this section is organized based on some materials available on Wikipedia.[1]

---

[1] https://www.wikipedia.org/.

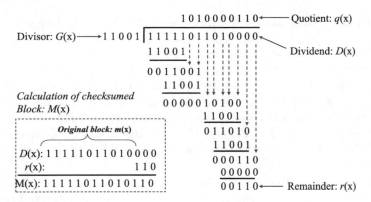

Fig. 2.1 Example on polynomial division and checksum calculation

2.1.2.1 Polynomial Arithmetic

The following polynomial arithmetic is defined for polynomial calculation used for CRC. (i) The calculation is done modulo 2. (ii) There are no carriers for addition and no borrows for subtraction, which means that the result of the addition of two bits equals that of their subtraction. The addition is also identical to the exclusive OR operation ($\oplus$). (iii) The polynomial multiplication is defined as follows: $1 \otimes 1 = 1$, $0 \otimes 0 = 0$ and $1 \otimes 0 = 0$.

The polynomial division is defined as follows: $\frac{D(x)}{G(x)} = q(x) + r(x)$, where $D(x)$ is the dividend, $G(x)$ is the divisor, $q(x)$ is the quotient and $r(x)$ is the remainder. Given $D(x)$ and $d(x)$, the $q(x)$ and $r(x)$ are calculated following the algorithm defined below, and an example is depicted in Fig. 2.1.

1. Start from the left of $D(x)$, take a part of $D(x)$, say $D'(x)$, such that the number of bits of $D'(x)$ is equal to that of $G(x)$.
2. If the first bit of $D'(x)$ is 0, take 0 for $q(x)$ and $r(x) = D'(x)$; otherwise, take 1 for $q(x)$ and $r(x) = D'(x) \ominus d(x)$.
3. Take the leftmost bit of the remaining part of $D(x)$, and append this bit at the end of $r(x)$ to form another $D'(x)$. Then go to step 2.
4. Continue the above manipulation for the remaining part of $D(x)$ bit-by-bit from the left to the right until the last bit is taken. Then the $r(x)$ and $q(x)$ are the final results of the calculation.

2.1.2.2 Principle of CRC

With CRC, both the transmitter and the receiver adopt the same generator polynomial $G(x)$ as the divisor. For $G(x)$, both its high-order bits and low-order bits are set to 1. Let polynomial $m(x)$ be a n-bit block to be transmitted, and $m(x)$ is longer than $G(x)$. A checksum is appended to the end of $m(x)$ to form a so-called checksummed block $M(x)$, which is divisible by $G(x)$. Then $M(x)$ is sent via the medium. Upon receiving it, the receiver divides the received block by $G(x)$. If there is a non-zero reminder, the received bit block is thought erroneous.

The checksumed $M(x)$ is calculated as follows. Given a k-bit $G(x)$, append $k - 1$ zero bits to the end of $m(x)$ to form a new polynomial, $D(x)$ with $n + k - 1$ bits. Then divide $D(x)$ by $G(x)$ using the module 2 division algorithm to calculate reminder $r(x)$. Finally, we can have $M(x) = D(x) \ominus r(x)$. An example illustrated in Fig. 2.1 is given for $m(x) = 1111101101$ with $G(x) = 11001$ (i.e., $k = 5$). Then, $D(x) = 11111011010000$, and using the above polynomial division, we have $r(x) = 110$. So $M(x) = 11111011010110$.

The major properties of CRC with a proper generator $G(x)$ are summarized below. It can detect (i) all single bit errors, (ii) all double-bit errors provided $G(x)$ contains term $x + 1$, (iii) all burst errors provided that the length of the burst is less than that of $G(x)$ and (iv) most larger burst errors.

2.1.2.3 Frame Check Sequence (FCS)

With FCS, the sender conducts a CRC calculation over the entire frame and appends the calculation result as a trailer to the data frame. The receiver recomputes the CRC on the frame using the same algorithm and parameters, and compares the results to the received FCS. If the two results are different, the received frame is considered erroneous.

2.2 Forward Error Correction (FEC)

Erasure codes [3] are often used in erasure channels, where transmitted signal may disappear, to support reliable transfer using redundancy. Bit-level erasure codes for channel coding [4] are often used as bit-level FEC. Basically, the sender adds a redundancy to the original information bit blocks to be sent for error control. The receiver will detect errors and correct them if possible. Several linear coding schemes, for which any linear combination of codewords is still a codeword, are used for this type of FEC, such as repetition codes, Hamming codes, Bose-Ray-Chaudhuri-Hocquenghem (BCH), Reed-Solomon (RS) and Low-Density Parity-Check Codes (LDPCs) etc. These linear coding schemes can provide more efficient encoding and decoding [5].

A simple scheme using redundancy is repetition codes, and a sophisticate one is Hamming code, which is named after its inventor, R.W. Hamming. Hamming code provides a simple bit error detection and error correction scheme. Actually, determining a redundant bit may be a complex function of many original information bits, which depends on coding schemes that are designed to cater for specific error patterns. In the following, only the above examples of FEC schemes are introduced

Table 2.1 Example of FEC using triplet modular redundancy

Triplet received	Interpreted as	Triplet received	Interpreted as
000	0	100	0
001	0	101	1
010	0	110	1
011	1	111	1

to illustrate the principles of FEC technologies. More readings on channel coding technology can be found in [4].

Packet-level erasure codes also use redundancy to support reliable data transfer over lossy channels, and are often called packet-level FEC. Such an FEC scheme encodes a message of k packets into a set of longer encoded messages each with n packets such that the original message can be recovered from a subset of the encoded messages. One such scheme requires a constant code rate $\frac{k}{n}$, such as tornado codes [6], which however is not suitable for highly dynamic channels [5]. Another type of such scheme, called rateless codes [7], releases such constraint without theoretical limit to the number of redundant packets to be transmitted to the receiver. The encoded packets can be sent as many as necessary until the packets are fully recovered, such as fountain codes [8, 9]. More discussions on their applications in UWANs can be found in Chap. 12.

2.2.1 Repetition Code

Table 2.1 lists samples using three bits to transmit one bit of the original data. If the three samples are zero, the original data is most likely a zero, while if three samples are one, the original data is most likely a one, and so on. With this triple modular redundancy, the result is processed by a voting system to produce a single output. Generally, a single data bit is repeatedly sent multiple times in a group, and the correct output is given by the most frequently occurring value in each group.

2.2.2 Hamming Code

It can locate erroneous bits in a bit block and correct them accordingly based on parity check.

Fig. 2.2 Example of
Hamming code and grouping

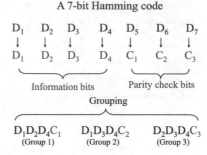

A 7-bit Hamming code

2.2.2.1 Error Location

An example similar to that in the MODICOM experiment manual [10] is used here to demonstrate the principle of Hamming code. As illustrated in Fig. 2.2, a 7-bit Hamming codeword consists of a 4-bit information block from D_1 to D_4 for the original signal, and a 3-bit redundancy for parity check from C_1 and C_3. This codeword is arranged in a form of $D_1 \sim D_4 C_1 \sim C_3$ artificially to simplify discussion, and how to construct Hamming code is discussed in Sect. 2.2.2.2.

A Hamming error control scheme for a single bit error operates as follows. At the transmitter, the four digits $D_1 \sim D_4$ are organized into three groups each consisting of 3 bits as follows: $D_1 D_2 D_3$, $D_1 D_2 D_4$ and $D_1 D_3 D_4$. A parity check digit bit (i.e., $C_1 \sim C_3$) is added to each group as follows: $D_1 D_2 D_3 C_1$, $D_1 D_2 D_4 C_2$ and $D_1 D_3 D_4 C_3$, each of which is a parity check group. If an error occurs in any of these digital bits in a group, the parity of this group will be lost, resulting in a parity check failure at the receiver.

Given $D_1 D_2 D_3 D_4 = 1010$, the following three parity check groups can be generated by using even parity check for each group, i.e., $D_1 D_2 D_3 C_1 = 1010$, $D_1 D_2 D_4 C_2 = 1001$ and $D_1 D_3 D_4 C_3 = 1100$, with the parity bits of $C_1 C_2 C_3 = 010$. Then the Hamming codeword is as follows: $D_1 D_2 D_3 D_4 C_1 C_2 C_3 = 1010010$, which is transmitted over the medium.

Suppose a node receives a codeword of 1011010. The following analysis is conducted to check whether there is a bit error in this coded word, and correct it if any, as demonstrated in Fig. 2.3. Here, we assume that maximally only one bit error may occur in the course of transmission.

- Group 1: $D_1 D_2 D_3 C_1 = 1010$ passes the parity check, which means that all bits in this group, i.e., D_1, D_2, D_3 and C_1, should be correct.
- Group 2: $D_1 D_2 D_4 C_2 = 1011$ fails in parity check, which means that one of the bits in this group should be in error. Since D_1 and D_2 therein already pass parity check in Group 1, the erroneous bit should be D_4 or C_2.
- Group 3: $D_1 D_3 D_4 C_3 = 1110$ fails in parity check too, which means that D_4 or C_3 may be erroneous. However, D_4 is the common possible erroneous bit in both Groups 2 and 3, which means that D_4 is most likely the erroneous bit.

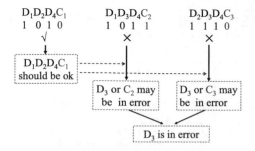

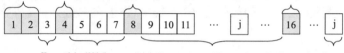

Fig. 2.3 Example of error correction with Hamming code

Parity bit (i.e., C bit) located at position 2^n (n = 0,1,2 ...).

Data bit (D) located at the rest positions, i.e., 3, 5 ..., equal to the sum of 2^n (n = 0,1,2 ...).

Fig. 2.4 Construction of the Hamming codeword

Since the original value of D_4 is 1, the correct one is 0, and the correct signal bits should be 1010. Note that, the above method can detect an erroneous bit from $D_1 \sim D_4 C_1 \sim C_3$ if there is only one bit in error.

2.2.2.2 Hamming Code Construction

An algorithm [1] is described here to construct the Hamming codeword consisting of the original data bits of N bits and the corresponding parity bits. As illustrated in Fig. 2.4, each bit in a Hamming codeword is numbered from the left to right, starting from 1, 2, 3 and so on. We first determine the positions of the parity bits and the data bits in the codeword, following the algorithms described below. (i) All the parity bits are located at the positions indexed by $2^n = 1, 2, 4, 8, 16$, where $n = 0, 1, 2......$, with $2^n \leq N + n$; (ii) The data bits occupies the rest positions of the codeword, i.e., 3, 5, 6, 7, 9, 10, 11,..., but not bigger than $N + n$ either.

Consider a bit block consisting of 7 bits $D_1 D_2 D_3 D_4 D_5 D_6 D_7$, which has the same size as that of the ASCII code. It can be shown that 4 parity bits (i.e., $n = 3$) are necessarily inserted at positions as follows: $C_1 = 2^0 = 1, C_2 = 2^1 = 2, C_3 = 2^2 = 4$ and $C_4 = 2^3 = 8$. Now a questions is whether more parity bits are needed in this case. Let us have the fifth parity bit C_5, i.e., $n = 4$, which means that its position should be at $2^4 = 16$th bit in the codeword. However, the total number of bits in this codeword is the sum of 7 (data bits) and 5 (parity bits), i.e., 12, which means that the position of C_5 will be out of the range of the codeword. Thus, C_5 is unnecessary

Table 2.2 Parity check group calculation of the Hamming codeword for ASCII code

Data bit	Data bit position	Parity bit and its position			
		$C_1(2^0 = 1)$	$C_2(2^1 = 2)$	$C_3(2^2 = 4)$	$C_4(2^3 = 8)$
D_1	$3 = 2^0 + 2^1$	✓	✓		
D_2	$5 = 2^0 + 2^2$	✓		✓	
D_3	$6 = 2^1 + 2^2$		✓	✓	
D_4	$7 = 2^0 + 2^1 + 2^2$	✓	✓	✓	
D_5	$9 = 2^0 + 2^3$	✓			✓
D_6	$10 = 2^1 + 2^3$		✓		✓
D_7	$11 = 2^0 + 2^1 + 2^3$	✓	✓		✓
Data bit of parity group		$D_1 D_2 D_4 D_5 D_7$	$D_1 D_3 D_4 D_6 D_7$	$D_2 D_3 D_4$	$D_5 D_6 D_7$

in this case. Finally, we have an 11-bit codeword (i.e., $7 + 4$) organized as follows: $C_1 C_2 D_1 C_3 D_2 D_3 D_4 C_4 D_5 D_6 D_7$.

Now we determine the grouping, i.e., which data bits of $D_1 \sim D_7$ should be associated with which parity bits to form a parity check group. To this end, we first determine which parity bits of $C_1 \sim C3$ are associated with a particular data bit. Then we take all data bits that use the same parity bit to form a group associated with this parity bit, whose value is determined by the value of each data bit in this group.

The following algorithm can be used to determine the parity bits associated with a particular data bit. For a data bit located at kth position, k can be rewritten as a sum of powers of 2, i.e., $k = 2^{j_1} + 2^{j_2} + \cdots 2^{j_i}$, where a 2^{j_i} just corresponds to the position of a parity bit. For example, for a data bit located on position 11, i.e., b_{11}, $11 = 2^0 + 2^1 + 2^3 = 1 + 2 + 8$, which means that this data bit is associated with parity bits located at positions 1, 2 and 8, i.e., C_1, C_2 and C_4, as illustrated in Fig. 2.4. Repeating the above process for all data bits, we can find the association relationship between all data bits and the parity bits.

Table 2.2 demonstrates how a parity check group is determined for the ASCII code, which is further explained below.

- D_1 is at position $3 = 2^0 + 2^1 = 1 + 2$, and covered by parity bits C_1 and C_2;
- D_2 is at position $5 = 2^0 + 2^2 = 1 + 4$, and covered by C_1 and C_3;
- D_3 is at position $6 = 2^1 + 2^2 = 2 + 4$, and covered by C_2 and C_3;
- D_4 is at position $7 = 2^0 + 2^1 + 2^2 = 1 + 2 + 4$, and covered by C_1, C_2 and C_3;
- D_5 is at position $9 = 2^0 + 2^3 = 1 + 8$, and covered by C_1 and C_4;
- D_6 is at position $10 = 2^1 + 2^3 = 2 + 8$, and covered by C_2 and C_4;
- D_7 is at position $11 = 2^0 + 2^1 + 2^3 = 1 + 2 + 8$, and covered by C_1, C_2 and C_4.

Then we can have the following groups to determine the value of each parity bit: (i) $D_1 D_2 D_4 D_5 D_7 C_1$, (ii) $D_1 D_3 D_4 D_6 D_7 C_2$; (iii) $D_2 D_3 D_4 C_3$ and (iv) $D_5 D_6 D_7 C_4$. For example, if $D_1 D_2 D_3 D_4 D_5 D_6 D_7 = 1001000$, with even parity check, we have:

- $D_1 D_2 D_4 D_5 D_7 C_1 = 100100$,
- $D_1 D_3 D_4 D_6 D_7 C_2 = 101000$,
- $D_2 D_3 D_4 C_3 = 0011$,
- $D_5 D_6 D_7 C_4 = 0000$.

Then the Hamming codeword $C_1 C_2 D_1 C_3 D_2 D_3 D_4 C_4 D_5 D_6 D_7$ is 00110010000.

Now we can check back the rationality of the Hamming code used in Fig. 2.2 from Table 2.2, i.e., C_1 covers D_1, D_2 and D_3, C_2 covers D_1, D_3 and D_4, while C_3 covers D_2, D_3 and D_4, which match the groups used over there.

2.3 Automatic Repeat ReQuest (ARQ)

With ARQ, the sender will retransmit unsuccessfully received frames to assure transmission reliability. To this end, a frame needs to include error detection bits along with data bits as redundancy. These bits are usually arranged in the form of CRC, and are used by the receiver to detect errors [11]. If a frame fails in CRC check, it is considered erroneous. Both acknowledgment (ACK) and timeout mechanisms are used to provide reliable data transmission. An ACK is a small frame sent by the receiver to indicate that it has successfully received the frame transmitted by the sender. If the sender cannot receive the ACK on what have sent after timeout, it retransmits the same frame until it receives the ACK subject to the maximum number of retrials. Typical ARQ protocols include Stop-and-Wait, Go-back-N and Selective Repeat ARQ, each of which is described below.

2.3.1 Stop-and-Wait

The protocol is described along with a performance analysis.

2.3.1.1 Protocol

After transmitting a data frame, the sender cannot transmit any other data frames until it has received the ACK on the previously transmitted frame. The receiver needs to transmit an ACK immediately to the sender after it has successfully received the frame from the sender. Once the sender receives the ACK, it transmits another frame in its buffer if any. If the sender has not received the expected ACK after timeout, it will re-transmit the same data frame [12].

As illustrated in Fig. 2.5, after the sender transmits a data frame F_1, it waits for the ACK from the receiver. Once it receives this ACK, it discards F_1 from the buffer and repeats the same process for the next data frame F_2. If the sender does not receive the ACK after timeout, it has to retransmit F_1.

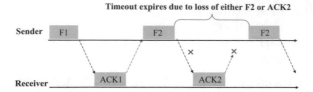

Fig. 2.5 Stop-and-Wait protocol

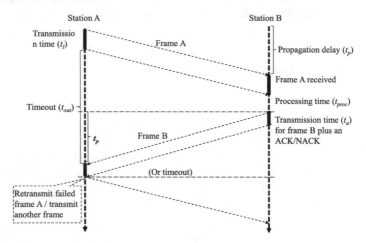

Fig. 2.6 Diagram of Stop-and-Wait protocol for analysis

The major weakness of this protocol is channel waste due to the blocking of transmitting other frames during waiting period for the ACK on the transmitted frame, and the following analysis provide a quantitative view on this issue.

2.3.1.2 Performance Analysis

This section introduces analytical results on the maximum throughput under a saturated state for this protocol reported in [2]. In a saturated state, the sender always has frames waiting for transmission.

The throughput analyzed here (Λ) is defined as the number of frames acknowledged positively by the receiver to the sender per time unit. Suppose that a frame is received in error with probability p, and the error in the reverse direction from the receiver to the sender is negligible. In this case, Λ is determined by the average minimum time interval between successive data frames positively acknowledged (t_v), i.e., $\Lambda = \frac{1}{t_v}$, which is calculated below.

With this protocol, after the sender has transmitted one data frame, it will not transmit any other data frames until it receives an ACK or the timeout happens. As illustrated in Fig. 2.6, the minimum time interval between successive data frames in

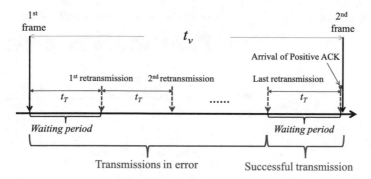

Fig. 2.7 Time interval between successive frames positively acknowledged (t_v) with Stop-and-Wait

this case, t_T, is given by

$$t_T = t_l + t_{out},\qquad(2.1)$$

where t_l is the transmission time of a data frame, and

$$t_{out} = 2t_p + t_{proc} + t_a,\qquad(2.2)$$

where t_p denotes the propagation delay between the sender and the receiver, t_{proc} is the processing time of a data frame upon it arrives at the receiver, and t_a is the transmission time of the ACK frame plus other information for the received data frame.

The calculation of t_v can be divided into two parts: one for the retransmission of the data frames received in error and the other for the final successful transmission of the data frame, as illustrated in Fig. 2.7. For the second part, the time interval between the last retransmission and the arrival of the positive ACK at the sender is just equal to t_T because the arrival of the positive ACK at the sender triggers the transmission of a new data frame.

For the average retransmission times of data frames received in error, the probability for having i number of retransmissions is equal to $p^i(1 - p)$, which means that i times of previous transmissions of a data frame (i.e., the first transmission plus the follow-up $i - 1$ retrials) fail with probability p^i, while the ith retrial succeeds with probability $(1 - p)$. Each trial takes an amount of time t_T. With an unlimited number of retransmissions, t_v is given by

$$t_v = t_T + t_T \sum_{i=1}^{\infty} i p^i (1 - p)$$

$$= \frac{t_T}{1 - p},\qquad(2.3)$$

Fig. 2.8 Go-back-N
protocol

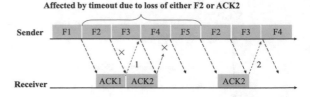

which yields

$$\Lambda = \frac{1 - p}{t_l + t_p + t_{proc} + t_a}.$$ (2.4)

2.3.2   Go-Back-N

This section also introduces analytical results on the maximum throughput under a
saturated state for this protocol reported in [2] after an introduction to it.

2.3.2.1   Protocol

Different from Stop-and-Wait, here a sender can continue to transmit data frames
even without receiving the ACKs on previously transmitted data frames. However,
the receiver keeps the track of sequence number of the next data frame that it expects
to receive, and drops any other data frames. Once a node successfully receives a data
frame destined for it, it acknowledges the sender of the frame reception. If the sender
finds that a transmitted data frame has not been received successfully, it goes back to
retransmit not only this frame but all the frames that have been transmitted following
this frame [12].

As illustrated in Fig. 2.8, the sender has transmitted data frames F_1 to F_5 without
waiting for the ACKs. After it has received the ACK for F_1, it discards F_1 from its
buffer. Since F_2 has not been acknowledged due to the loss of either the frame itself
or its ACK, F_2 along with F_3 and F_4 has been retransmitted after timeout.

Note that, in order to control the number of standing frames at the sender, a sliding
window is used to control the maximum number of frames stored in the buffer, which
includes those to be transmitted and those transmitted but not yet acknowledged.
More detail about sliding window will be described later for TCP flow control in
Chap. 6.

2.3.2.2   Performance Analysis

The sender continues to transmit other data frames after the transmission of the first
data frame as illustrated in Fig. 2.9. However, if the first data frame is not positively

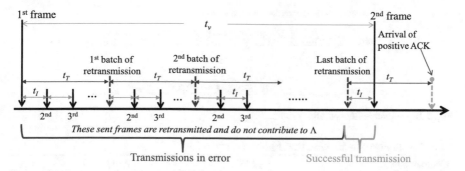

Fig. 2.9 Time interval between successive frames positively acknowledged (t_v) with Go-back-N

acknowledged before timeout, this frame along with all other frames transmitted after it has to be retransmitted. However, among the retransmitted data frames other than the first one, the successfully received data frames do not contribute to the throughput. Accordingly, we can have the t_v as illustrated in Fig. 2.9 for the calculation of t_v. The only difference from Stop-and-Wait is the time interval between the last retransmission of the first data frame and the beginning of the transmission of the second frame. For Stop-and-Wait, it is t_T as illustrated in Fig. 2.7, while it is t_I for Go-back-N because the sender does not stop transmission. In this case, the interval between successive frame transmissions is always equal to t_I. After the last retrial leading to a successful transmission, the cycle for another data frame just starts as depicted in this figure. Thus, with an unlimited number of retransmissions, we have

$$t_v = t_I + t_T \sum_{i=1}^{\infty} i p^i (1 - p)$$

$$= \frac{t_I + t_{out}/p}{1 - p}, \tag{2.5}$$

which yields

$$\Lambda = \frac{p(1 - p)}{p t_I + t_p + t_{proc} + t_a}. \tag{2.6}$$

2.3.3 Selective Repeat ARQ

The major difference between this protocol and Go-back-N is that here the receiver continues to accept and acknowledge each data frame that has been received successfully even after a data frame is received in error, and the sender retransmits only the data frames that have not been acknowledged positively by the receiver [12].

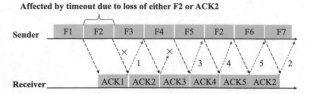

Fig. 2.10 Selective Repeat ARQ protocol

Table 2.3 Comparison between FEC and ARQ [12]

Compared items	FEC	ARQ
Information redundancy	Larger	Smaller
Error control delay	Shorter	Longer
Necessity of feedback channel	No	Yes
Protocol complexity	Lower	Higher
Computation complexity	Higher	Lower
Bandwidth consumption	Larger	Smaller
Reliability guarantee	No	Yes

As illustrated in Fig. 2.10, the sender has transmitted out $F_1 \sim F_5$ without waiting for the ACK on the early transmitted data frames. When it finds that F_2 has not been acknowledged positively after timeout, it retransmits only F_2 but not $F_3 \sim F_5$, and then continues to transmit $F_6 \sim F_7$.

2.4   FEC Versus ARQ

Table 2.3 lists some primary characteristics of FEC and ARQ. Generally, FEC, which is usually implemented in hardware, can yield a short delay of error control because it does not require retransmission of erroneous frames for error correction. There, no feedback channel is required so that it can be used in situations where retransmission is costly or impossible, such as massive data storage devices to protect against damage. However, the transmission of redundancy consumes extra bandwidth, and sophisticated coding schemes are required to achieve optimal tradeoff between error control efficiency and channel utilization.

With ARQ, which is usually in software, retransmission is used to correct transmission error. It jointly uses acknowledgement schemes and timeout mechanism. Therefore, a feedback channel is required for acknowledgement, and error control delay is longer due to retransmission. However, it requires neither complex coding nor large redundancy for error control. Note that FEC can only optimize transmission reliability, while ARQ can guarantee transmission reliability.

2.5 Summary

This chapter briefly describes typical error control approaches used for communication over wireless and wired links. Error control needs to consider error patterns, which do present in wireless channels such as random error and burst error due to different types of interference sources. This issue is primarily considered by FEC on the physical layer with channel coding. On the other hand, ARQ is widely used by the data link layer to handle residual errors of the lower layer, and also adopted by higher layers such as transport layer to recover data lost during its journey to the destination. For both FEC and ARQ, they need to address challenging issues raised by limited channel capacity and low channel quality, both of which varies with time, especially in mobile networks and underwater environments. It is necessary to design smart error control protocols that can exploit more information on realtime channel quality, adaptively configure FEC and ARQ settings with subject to constrains in energy and channel capacity. Please refer to [13–16] for more discussions on these issues.

References

1. Tanenbaum, A.S.: Computer Networks, 4th edn. Pearson Education International (2003)
2. Schwartz, M.: Telecommunication Networks, Protocols. Modeling and Analysis. Addison-Wesley Publishing Company, Massachusetts, USA (1987)
3. Luby, M.G., Mitzenmacher, M., Shokrollahi, M.A., Spielman, D.A.: Efficient erasure correcting codes. IEEE Trans. Inform. Theory 47(2), 569–584 (2001). Feb
4. Declercq, D., Fossorier, M., Biglieri, E.: Channel Coding Theory, Algorithms, and Applications. Academic Press, Cambridge (2014)
5. Jiang, J.F., Han, G.J., Zhu, C.S., Chan, S., Rodrigues, J.J.P.C.: A trust cloud model for underwater wireless sensor networks. IEEE Commun. Mag. 110–116 (2017)
6. Luby, M., Mitzenmacher, M., Shokrollahi, A.., Spielman, D.., Stemann, V.: Practical loss-resilient codes. In: Proceedings of Annual ACM Symposium Theory of Computing (STOC), pp. 150–159. El Paso, TX, USA (1997)
7. Maymounkov, P., Mazieres, D.: Rateless codes and big downloads. In: Proceedings of International WS Peer-to-Peer System. Berkley, USA (2003)
8. Luby, M.: LT codes. In: Proceedings of Annual IEEE Symosium Foundations of Computer Science (FOCS), pp. 271–280. Washington, DC, USA (2002)
9. Shokrollahi, A.: Raptor codes. IEEE Trans. Inform. Theory 52(6), 2551–2567 (2006). Jun
10. LJ Technical Systems, *CT02 Curriculum Manual*
11. Stojanovic, M.: Optimization of a data link protocol for an underwater acoustic channel. In: Proceedings of MTS/IEEE OCEANS. Washington, DC, USA (2005)
12. Jiang, S.M.: On reliable data transfer in underwater acoustic networks: a survey from networking perspective. IEEE Commun. Surv. Tutor. vol. P(99) (2018)
13. Zorzi, M., Rao, R.R.: Energy-constrained error control for wireless channels. IEEE Personal Commun. Mag. 4(6), 27–33 (1997). Dec
14. Akyildiz, I.F., Su, W., Sankarasubramaniam, Y., Cayirci, E.: Wireless sensor networks: a survey. Computer Net. 38(4), 393–422 (2002). Mar

15. Sen, J., Bhattacharya, S.: A survey on cross-layer design frameworks for multimedia applications over wireless networks. Int. J. of Comput. Sci. Inf. Tech. **1**(1), 29–42 (2008)
16. Ramis, J., Femenias, G., Carrasco, L.: Cross-layer design of multi-rate wireless networks based on link layer truncated ARQ. In: Proceedings of IEEE International WS Cross Layer Design (IWCLD), pp. 1–5. Palma de Mallorca (2009)

Chapter 3
Medium Access Control (MAC)

Abstract In a wireless network, the communication medium is usually shared simultaneously by multiple users, and a Medium Access Control (MAC) protocol is used to arbitrate the sharing. This chapter focuses on the fundamental design issues of MAC protocols, and introduces a reference model used to describe various MAC protocols. The major references for this chapter include (Chandra et al., Wireless medium access control - a review, 1999, [1], Walke, Mobile radio networks: networking, protocols and traffic performance. Wiley, New York, 2002, [2], Jiang, Proceedings ACM international conference underwater networks & systems (WUWNet), 2015, [3]).

3.1 Fundamental Issues

MAC protocol design for wireless networks should take into account application requirements, network topologies and implementation complexity, with focus on the following performance criteria: medium access delay, throughput, fairness, security, robustness and energy efficiency, each of which defined below.

- Medium access delay is the time interval between when a frame arrives at the sender to when the frame is successfully received by the receiver. This delay affects MAC performance for QoS support.
- Throughput is defined as the amount of data that is received successfully per time unit.
- Fairness refers to the opportunity for a node to access the shared medium, and the nodes of the same type should have an equal opportunity to use the shared medium.
- Security aims to protect the integrity and confidentiality of the frames transmitted over a wireless medium.
- Robustness reflects the capability of maintaining the normal MAC operation in the case of interference and nodal failure.
- Energy efficiency denotes the amount of energy consumed per successful transmission/reception.

© Springer Nature Singapore Pte Ltd. 2018

S. Jiang, *Wireless Networking Principles: From Terrestrial to Underwater Acoustic*,

https://doi.org/10.1007/978-981-10-7775-3_3

Fig. 3.1 Signal collision

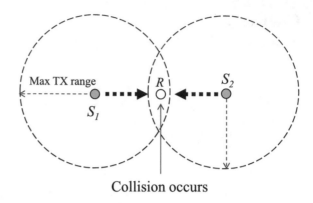

Collision occurs

As discussed below, MAC protocol design faces many challenges for broadcast wireless media, and some of them are caused by the carrier sensing mechanism, which is adopted by many MAC protocols.

3.1.1   Signal Collision

The objective of a MAC protocol is to allow multiple users to use efficiently a shared medium in a fair manner. This endeavor is meaningful only if the frame reception is successful. The major event leading to reception failure is the signal collision occurring at the receiver. As illustrated in Fig. 3.1, if node R is close to both S_2 and S_2, the signals from them will collide at R, which may cause R unable to decode successfully the expected signal. Therefore, avoiding such collision is the primary task of each MAC protocol.

To avoid effectively the collision at a receiver, ideally a MAC transmission decision should be made by the receiver, which is called receiver-centric MAC protocol henceforth. This is because that only the receiver can know exactly and immediately the real situation about whether a new transmission should be allowed without any impact on its collision-free reception. Actually, it is the sender rather than the receiver that controls the traffic source and triggers each transmission. Thus, certain coordination between the sender and receiver is necessary to make and enforce a proper MAC transmission decision. However, this comes at a cost of more protocol overhead for information exchange, consuming more bandwidth and energy. When propagation delay is negligible, such as in RF wireless networks discussed here, a sender can also make MAC transmission decisions, and the above mentioned protocol overhead can be avoided to simplify MAC protocol design.

3.1.2   Capture Effect

When multiple signals overlap at the same node simultaneously, the stronger signal can be decoded without error but may not for the weaker one. This phenomenon is

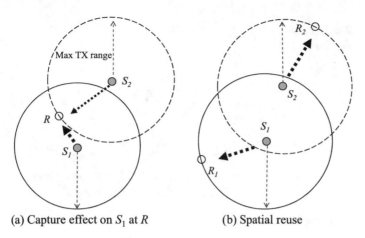

Fig. 3.2 Capture effect and spatial reuse

called capture effect. As illustrated in Fig. 3.2a, the receiver R can receive the signals from both S_1 and S_2. Since the distance between R and S_1 is much shorter than that between R and S_2, the strength of the signal from S_1 is much stronger than that from S_2. In this case, the signal from S_1 can be still decoded successfully at R but not for the signal from S_2. This effect may cause unfairness among transmitters, and is the cause of the near-far problem, with which, nodes closer to the base station have more opportunities than those farther far away from the base station if their transmission powers are at the same level.

3.1.3 Spatial Reuse

As illustrated in Fig. 3.2b, if R_1 is far away from S_2 and R_2 is also far away from S_1, then both R_1 and R_2 can successfully receive the signals from their senders S_1 and S_2, respectively, no matter how S_1 and S_2 are close to each other. This feature is called spatial reuse, which should be exploited as much as possible to maximize channel utilization because it allows S_1 and S_2 to transmit simultaneously to R_1 and R_2 without harm.

3.1.4 Hidden Terminals

To avoid collision with ongoing transmissions, carrier sensing is a mechanism that is often used by a MAC protocol to detect whether there is an ongoing one, e.g., IEEE 802.3 for Ethernet, and IEEE 802.11 for WiFi. With this mechanism, a node senses the medium to detect carrier first. If it hears nothing, it infers that there is no

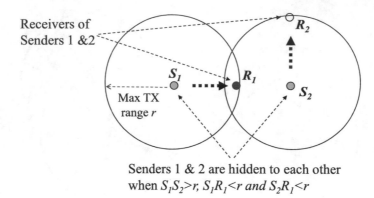

Senders 1 & 2 are hidden to each other
when $S_1S_2>r$, $S_1R_1<r$ and $S_2R_1<r$

Fig. 3.3 Hidden terminal problem

transmission undergoing. However this inference may be not correct in the presence of hidden terminals as discussed below.

As illustrated in Fig. 3.3, since the distance between senders S_1 and S_2 is so long that they cannot hear each other. Suppose S_2 is transmitting to node R_1. With carrier sensing, S_1 cannot detect the ongoing transmission of S_2. In this case, S_1 infers that the channel is idle, and will start its transmission while S_2 is transmitting, causing the signals from S_1 and S_2 to collide at R_1. This phenomenon is called hidden terminal problem, by which, S_2 is a hidden terminal of S_1, and vice versa.

The receiver plays a key role in handling this problem because only it can know whether it is covered by any hidden terminals. To prevent other nodes from transmitting simultaneously to the same receiver that is going to receive a transmission, the receiver has to block all unexpected transmissions that may interfere with reception. To this end, the receiver should inform all its neighbors of the scheduled reception so that they can avoid any transmissions during the scheduled period. A node wishing to transmit to the receiver should get a permission first from the receiver before starting its transmission.

The above solution can be implemented through using a well-known handshake protocol: Request-to-Send (RTS)/Clear-to-Send (CTS) . The RTS/CTS protocol was originally used in a wired local area network (LAN) to link Apple computers, and was applied in RF wireless networks to design a MAC protocol, called Multiple Access Collision Avoidance (MACA) [4]. It was then adopted by the Floor Acquisition Multiple Access (FAMA) protocol [5] to eliminate the hidden terminal problem. In the IEEE 802.11 MAC protocol standard, the RTS/CTS scheme is slightly modified as an optional solution to the hidden terminal problem in wireless local area networks (WLANs). The principles of both MACA and FAMA are introduced in the following subsections, and the RTS/CTS of IEEE 802.11 will be described in Sect. 4.7.

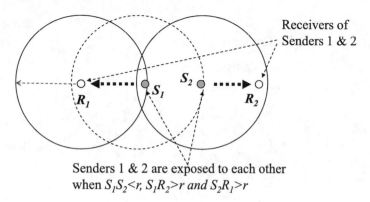

Fig. 3.4 Exposed terminal problem

3.1.5 *Exposed Terminals*

As illustrated in Fig. 3.4, if the distance between the receiver (R_2) of a sender (S_2) and the receiver (R_1) of another active sender (S_1) is so large that the transmission of S_2 will not collide with the ongoing transmission of S_1 at R_1, and vice versa. In this case, even if S_1 is transmitting to R_1, S_2 can still transmit to R_2 in parallel with S_1 without harm to each other with spatial reuse as discussed earlier. However, simply using carrier sensing will forbid a simultaneous transmission from S_2 since it can detect the ongoing transmission of S_1. This phenomenon is called the exposed terminal problem.

3.1.6 *Energy Efficiency*

Mobile terminals are usually battery-operated with a limited energy capacity, which determines the operational time of network nodes, and eventually affects the network lifetime. Whether an exhausted node can be recharged depends on types of wireless networks. In mobile cellular networks and WLANs, mobile phones and notebooks are rechargeable, whereas in wireless sensor networks (WSNs), it is not cost-effective to recharge exhausted sensor nodes, especially for densely deployed nodes with low cost. In both cases, energy consumption control is essential to maintain reliable network performance and prolong network lifetime. Particularly for MAC protocols, reducing protocol overhead and communication activities are among the major measures to improve energy efficiency because it is demonstrated empirically that energy consumption for communication is much higher than that for computing [6, 7].

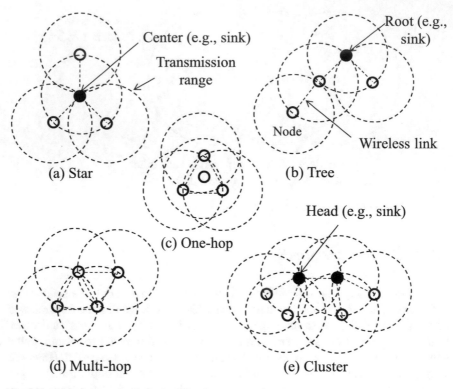

Fig. 3.5 Typical network topologies [8]

3.1.7 Network Topologies

The complexity of a MAC protocol depends on the network topology that the protocol is designed for. Topologies can be static or dynamic. A dynamic topology changes with time due to terminal mobility, disruptive changes in transmission power, resulting in the lack of a fixed long-term structure. However, a static topology does not change with time. Both topologies can be classified into centralized and distributed topologies as well as cluster that combine these both as depicted in Fig. 3.5, and are discussed below.

3.1.7.1 Centralized Topology (CT)

There is a central node used to coordinate communication, especially between itself and other nodes within its communication range. It may also function as a data sink, bridge or gateway to interface with other networks, similar to a base station or an access point (AP). The central node may become one point failure of the entire network, whereas a network with this topology has a limited coverage. As discussed in [3, 8], such topology can be further divided into:

- Star (ST): The central node is in the mutual range with every node, and can fully coordinate MAC operation as illustrated in Fig. 3.5a. In this case, the MAC protocol can be simple and efficient to support QoS.
- Tree (TR): It is a hierarchical structure of star topologies to cover a large area as illustrated in Fig. 3.5b. A special case is that multiple strings are linked to one root. The parent node may also become one point failure of its covered subtree.

3.1.7.2 Distributed Topology (DT)

There is no central node as used in the centralized topology, and nodes are allowed to communicate each other directly. It can be further divided into [3, 8]:

- One-hop (OH), also called mesh network: The distance between any nodes is one hop so that they can hear each other as illustrated in Fig. 3.5c. This kind of network has neither hidden terminals nor exposed terminals, but with a limited network coverage.
- Multi-hop (MH): The distance between some nodes is larger than one hop as illustrated in Fig. 3.5d, and data destined for some nodes needs to be relayed by intermediate nodes. It can cover a larger area, whereas end-to-end network performance degrades quickly as the number of hops increases.

3.1.7.3 Cluster Topology (CL)

It combines the centralized and distributed topologies by organizing the nodes into groups as illustrated in Fig. 3.5e. A group may be either a star or a tree topology, which is called cluster. The cluster head can form a multi-hop network to cover a large area, such as cellular networka for mobile personal communication [3, 8].

3.2 A MAC Reference Model

As discussed in [8], this reference model[1] decomposes a MAC protocol into three components as illustrated in Fig. 3.6: operation cycle (OC), medium access unit (MAU) and MAC mechanism (MM), and the relationship between them is illustrated in Fig. 3.7. At the beginning of one operation cycle, a node runs some MAC mechanisms such as carrier sensing, and then accesses the medium with certain medium access units like time slots. Adopted MAC mechanisms, their running sequence, the content of messages exchanged and the number of medium access units available per

---

[1]This MAC reference model is mainly used to describe various MAC protocols proposed for UWANs, and the readers can skip this part to learn MAC protocols proposed for RF wireless networks.

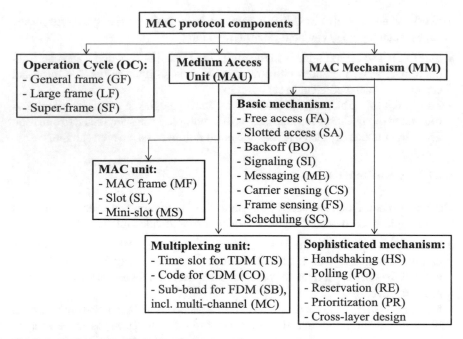

Fig. 3.6   A reference model for MAC protocols [8]

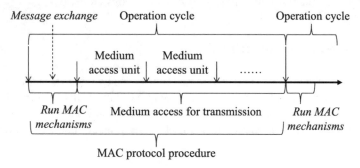

Fig. 3.7   Relationship between the MAC protocol components [8]

operation cycle make up the MAC protocol procedure, which vary with the particular protocol design [8].

3.2.1   Operation Cycle

A MAC cycle is a repeated time epoch, during which each node follows the same MAC procedure to access the medium. It can be either a time interval with a fixed format, or a random time interval depending on transmission lengths. The time interval for MAC operation in a cycle is usually organized into either a general frame, a large frame or a superframe as discussed below [8].

- General frame (GF): It primarily consists of a frame transmission time and the other time used for MAC operation, such as carrier sensing or backoff. It is often adopted by simple MAC protocols, such as ALOHA and Carrier Sensing Multiple Access (CSMA).
- Large frame (LF): The time is divided into small units such as slots, each of which is allocated to one node for medium access following the same access scheme.
- Superframe (SF): The time is divided into several segments, each of which adopts different access schemes, such as random access or reservation access.

3.2.2 Medium Access Units

As discussed in [8], a medium access unit (MAU) is the basic unit for each node to share the medium, and can be defined by either a MAC protocol of the data link layer (called MAC unit) or a multiplexing scheme of the physical layer (called multiplexing unit), as discussed below.

3.2.2.1 MAC Units

Typical MAC units include MAC frame, slot and mini-slot as discussed in [3, 8].

MAC Frame (MF)

A MAC frame (MF) is also called "MAC protocol unit", which is a bit block corresponding to once MAC transmission with or without a pre-defined frame delimitation. Without a frame delimitation, the frame boundary is detected through transmission activity.

Slot (SL)

A slot (SL) (Note that "time slot" is reserved for the use in multiplexing-based MAC protocols) is a medium allocation unit that can be occupied by a node for once transmission. Collision with an ongoing transmission can be reduced by regulating each transmission to start only at the beginning of each slot. To this end, a network level time synchronization is needed to make all nodes to have an identical time reference. Furthermore, to avoid collision between consecutive slots due to uncertainties, a guard time has to be inserted between them (Refer to Sect. 3.2.2.2 for more details).

Mini-Slot (MS)

Similar to SL, a mini-slot (MS) is also an allocation unit used by a node for control message transmission with a smaller size than that of the SL. For example, an MS is used to submit reservation requests in the Packet Reservation Multiple Access (PRMA) protocol [9]. MSs are usually accessed through random access without using time synchronization.

3.2.2.2   Multiplexing Units

As discussed in [8], signal multiplexing of the physical layer is used to enable multiple signals to be transmitted over a shared medium to improve medium utilization. It divides the medium into multiple subchannels each transmitting one signal in form of either sub-frequencies, time slot or codes, which are called multiplexing allocation units here. Typical multiplexing schemes include Frequency Division Multiplexing (FDM), Time Division Multiplexing (TDM) and Code Division Multiplexing (CDM). Their multiplexing allocation units are described below, and more relevant issues will be discussed in multiplexing-based multiple access schemes in Sect. 3.3.1.

Subband (SB) for FDM

With FDM, a shared frequency spectrum is divided into multiple fixed-size subbands each to be occupied by one signal. A guard band between adjacent subbands is inserted to tolerate transmission uncertainty. Orthogonal FDM (OFDM) allows an overlapping of adjacent subbands in a mutually orthogonal mode to improve spectral efficiency.

Time Slot (TS) for TDM

With TDM, the time is divided into time slots (TSs), each of which is allocated to transmit one signal. A guard time needs to be inserted between two adjacent time slots to guarantee collision-free reception.

Code (CO) for CDM

With CDM, spread spectrum (SS) techniques are applied to allow multiple signals to use a shared frequency spectrum concurrently, typically including frequency-hopping SS (FHSS) and direct-sequence SS (DSSS). FHSS is a combination of TDM and FDM. It allows each signal to occupy a different subband for transmission at a particular time point, and hop to another subband following a hopping sequence at the next time point. The hopping sequence is different for each signal transmission such that each signal is transmitted in different subbands at any time points.

With DSSS, each signal is assigned a unique orthogonal spreading code, which is used to spread the signal over the entire spectrum for transmission. The physical sharing unit is the subband for FHSS and the entire spectrum for DSSS, respectively. They are shared according to the hopping sequence and the spreading code, which are collectively called "code" for the allocation unit.

3.2.3 MAC Mechanisms

As discussed in [8], when no multiplexing is conducted on the physical layer, certain MAC mechanisms should be used to prevent signal collision. A MAC mechanism refers to an action defined by a MAC protocol and used by a node to access the medium. Such mechanisms can be divided into basic mechanisms, combinations of basic ones and those based on cross-layer design. Note that an acknowledgement scheme implemented on the Logical Link Control (LLC) sub-layer is often used jointly with a MAC protocol, which stipulates that the receiver of a MAC frame should acknowledge the sender of the reception. Table 3.1 summarizes the major characteristics of the MAC mechanisms, each of which is described below.

3.2.3.1 Basic Mechanisms

Some popular basic mechanisms are discussed below [3, 8].

Free Access (FA)

Free access (FA) schemes include random access and probabilistic access. With random access, a node accesses a medium at its will without using any other MAC mechanisms like ALOHA [10]. With probabilistic access, a node accesses a medium with probability ω, and does not with probability $1 - \omega$. Random access is simple, robust and applicable to any topologies, but its bandwidth utilization is very low due to highly frequent collisions.

Slot Access (SA)

A slot access (SA) scheme divides a medium logically into a series of slots along the time axis, stipulating that any new medium access is allowed only at the beginning of each slot. Although it can prevent an ongoing transmission from being interfered by a new attempt, it cannot avoid collision completely because the beginning of each slot is just a synchronization point that invites multiple nodes to transmit simultaneously.

Backoff (BO)

A backoff (BO) scheme enforces a node to wait a time period before accessing the medium. A channel access prioritization can be implemented by assigning a shorter backoff time to higher priority nodes.

Signaling (SI)

With a signaling (SI) scheme, a node deliberately sends a short signal to either block ongoing transmissions (e.g., CSMA/CD), jam competitors (e.g., HIPERLAN [11])

Table 3.1 Characteristics of MAC mechanisms and multiplexing-based access schemes [8]

Mechanism/scheme	Basic operation condition	Main strength	Main weakness
MAC mechanisms			
Free access (FA)		Very simple and robust	Frequent collision
Slotted access (SA)	Time synchronization (SYN)	Less collision than free access	Same as TDMA
Backoff (BO)		Useful for collision avoid-ance/prioritization	Bandwidth waste due to backoff time
Signaling (SI)		Invulnerable to channel quality without decoding	No much information available
Carrier sensing (CS)		Simple and robust	Hidden and exposed terminal problems
Messaging (ME)	Channel quality can satisfy decoding of received signal for successful reception of message	Availability of explicit information. It can provide as much information as necessary	Vulnerable to channel quality, susceptive to propagation delay. More protocol overhead for messaging
Frame sensing (FS)		Efficient for collision avoidance at receiver. QoS is supported with polling and reservation	
Handshaking (HS)			
Polling (PO)*			
Reservation (RE)			
Scheduling (SC)	Time synchronization for TDMA-based schemes to support simultaneous transmissions	Able to leverage long propagation delay for higher bandwidth utilization	Affected by avail-ability/accuracy of information for scheduling calculation
Cross-layer design	Depend on schemes jointly used in MAC protocol design	Further improved performance, bandwidth utilization implementation and inter-operability issue and energy efficiency	More complex protocol, implementation and inter-operability issue

(continued)

Table 3.1 (continued)

Mechanism/scheme	Basic operation condition	Main strength	Main weakness
Multiplexing-based access schemes			
TDMA	Time synchronization	Relatively simple	Bandwidth waste due to guard-time proportional to propagation delay
CDMA	Orthogonal/uncorrelated code, SYN may be used for optimum multiuser detection	Collision-free simultaneous transmission, specially for data/control frame transmission for multi-channel. High bandwidth utilization and communication security for CDMA	Complex implementation, affected by nonorthogonality and near-far effect
FDMA, incl Multi-channel	Sender-receiver frequency synchronization		Bandwidth waste due to guard-band
			Control channel is a bottle-neck

*Suitable for star-topology

or indicate contest intent (e.g., T-Lohi [12]). The receiver does not need to decode the received signal. This scheme can prioritize nodes in medium access with different signaling lengths, and is applicable to any topologies.

Carrier Sensing (CS)

With carrier sensing (CS), whether a channel is busy or idle is judged in terms of the received signal strength without decoding. It can be used by a node to prevent its attempt from colliding with an ongoing transmission for any topologies. Because it relies on the received signal, its efficiency is affected by signal propagation delay.

CS has been widely used to design MAC protocols, but may not work well in the presence of hidden and exposed terminals as discussed in Sects. 3.1.4 and 3.1.5. Actually, similar consequences can also be caused by the spatio-temporal uncertainty to be discussed for UWANs in Sect. 9.2.1.3.

Messaging (ME)

Messaging (ME) is an action taken by a node to send messages to the relevant nodes. A message is much more informative than signaling provided that the message is successfully received after decoding the received signal. A piggyback scheme can allow the information to be carried incidentally by data frames.

Frame Sensing (FS)

Different from the above carrier sensing, with frame sensing (FS), a node obtains information through decoding the received signal [13]. FS is much more informative than CS, and can be applicable to any topologies. But it is susceptive to propagation delay, message transmission time and channel quality.

Scheduling (SC)

Scheduling (SC) allows a node itself to calculate its transmission time to assure collision-free transmission and/or successful reception as well as its sleeping and wakeup time for energy efficiency. The performance of a scheduling scheme depends on the accuracy of the information used in computing schedules and time synchronization used to enforce the schedule.

3.2.3.2   Sophisticated Mechanisms

There are some sophisticated MAC mechanisms, which combine the basic mechanisms as discussed below [8].

Handshaking (HS)

A handshaking (HS) scheme is a combination of messaging and frame sensing, which allows nodes to exchange messages according to a predefined protocol.

Polling (PO)

A polling (PO) scheme also combines messaging and frame sensing, and only the polled node can access the medium. It can efficiently support QoS and is suitable for a star topology, in which a central node can poll the other nodes.

Reservation (RE)

With reservation (RE), a node submits a request usually via random access for any data transmission. Note that if request submission is scheduled locally, then such MAC protocol is categorized as scheduling. Different from scheduling, the transmission decision here is not made by the node itself. When the decision is made by the receiver, it is a receiver-centric MAC protocol. Reservation can reduce message exchange overhead and delay for large and periodic data transmission.

Prioritization (PR)

A prioritization (PR) scheme aims to prioritize nodes for medium access through, for example, setting either a longer jamming time or a shorter backoff time for higher priority nodes.

Cross-Layer Design

Some communication techniques of the physical layer and information available on other layers such as the network layer and the transport layer are used jointly in MAC protocol design to improve further protocol performance, bandwidth utilization and energy efficiency.

3.3   Categories of MAC Protocols

There are many taxonomies for MAC protocols, and here we categorize them into the multiplexing based and the non-multiplexing based. The former can be further classified according to particular multiplexing schemes, while the latter can be further divided into the coordination-based and contention-based.

3.3.1   Multiplexing-Based Multiple Access Schemes

As mentioned earlier, typical multiplexing technologies include frequency division multiplexing (FDM), time division multiplexing (TDM) and code division multiplexing (CDM). The corresponding multiple access schemes for users to share a medium include Frequency Division Multiple Access (FDMA), Time Division Multiple Access (TDMA) and Code Division Multiple Access (CDMA). A MAC protocol based on multiple access schemes can simplify its design because they can facilitate the solution to avoid collision caused by signals sent from different users. To this end, each user has to submit a request for a subchannel allocation, and the primary MAC issues include request submission and allocation notice. Table 3.1 summarizes the major characteristics of these multiplexing-based multiple access schemes, each of which is described below [3, 8].

3.3.1.1   Frequency Division Multiple Access (FDMA)

As discussed in [8], with FDMA, each node is assigned with a different subband, which allows interference-free concurrent transmissions without using time synchronization. However, receiver-transmitter frequency synchronization is needed to avoid inter-carrier interference (ICI). A guard band between adjacent subbands is used to tolerate transmission uncertainty as mentioned earlier. With Orthogonal FDMA (OFDMA), the adjacent subbands allocated to different users for transmission overlap in an mutually orthogonal manner similar to OFDM. For both FDMA and OFDMA in UWANs, a small bandwidth channel is more vulnerable than a wide one to frequency-selective fading and multipath propagation effect [14–16]. Furthermore, in a mobile UWAN, non-negligible Doppler effect may cause a large frequency shift at the receiver [17]. All these factors may severely degrade communication quality. More discussion on these issues can be found in Sect. 9.1.

Multichannel

Multichannel is a special case of FDMA. It also divides a channel physically into multiple subchannels for MAC operation, and a guard-band is inserted between each pair of adjacent subchannels. A primary issue is to synchronize a sender-receiver pair to the same subchannel for communication between them. One solution is to use a dedicated control channel (DCC) for the nodes to negotiate the allocation of data subchannels for collision-free communication between them. However, a DCC may become the bottleneck and even one-point-failure of the network. Furthermore, since a DCC cannot be used to transmit data, resulting in low channel utilization. Thus, several non-DCC multichannel schemes have been investigated in the literature such as [18].

Triple Hidden-Terminal

With a single-transceiver, a so-called triple hidden-terminal problem may arise [19]. In this case, a node can work either on the DCC or a DCH but not on both simultaneously. Thus, a node communicating in a DCH cannot learn about the channel assignment undergoing in the DCC. Consequently, it may select a DCH already allocated to another node for new transmission, leading to collision. Similarly, in the case of long propagation delay, a CTS frame carrying a DCH assignment may arrive at a node just after it sent out another CTS frame carrying the same DCH assignment to another node [8].

3.3.1.2   Time Division Multiplex Access (TDMA)

A TDMA-based MAC protocol is mainly responsible for allocating different time slots to each requesting node, and informing them of the allocation results. A star network topology is preferable since the central node can be in charge of request collection, time slot allocation and result notification.

Precise time synchronization is needed to establish a common timing reference for every node in the network to locate the time slots allocated to it. A guard time should be inserted between two adjacent time slots to guarantee collision-free reception. The length of the guard time depends on the propagation delay [20]. For the worst-case consideration, the maximum propagation delay has to be considered. Furthermore, a time margin is further needed between two consecutive time slots to handle the shift in the clocks used by different nodes, which also increases with propagation delay. This time margin can be considered together in setting the guard time [8].

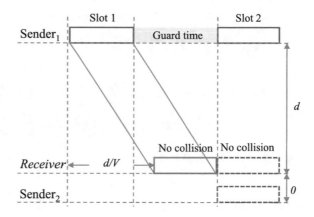

Fig. 3.8 Guard time setting for collision-free reception with TDMA [8]

For guard time setting, as illustrated in Fig. 3.8, the size of the guard time (t_g) is set subject to the propagation delay, where sender 1 transmits at time slot 1, and sender 2 at time slot 2, both to the same receiver. The distance between sender 1 and the receiver is d, while sender 2 is very close to the receiver, and can receive the signal from sender 2 almost immediately. To prevent the signal of sender 2 from interfering with that of sender 1, t_g must satisfy the following condition: $t_g \geq \frac{d}{V}$, which means that the next time slot cannot start until the data transmitted in time slot 1 has reached all nodes to avoid any possible overlaps [20]. For the worst-case consideration, d should be set to the maximum distance between any nodes (d_{max}). Thus, the channel utilization with TDMA will decrease with the network size, and the maximum channel utilization can be calculated by

$$\mu = \frac{t}{t + \frac{d_{max}}{v}}, \tag{3.1}$$

where t is the size of the time slot. Obviously, a large propagation delay (i.e., $\frac{d_{max}}{v}$) yields poor channel utilization [15].

3.3.1.3 Code Division Multiple Access (CDMA)

As mentioned earlier, spread spectrum (SS) technology include frequency-hopping SS (FHSS) and direct-sequence SS (DSSS). Accordingly, CDMA is also classified as FHSS-CDMA and DSSS-CDMA, each of which is discussed more below [3, 8].

As discussed in [21], the wide bandwidth of spread spectrum is robust to frequency-selective fading, resilient to Doppler effect, and can compensate for the multipath effect by using Rake filters. DSSS CDMA with multicarrier transmissions may offer higher spectral efficiency than that with single-carrier. It is shown that in underwater environments, the raw bit error rate (BER) of FHSS is higher than that

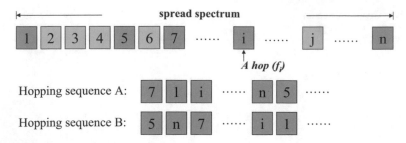

Fig. 3.9 Example of frequency hopping

of DSSS with the same signal-to-noise ratio (SNR) [21].

FHSS-CDMA

Actually, FHSS-CDMA can be regarded as a combination of FDMA and TDMA, i.e., at a particular time slot, every node occupies a different subband, and hops to another subband at the end of the time slot according to a pseudo-random hopping sequence [22]. This hopping sequence of each node is different from each other to prevent them from hopping to the same subband at the same time slot.

Figure 3.9 depicts an example of the frequency hopping scheme. The whole spectrum is divided into n subbands and indexed sequentially. A hopping sequence is just a serial of these subbands. Two hopping sequences need to be orthogonal so that the transmissions following these two sequences never occur in the same frequency and the same slot simultaneously. For example, node A's hopping sequence is "7,1,i......n,5", while that for node B is "5,n,7......i,1", following the same time slot sequence. On time slot 1, node A transmits at subband 7, while node B at subband 5, and so on.

Frequency hopping has the following disadvantages that limit its performance. (i) Guard bands between adjacent subbands wastes bandwidth like FDMA. (ii) Hopping time for subband switching affects network throughput. (iii) If the number of nodes changes frequently, hopping sequence also need to be adjusted accordingly, which is not scalable in the case of a large number of dynamic nodes.

DSSS-CDMA

A spreading code can be either an orthogonal code or a pseudo-noise. With the former, the codes assigned to different users are orthogonal to each other, and the mathematical properties of such codes are exploited as described below. However, the users in such systems need to be coordinated with each other, while uncorrelated to each other with the latter. The spread signals can be decoded with orthogonal codes if the orthogonality of the codes holds for the received signal, and with pseudo-noise if the received signal from each node is at the same level. Thus, power control is used to avoid the near-far effect, with which strong signals cause weak signals to fail in decoding. In a multiuser system, SYN is for optimum multiuser detection [23, 24].

A spreading code (or chip sequence) is an m-bit series. It is used by a node to transmit bit 1, while its negative is used to transmit bit 0 in multiplexing. For example, given an 8-bit chip sequence 01011001, its negative is 10100110. Then bit "1" is transmitted by this 8-bit chip sequence, i.e., 01011001, while 10100110 for the transmission of bit "0". To facilitate mathematical manipulation, the chip sequence is often expressed in form of a vector using "-1" to represent "0". For example, the above chip sequence and its negative can be expressed as follows: $\mathbf{A} = [-1, 1, -1, 1, 1, -1, -1, 1]$ and $\bar{\mathbf{A}} = [1, -1, 1, -1, -1, 1, 1, -1]$, respectively.

With CDMA, the spreading codes for different nodes have to be orthogonal to each other so that the received multiplexed signal can be successfully decoded. Given two m-bit codes in form of vector: $\mathbf{A} = [a_1, a_2, \ldots, a_m]$ and $\mathbf{B} = [b_1, b_2, \ldots, b_m]$, $\mathbf{A}$ and $\mathbf{B}$ are said to be orthogonal to each other if their normalized inner product is equal to zero as formalized as follows:

$$\mathbf{A} \bullet \mathbf{B} = \frac{1}{m} \sum_{i=1}^{m} a_i b_i = 0. \tag{3.2}$$

The following approved properties of the normalized inner product can be used to decode the multiplexed signal if the receiver knows the spreading code used to modulate the signal.

- The normalized inner product of a chip sequence and itself is equal to 1, i.e.,

$$\mathbf{A} \bullet \mathbf{A} = \frac{1}{m} \sum_{i=1}^{m} a_i^2 = 1. \tag{3.3}$$

- The normalized inner product of a chip sequence and its negative is equal to -1, i.e.,

$$\mathbf{A} \bullet \bar{\mathbf{A}} = \frac{1}{m} \sum_{i=1}^{m} (-1) \times 1 = -1. \tag{3.4}$$

- If $\mathbf{A}$ is orthogonal to $\mathbf{B}$, then $\mathbf{A}$ is also orthogonal to the negative of $\mathbf{B}$ ($\bar{\mathbf{B}}$), i.e.,

$$\mathbf{A} \bullet \bar{\mathbf{B}} = \frac{1}{m} \sum_{i=1}^{m} a_i \times (-b_i) = -\frac{1}{m} \sum_{i=1}^{m} a_i b_i = 0. \tag{3.5}$$

The decoding operation is carried out by the receiver through multiplying the multiplexed signal with the spreading code assigned to the sender. Table 3.2 demonstrates how the above properties can be used by a receiver to decode a multiplexed signal, which is sent from nodes A and B using a spreading code $\mathbf{A}$ and $\mathbf{B}$, respectively.

Table 3.2 Decoding of node A's signal from the multiplexing of A and B

Input signals		Process for decoding signal A from multiplexed signals	Decoded signal A
A	B		
1	1	$\mathbf{A} \bullet (\mathbf{A} + \mathbf{B}) = \mathbf{A} \bullet \mathbf{A} + \mathbf{A} \bullet \mathbf{B}$	$1 + 0 = 1$
1	−1	$\mathbf{A} \bullet (\mathbf{A} + \bar{\mathbf{B}}) = \mathbf{A} \bullet \mathbf{A} + \mathbf{A} \bullet \bar{\mathbf{B}}$	$1 + 0 = 1$
−1	1	$\mathbf{A} \bullet (\bar{\mathbf{A}} + \mathbf{B}) = \mathbf{A} \bullet \bar{\mathbf{A}} + \bar{\mathbf{A}} \bullet \mathbf{B}$	$-1 + 0 = -1$
−1	−1	$\mathbf{A} \bullet (\bar{\mathbf{A}} + \bar{\mathbf{B}}) = \mathbf{A} \bullet \bar{\mathbf{A}} + \mathbf{A} \bullet \bar{\mathbf{B}}$	$-1 + 0 = -1$

3.3.2   Contention-Based Protocols

With a contention-based MAC protocol, neither a central coordinator nor an inter-node coordination is available for medium sharing. Each node should learn how to coordinate each other, and decide locally for medium access in a way similar to the random access and probabilistic access MAC mechanisms discussed above. The major advantages of such protocols (e.g., ALOHA and CSMA as well as their variants) include simple implementation, adaptability and robustness to dynamic networking environments. However, they are inefficient to support QoS with poor protocol performance and low bandwidth utilization.

3.3.3   Coordination-Based Protocols

With a coordination-based MAC protocol, there is a central coordinator or a distributed coordination scheme used to arbitrate medium access. Such kind of MAC protocol can effectively ensure only one transmitter present at one time, and efficiently support QoS, because the medium is fully under control and resource reservation can be made if necessary. However, the MAC coordination operation is costly, and it is difficult to apply such kind of MAC protocol in dynamic topology networks. With a central coordinator, the coordination is simply carried out by the coordinator through polling each node in turn. With a distributed coordination, each node is equally involved in the coordination process for medium sharing. More discussions are given below.

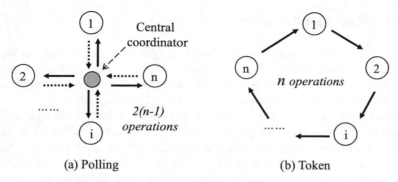

(a) Polling   (b) Token

Fig. 3.10   Polling versus token-passing

3.3.3.1   Central Coordination

With central coordination, a fixed node coordinates medium sharing among all competing nodes. To this end, a node needs to obtain a medium allocation for its transmission, which is granted by the coordinator. This grant contains information such
as when and for how long a node can use the medium. This kind of protocol is simple
and efficient for QoS support, and more suitable for a star or a tree topology network.
However, the coordinator may become the bottleneck or even an one-point-failure
of the whole network.

The coordination can be carried out based on rotation or on-demand. With rotation,
the coordinator grants medium access to each node one-by-one according to a pre-
defined sequence, e.g., polling. This kind of coordination is simple, and can provide
equal opportunity to each node to access the medium. However, it may cause waste
when a polled node has no traffic to send. As illustrated in Fig. 3.10, polling may
cause lots of ping-pong operation between the coordinator and nodes. For one round
polling, the number of such operation is $2(n - 1)$, where n is the number of nodes
in the network. Therefore, the coordinator works much more than the other nodes.

With an on-demand scheme, a node needs to submit a request first to the coordinator that will grant the request if there is enough bandwidth to satisfy the request.
Note that a small portion of bandwidth needs to be used by each node to submit
its request, which is usually conducted through random access. An example is the
Packet Reservation Multiple Access (PRMA) protocol [9], which is a centralized and
slotted MAC scheme, providing a mechanism for slot reservation at the "talkspurt"
level for voice and data applications. It stipulates that the first frame of the contention
winner makes the reservation for the subsequent frames. PRMA was originally proposed for WLANs using a base station to coordinate MAC operation. It is extended
for MANETs in [25].

3.3.3.2   Distributed Coordination

With distributed coordination, every node is equally involved in coordination process for medium sharing without the bottleneck problem of central coordination. It is suitable for distributed topology networks. The coordination can be carried out by using a token or a pre-defined sharing sequence as described below.

With token-passing, a token circulates among the nodes that are sharing a medium, which can only be accessed by the token holder. As illustrated in Fig. 3.10b, token-passing operates in a fully distributed manner following a token-passing structure such as token-ring or token-bus, in which all nodes function identically. However, one node fails in operation will affect the whole network. Furthermore, token loss severely degrades network performance, and may happen frequently in unreliable wireless networks especially with highly dynamic topologies such as MANETs, because node mobility may cause more token losses.

Typical examples include IEEE 802.4 for a ring topology LAN (called token-ring) and IEEE 802.5 for a bus topology LAN (called token-bus). The token-passing sequence determines the sequence for the nodes to access a medium, and can be changed if necessary, but frequent changes will take time.

3.4   Summary

This chapter discusses the primary issues on MAC protocol design for terrestrial wireless networks. Since the MAC protocol is one of the key networking primitives for shared-medium networks, so many protocols have been proposed for different types of wireless networks in the literature. An abstracted structure of various MAC protocols, called MAC reference model, is defined to facilitate understanding of MAC protocols for terrestrial wireless networks and especially for UWANs. This will be described in later chapters. Please also refer to Refs. [26–29] for more reading on these MAC design issues.

References

1. Chandra, A., Gummalla, V., Limb, J.O.: Wireless medium access control - a review (1999). http://www.gatech.edu/fac/John.Limb/
2. Walke, B.H.: Mobile Radio Networks: Networking, Protocols and Traffic Performance, 2nd edn. Wiley, New York (2002). ISBN 0-471-49902-1
3. Jiang, S.M.: A Reference model for MAC protocols in underwater acoustic networks (extended abstract). In: Proceedings ACM International Conference Underwater Networks & Systems (WUWNet). Arlington, VA, USA (2015)
4. Karn, P.: MACA - a new channel access method for packet radio. In: Proceedings of the 9th Computer Networking Conference. London, Ontario Canada (1990)
5. Fullmer, C.L., Garcia-Luna-Aceves, J.J.: Floor acquisition multiple access (FAMA) for packet-radio networks. In: Proceedings of ACM SIGCOMM. Cambridge, USA (1995)

6. Etoh, M., Ohya, T., Nakayama, Y.: Energy consumption issues on mobile network systems. In: Proceedings of IEEE International Symposium on Application & the Internet (SAINT), pp. 365–368. Turku, Finland (2008)
7. Baliga, J., Ayre, R., Hinton, K., Tucker, R.S.: Energy consumption in wired and wireless access networks. IEEE Commun. Mag. **49**(6), 70–77 (2011)
8. Jiang, S.M.: State-of-the-art medium access control (MAC) protocols for underwater acoustic networks: a survey based on a mac reference model. IEEE Commun. Surv. Tutor. **20**(1) (2018)
9. Goodman, D.J., Valenzuela, R.A., Gayliard, K.T., Ramamurthi, B.: Packet reservation multiple access for local wireless communications. IEEE Trans. Commun. **37**(8), 885–890 (1989)
10. Yu, Y., Giannakis, G.B.: High-throughput random access using successive interference cancellation in a tree algorithm. IEEE Trans. Inform. Theory **53**(12), 4628–4639 (2007)
11. European Telecommunication System Inst (ETSI), *Radio Equipment and Systems (RES): High Performance Radio Local Area Network (HIPERLAN) Type 1; Functional Specification ETS 350 652 ed* (1996)
12. Syed, A.A., Ye, W., Heidemann, J.: T-Lohi: A new class of MAC protocols for underwater acoustic sensor networks. In: Proceedings of IEEE INFOCOM, pp. 789–797. Phoenix, Arizona, USA (2008)
13. Fullmer, C.L., Garcia-Luna-Aceves, J.J.: Solutions to hidden terminal problems in wireless networks. In: Proceedings of ACM SIGCOMM. Cannes, France (1997)
14. Sozer, E.M., Stojanovic, M., Proakis, J.G.: Underwater acoustic networks. IEEE J. Ocean. Eng. **25**(1), 72–83 (2000)
15. Pompili, D., Akyildiz, I.F.: Overview of networking protocols for underwater wireless communications. IEEE Commun. Mag. 97–102 (2009)
16. Melodia, T., Kulhandjian, H., Kuo, L.-C., Demirors, E.: Advances in underwater acoustic networking. In: Basagni, S., Conti, M., Giordano, S., Stojmenovic, I. (eds.) Mobile Ad Hoc Networking: The Cutting Edge Directions, chapter 23, pp. 804–852. Wiley-IEEE Press (2013)
17. Stojanovic, M.: Underwater acoustic communications: design considerations on the physical layer. In: Proceedings of Annual Conference Wireless on Demand Network System & Services (WONS). Garmisch-Partenkirchen (2008)
18. You, L.N., Jiang, S.M., Wei, G.: A multi-channel mac using no dedicated control channels for wireless mesh networks. In: Prof. International Conference on Wireless Communication and Signaling Processing (WCSP), pp. 2996–3000. Nanjing, China (2009)
19. So, J., Vaidya, N.: Multi-channel mac for ad hoc networks: handling multi-channel hidden terminals using a single transceiver. In: Proceedings of Annual ACM International Conference on Mobile Computing & Network (MobiCom), pp. 222–233. Philadelphia, USA (2004)
20. Proakis, J.G., Sozer, E.M., Rice, J.A., Stojanovic, M.: Shallow water acoustic networks. IEEE Commun. Mag. **39**(11), 114–119 (2001)
21. Freitag, L., Stojanovic, M., Singh, S., Johnson, M.: Analysis of channel effects on direct-sequence and frequency-hopped spread-spectrum acoustic communication. IEEE J. Ocean. Eng. **26**(4), 586–93 (2001)
22. Sklar, B.: Digital Communications: Fundamentals and Applications, 2nd edn. Prentice-Hall, Upper Saddle River (2002). ISBN 7-5053-7870-8
23. Chang, C.M., Chen, K.: Multiuser synchronization. In: Proceedings of IEEE Symposium Personal, Indoor & Mobile Radio Communication (PIMRC), vol. 2, pp. 1090–1095. London, UK (2002)
24. Kodithuwakku, J., Letzepis, N., McKilliam, R., Grant, A.J.: Decoder-assisted timing synchronization in multiuser CDMA systems. IEEE Trans. Commun. **62**(6), 2061–2071 (2014)
25. Jiang, S.M., Rao, J.Q., He, D.J., Ling, X.H., Ko, C.C.: A simple distributed PRMA for MANETs. IEEE Trans. Veh. Tech. **51**(2), 293–305 (2002)
26. Kumara, S., Raghavanb, V.S., Deng, J.: Medium access control protocols for ad hoc wireless networks: a survey. Ad Hoc Netw. **4**(3), 326–358 (2006)
27. Demirkol, I., Ersoy, C., Alagoz, F.: MAC protocols for wireless sensor networks: a survey. IEEE Commun. Mag. **44**(4), 115–121 (2006)

28. Menouar, H., Filali, F., Lenardi, M.: A survey and qualitative analysis of MAC protocols for vehicular ad hoc networks. IEEE Wirel. Commun. Mag. **13**(5), 30–35 (2006)
29. Kuntz, R., Gallais, A., Noel, T.: Medium access control facing the reality of WSN deployments. ACM SIGCOMM Comp. Commun. Rev. (CCR) **39**, 22–27 (2009)

Chapter 4
MAC Protocols for RWNs

Abstract This chapter focuses on the principles of typical MAC protocols and standards for RF-based wireless terrestrial networks.

4.1 Overview

The protocols to be discussed are listed in Table 4.1, which describes them in the context of the MAC reference model described in Chap. 3. Most of these protocols are not multiplexing-based. As mentioned earlier, multiplexing-based MAC protocols are relatively simple, and typically applied in mobile cellular networks as access networks. Their particulars are well documented in the literature [1].

4.2 ALOHA

ALOHA is a well-known MAC protocol invented in 1968. In the following, we describe this protocol and an analysis on protocol performance for both ALOHA and slotted ALOHA.

4.2.1 Protocols

As illustrated in Fig. 4.1, with ALOHA, a node can send a MAC frame whenever it is available. If a collision happens, the collided frame will be retransmitted after a random backoff period in order to avoid a second collision. Different from wired networks, in wireless networks, a sender cannot detect whether its transmission is collided in the course of transmission. The sender can only infer a collision situation after the transmission with an acknowledgement (ACK) scheme. That is, if a node successfully receives a MAC frame destined for it, it returns an ACK frame to the

© Springer Nature Singapore Pte Ltd. 2018

S. Jiang, *Wireless Networking Principles: From Terrestrial to Underwater Acoustic*,

https://doi.org/10.1007/978-981-10-7775-3_4

Table 4.1 Typical MAC protocols in the context of the MAC reference model (Fig. 3.6) [2]

Typical MAC	Topology	MAC cycle			Access units	Basic MAC mechanisms							
		GF	LF	SF		FA	SA	SI	CS	ME	FS	SC	BO
ALOHA		✓			MF	✓							
S-ALOHA		✓			SL		✓						
CSMA		✓			MF				✓				
CSMA/CD†		✓			MF			✓	✓				
CSMA/CA		✓			MF				✓				✓
BTMA‡		✓			MF			✓	✓				✓
MACA		✓			MF	✓			✓	✓			✓
FAMA		✓			MF	✓			✓	✓			✓
802.11 DCF		✓			MF			✓	✓	✓			✓
802.11 PCF	ST			✓	MF			✓	✓	✓			✓
802.15.4	ST			✓	MF, SL			✓	✓	✓			✓
HIPERLAN		✓			MF			✓	✓	✓	✓		✓

†CSMA/CD is not suitable for wireless network environments
‡BTMA is a multichannel protocol

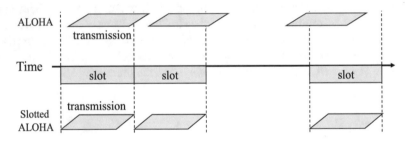

Fig. 4.1 ALOHA versus slotted ALOHA

sender. If the sender has not received the ACK after a timeout, it will retransmit the same frame, subject to a predefined maximum number of retrials.

The major advantage of ALOHA is simplicity, whereas its bandwidth utilization is low because a new transmission attempt may easily collide with an ongoing transmission. To reduce such collision, a slotted ALOHA is proposed. It divides the time into equal-sized slots as illustrated in Fig. 4.1. In this case, a new transmission is allowed to start only at the beginning of a new slot in order to prevent an ongoing transmission from being interfered by a new transmission attempt. It can be shown analytically below that slotted-ALOHA can double the throughput of ALOHA. However, collision may still happen if multiple nodes start their transmissions simultaneously at the beginning of the same new slot.

Fig. 4.2 Diagram for
ALOHA modeling [1]

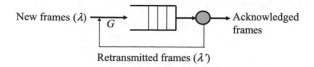

4.2.2 Performance Analysis

The major parameters for the MAC protocol analysis [3, 4] are defined below.

- Average frame arrival rate (G) is measured by the number of frames arriving to a node per time unit, and equal to the sum of new frame arrival rate (λ) and the frame retransmission rate (λ'), i.e., $G = \lambda + \lambda'$, as illustrated in Fig. 4.2.
- Frame length (l) is the number of bits in a MAC frame.

The following assumptions are adopted in the analysis:

- Propagation delay (a) is negligible in comparison with the frame transmission time (f). If the time unit is normalized with per-frame transmission time, then $f = 1$.
- MAC frames sent by different nodes have the same length. For slotted ALOHA, the frame length is equal to the slot length.
- An infinite number of nodes generate MAC frames following a Poisson process with mean λ. This assumption guarantees that there are always frames to be transmitted. If $\lambda \geq 1$, almost every sent frame will collide with each other. Therefore, we should have $\lambda < 1$ if not specified otherwise.
- Probability for there are k transmission attempts for both new frames and retransmitted frames during a time period t also follows a Poisson process with mean tG, i.e.,

$$p(k) = \frac{(tG)^k e^{-tG}}{k!}, \tag{4.1}$$

In the case of low traffic load, almost no retransmission occurs, then $G \approx \lambda$.

4.2.2.1 System Throughput

System throughput (S) is the average number of MAC frames transmitted successfully per time unit in the network. To determine G, we need to find the situation in which an ongoing frame transmission will not suffer from any collision. As illustrated in Fig. 4.3, this situation should satisfy the following two conditions: (i) during the whole period of the current frame transmission, there is not any new frame transmission attempt; and (ii) there should be no other frame transmissions initiated during the frame period preceding the current frame; otherwise, the end part of the preceding transmission collides with the beginning of the current one. In this case, the total vulnerable period for a successful frame transmission is two frames long. In other words, if there are no frame arrivals during this vulnerable period, the current

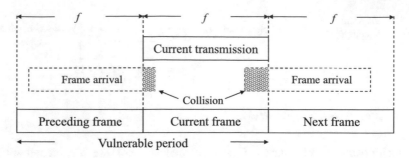

Fig. 4.3 Condition for a successful frame transmission with ALOHA

frame transmission will succeed. Following (4.1), the probability for this successful transmission is given by $p(0) = e^{-2fG}$. Then

$$S = Gp(0) = Ge^{-2fG}. \tag{4.2}$$

Letting $\frac{dS}{dG} = 0$, we can get the peak of S at $G = 0.5$, resulting in $S = 0.184$. This shows that the channel utilization of ALOHA is very low, and approaches to zero when G is very large.

With slotted ALOHA, a frame arriving during the current slot cannot be transmitted until the next slot begins. This stipulation eliminates collisions between an ongoing transmission and any new attempts arriving during the current slot. However, if there is at least one other frame arriving during the preceding slot, their transmissions starting from the beginning of the current slot will collide with the current frame transmission. Thus, this period of one-slot length is just the vulnerable period for slotted ALOHA, which is a half of that for ALOHA. Therefore, $p(0) = e^{-fG}$, and accordingly,

$$S = Ge^{-fG}. \tag{4.3}$$

Similarly, letting $\frac{dS}{dG} = 0$, we can have the peak of S at $G = 1$, resulting in $S = 0.368$, which is double that of ALOHA. However, the channel utilization of slotted ALOHA is still low, and also approaches to zero too when G is very large.

4.2.2.2   Medium Access Delay

Medium access delay here (D) is the time elapsed from when a MAC frame is transmitted to when it arrives successfully at the receiver. This delay can be divided into two parts for calculation: average retransmission delay and the last successful transmission plus the propagation delay, as discussed below.

Given S and G, the average number of retransmissions for a successfully acknowledged frame, n, is given by

$$n = \frac{G}{S} - 1. \tag{4.4}$$

The delay for once retransmission, R, consists of the transmission time of a frame (f), a round-trip time (RTT) ($2a$: one for the transmitted frame and the other for the acknowledgement), the transmission time of an acknowledgement frame (α), and the average retransmission processing delay (i.e., mean backoff time, δ). Then we have

$$R = f + 2a + \alpha + \delta, \qquad (4.5)$$

where the time for the last successful transmission plus the propagation delay is $f + a$ without considering the acknowledgement delay. Thus, D is given by

$$D = nR + f + a. \qquad (4.6)$$

Given D, we can have per-node effective throughput as $\frac{l}{D}$, where l denotes the frame length in bits.

4.3 Carrier Sensing Multiple Access (CSMA)

CSMA is a well-known MAC protocol, and widely applied in practice such as Ethernet and WiFi. In the following, we describe the protocol and then introduce an analysis on protocol performance for both non-persistent CSMA and slotted non-persistent CSMA.

4.3.1 Protocols

Carrier sensing has been adopted by many MAC protocols to detect ongoing transmission to prevent collision between a new transmission and an ongoing transmission, and CSMA is a well-known example. As illustrated in Fig. 4.4, a node planning to transmit first checks the channel to find whether there is an ongoing transmission, which is also called "listen-before-transmit". If a carrier is detected, the planned transmission is delayed. In this case, what a node to do create several variants of CSMA as follows:

- 1-persistent CSMA: The node continues detecting carrier until the channel becomes idle.
- Non-persistent CSMA: The node reschedules a late carrier sensing operation.
- p-persistent CSMA: The node continues detecting carrier like 1-persistent CSMA with probability p, and does not like non-persistent CSMA with probability $1 - p$.

When no carrier is detected, according to the behavior of the node at this moment, there are also two variants: transmitting immediately or conducting a random backoff. With the latter, the node has to wait a random time before starting the transmission.

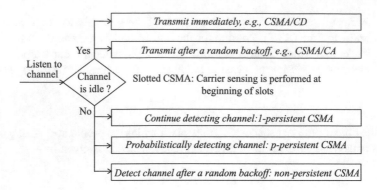

Fig. 4.4 CSMA variants

The time can also be slotted, by which a node can start carrier sensing only at the beginning of each slot. This variant of CSMA is called slotted CSMA.

Note that even if a channel is sensed idle, it is possible for multiple nodes finding the same situation simultaneously, and start their transmissions concurrently, probably leading to collisions. Therefore, collision detection (CD) and collision avoidance (CA) schemes are often jointly used with CSMA, which are called CSMA/CD and CSMA/CA as to be described below.

4.3.1.1   CSMA with Collision Detection (CSMA/CD)

CSMA/CD is an 1-persistent CSMA with immediate transmission when no carrier is sensed, jointly using CD, i.e., "listen-while-talk". It is adopted by the IEEE 802.3 MAC protocol for Ethernet. The CD operation is usually carried out as follows: A sender keeps checking whether there is a collision in the course of transmission. Once a collision is detected, it abandons its transmission, and sends a jamming signal to notify other nodes to stop transmissions. A retransmission is scheduled after a random backoff. For an electrical bus structure, a collision can be detected by comparing the original signal with its echo, and any difference between them means a collision. However, in wireless environments, a transceiver often operates in a half duplex mode, i.e., either transmitting or receiving but not both simultaneously because the transmitted signal is usually much stronger than the received signal. Furthermore, in wireless environments, it is almost impossible to regularly receive echoed signals in an open space, and multipath propagation may complicate the situation. Therefore, it is difficult to apply CSMA/CD in wireless environments.

4.3.1.2   CSMA with Collision Avoidance (CSMA/CA)

CSMA/CA is a popular MAC protocol adopted by many wireless networks. It is a non-persistent CSMA, jointly using CA. To avoid collision, when a channel is

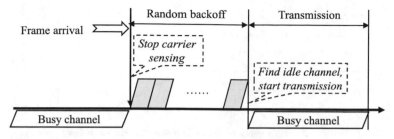

Fig. 4.5 Principle of CSMA/CA (non-persistent CSMA)

detected idle, the node will conduct a random backoff to minimize the probability for two or more nodes to transmit concurrently. Only if the channel is still sensed idle at the end of the backoff period, the node can start its transmission, as illustrated in Fig. 4.5. CSMA/CA is adopted by the IEEE 802.11 MAC protocol as described in Sect. 4.7.1.

4.3.2 Performance Analysis

A detailed analysis on the performances of the above mentioned CSMA variants can be found in [3]. Here, only a brief analysis of non-persistent CSMA is introduced because it is adopted by many wireless networks. The same set of parameter definition and assumptions as used by ALOHA is adopted here.

4.3.2.1 Non-persistent CSMA

As illustrated in Fig. 4.5, if the channel is sensed idle, the node transmits a frame; otherwise, the node defers sensing by a random backoff. At the end of the backoff period, the above procedure is repeated. As illustrated in Fig. 4.6, the time axis of MAC operation can be divided into cycles, each of which consists of a busy period (B) and an idle period (I). According to renewal theory [5], the system throughput (S) can be calculated in terms of channel utilization on average as follows:

$$S = \frac{\bar{U}}{\bar{B} + \bar{I}}, \tag{4.7}$$

where $\bar{B}$ and $\bar{I}$ indicate the average busy period and the average idle period, respectively, and $\bar{U}$ is the average period for a successful frame transmission. Apparently, $\bar{I} = \frac{1}{G}$.

For $\bar{U}$, it is calculated by the production of one frame transmission time (f) and the probability for successful frame transmission. This probability is equal to that

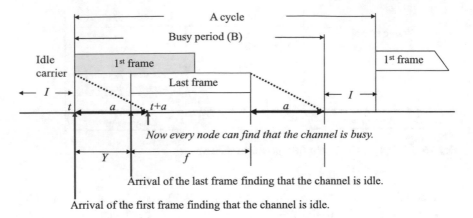

Fig. 4.6 Time division of MAC operation for non-persistent CSMA

for no other transmission to occur during the period between the arrival of the first frame that finds that the channel is idle and when this frame signal can be detected by the others. This is due to that once a node has detected the signal, it will not transmit during a certain period according to the protocol. This period is just equal to the maximum propagation delay between nodes (a). Therefore, this probability is equal to e^{-aG} following Poisson arrival process. So,

$$\bar{U} = fe^{-aG}. \tag{4.8}$$

For the busy period B of the system as illustrated in Fig. 4.6, it starts from the arrival of the first frame that finds the channel is idle (t) to when the last transmitted bit of the last frame also finding the idle channel reaches other nodes. Note that, once a node finds an idle channel, it starts transmission immediately. The delay from the transmission of the last bit and its arrival to other nodes is a maximum propagation delay (a). Thus, $B = Y + f + a$, where Y is the time interval between the arrivals of the first and last frames mentioned above.

For $\bar{B}$, since both a and f are constant, we only need to calculate the average Y. To this end, we use the same symbol Y to indicate the random variable for the arrival time of the last frame between t and $t + a$. After $t + a$, every node will find that the channel is busy so that no more new transmission is initiated according to the protocol. To calculate the expectation of Y, let $F_Y(y)$ be its distribution function, which can be determined as follows:

$$\begin{aligned} F_Y(y) &= \mathbb{P}\{Y \leq y\} \\ &= \mathbb{P}\{\text{No frames arrive between } t + y \text{ and } t + a\}, \end{aligned} \tag{4.9}$$

which just means that no new frames arrive during the remaining period of $a - y$. Thus,

$$F_Y(y) = e^{-(a-y)G}.\tag{4.10}$$

Then we can have

$$\bar{Y} = \int_0^a y e^{-(a-y)G} dy$$

$$= a - \frac{1 - e^{-aG}}{G}.\tag{4.11}$$

So,

$$\bar{B} = \bar{Y} + f + a$$

$$= 2a - \frac{1 - e^{-aG}}{G} + f.\tag{4.12}$$

Then with (4.7) and the formulas found for $\bar{U}$ and $\bar{I}$ discussed earlier, we have

$$S = \frac{f G e^{-aG}}{G(f + 2a) + e^{-aG}}.\tag{4.13}$$

4.3.2.2　Slotted Non-persistent CSMA

In this case, a node can sense a channel only at the beginning of each slot. The node starts its frame transmission if the channel is idle; otherwise, it needs to conduct a random backoff in a number of slots. Here the slot length is set equal to the propagation delay (a) so that the first bit of a transmitted frame will be detected at the beginning of the next slot, which assures no transmission occurring during the next slot. In the case of an infinity number of users generating frames, it is reasonable to have $a > \frac{1}{G}$, and then we can have $\bar{I} = a$.

As illustrated in Fig. 4.7, the maximum B is one slot although the actual busy period is just one frame transmission time, since the remaining period (roughly equal to $a - f$) albeit idle cannot be exploited by any other nodes as regulated the protocol. Thus the maximum cycle in this case consists of two slots.

The average length of the busy period is given by $a(1 - e^{-aG})$ because there should be at least one arrival during the preceding slot; otherwise the current slot will be idle again. The probability for at least one arrival during a slot period is just equal to $1 - e^{-aG}$. Consider $\bar{I} = a$ mentioned earlier, the average cycle length is equal to $a(2 - e^{-aG})$. To have a successful frame transmission (f) in the current slot, it requires that only one frame arrives during the preceding slot, which occurs with probability aGe^{-aG}, hence $\bar{U} = f G e^{-aG}$. Therefore, we have

$$S = \frac{f G e^{-aG}}{a(2 - e^{-aG})}.\tag{4.14}$$

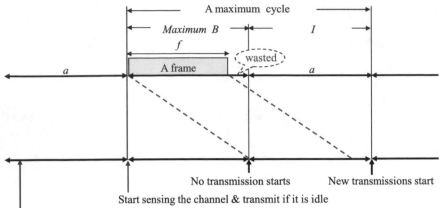

Fig. 4.7 Time division for slotted non-persistent CSMA

4.4 Busy Tone Multiple Access (BTMA)

BTMA is one of the earliest MAC protocols proposed in 1975 [6] to address the hidden terminal problem in wireless networks. It is a multichannel protocol and jointly uses signaling called busy tone. Basically, a channel is divided into two subchannels: one data channel (DCH) and one control channel (CCH), and the protocol operates as follows:

- Any node that hears an ongoing transmission on the DCH will transmit a busy tone via the CCH.
- Any node that hears a busy tone will not initiate any transmission.

As illustrated in Fig. 4.8, a sender (S) is transmitting to a receiver (R) over the DCH. However, this transmission may collide with other transmissions from any neighbors of R, which are often hidden to S. To avoid thus problem, a busy tone transmitted by R via the CCH can tell each node overhearing this tone not to transmit. Accordingly, a large inhibited area can be created by the busy tones sent by the neighbors of S that overhear its transmission over the DCH.

Actually, only the nodes in the inhibited area of node R (i.e., the shadowed area) should not initiate any transmission, while the transmissions of the other nodes are not harmful to the reception at R. Thus, BTMA causes the exposed terminal problem although it can resolve the hidden terminal problem. To solve this problem, a Receiver Initiated BTMA (RI-BTMA) protocol is proposed in [7], with which, a node will transmit a busy tone only after it has identified itself as the intended receiver.

Fig. 4.8 Principle of busy tone multiple access (BTMA) protocol

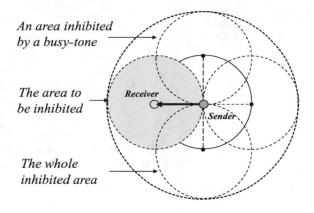

4.5 Multiple Access Collision Avoidance (MACA)

MACA is originally proposed for terrestrial wireless networks in [8], and operates as follows:

- A node sends an RTS frame without carrier sensing.
- The receiver will return a CTS frame if it is ready to receive from the RTS sender.
- A node overhearing an RTS defers long enough so that the RTS sender can receive the returning CTS frame.
- Upon receiving the CTS frame corresponding to the sent RTS frame, this CTS receiver starts transmission.
- Any node overhearing a CTS frame cannot initiate transmission to avoid collision with the returning data transmission.
- Any node overhearing an RTS frame but not a CTS frame can commerce transmission.

The working condition and collision scenarios of MACA are further introduced below.

4.5.1 Operational Conditions

The basic working conditions for MACA are depicted in Fig. 4.9, and explained below. (i) The propagation delay should be negligible. (ii) The sizes of RTS and CTS frames should be small in order to minimize the collision between them. (iii) Both the sent RTS and CTS frames should be overheard by all nodes in the corresponding areas. It is not difficult to understand conditions ii and iii, which can be satisfied. The following two examples show that if condition i cannot be satisfied, collision between RTS frames and data frames may still happen.

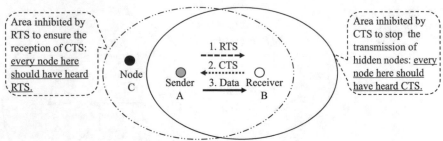

Working conditions: (i) Propagation delay should be negligible, (ii) RTS and CTS
frames should be short enough to minimize collision between them

Fig. 4.9   Basic operational conditions of MACA

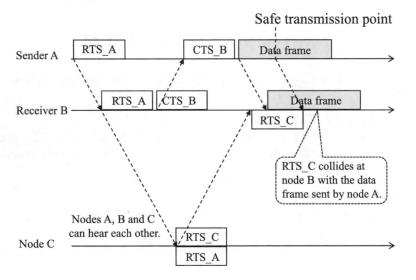

Fig. 4.10   Collision without hidden terminal

4.5.2   Collision Scenarios

Let us first look at the scenario in which all nodes can hear each other, resulting in
no hidden terminals present. However, there is still possible collision as illustrated
in Fig. 4.10, where nodes A, B and C can hear each other. Suppose node A sends
an RTS_A frame to the intended receiver node B, which replies node A an CTS_B
frame. Node A successfully receives CTS_B, and then transmits a data frame to
node B. Due to the propagation delay, it takes some time for RTS_A to reach node C.
Before the arrival of RTS_A at node C, node C may also send an RTS frame to node
B. The extreme case is that an RTS_C is sent out just before the arrival of RTS_A,
as depicted in the figure. This RTS_C will collide at node B with the data frame sent

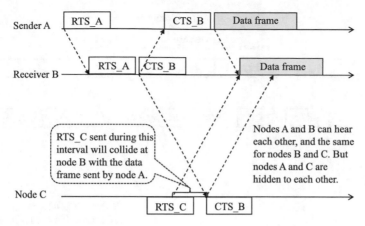

Fig. 4.11 Collision with hidden terminal

by node A. To avoid this situation, node A should defer its data transmission to the safe transmission point as illustrated in Fig. 4.10.

Figure 4.11 depicts another collision scenario with only one difference from Fig. 4.10, i.e., nodes A and C are hidden to each other. In this case, node C cannot overhear RTS_A but CTS_B, so it might send an RTS_C any time before the arrival of CTS_B at node C. However, if an RTS_C frame is sent during the interval as depicted in the figure, RTS_C will collide at node B with the data frame sent by node A.

4.6   Floor Acquisition Multiple Access (FAMA)

FAMA [9] aims to eliminate the hidden terminal problem by guaranteeing that a node having won the channel access will not suffer from collision during its transmission and retransmission, working as follows:

- A node needs to listen to the channel before an RTS transmission. Only when the channel is clear, the node can send the RTS frame. This excludes any RTS transmission during an ongoing transmission.
- The sent CTS frame lasts long enough to jam any transmission from the hidden terminals that have not overheard the sent RTS frame corresponding to this CTS.

As suggested in [9], the length of the CTS is set as long as one round-trip time (RTT) plus one RTS transmission time, as illustrated in Fig. 4.12, where node A first sends an RTS frame (RTS_A) to node B, which replies a CTS frame (CTS_B) accordingly. For other nodes that have not overheard the RTS or CTS, they might transmit RTS frames, such as node C here. The latest transmission of node C, i.e., RTS_C, is sent just before the arrival of CTS_B. The delay between when CTS_B is sent and when

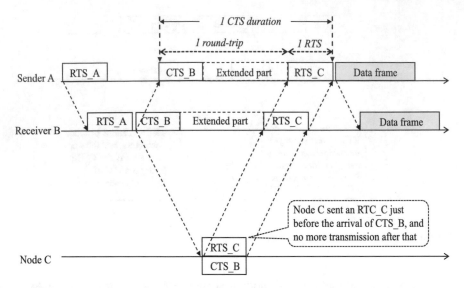

Fig. 4.12 CST setting with FAMA [9]

RTS_C arrives at node B is just one RTT. Such long CTS_B can also assure that node A receives CTS_B successfully. Here, the transition time between receiving and transmitting states in the transceiver is not considered.

4.7 MAC Protocol Standards

This section introduces some MAC protocol standards for wireless personal area networks (WPANs), wireless local area networks (WLANs) and Mobile Ad Hoc Networks (MANETs), which include IEEE 802.11, IEEE 802.15 and HIPERLAN as well as ECMA-368. These standards are based on the MAC protocols discussed earlier, which primarily include CSMA/CA, polling and frequency hopping as well as RTS/CTS handshake.

4.7.1 IEEE 802.11

IEEE 802.11 [10] is a popular WLAN MAC protocol standard used in practice. As illustrated in Fig. 4.13, the MAC protocol stack consists of two functions: distributed coordination function (DCF) and point coordination function (PCF). DCF is the basis of the MAC protocol block, providing contention-based medium access, while PCF provides coordinated medium access. Actually, DCF consists of the CSMA/CA and RTS/CTS handshake protocols discussed earlier. PCF requires an infrastructure

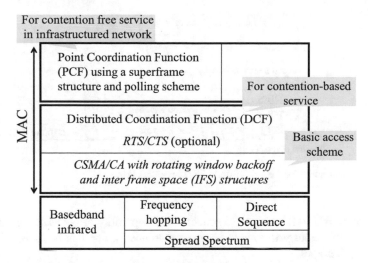

Fig. 4.13 Overview of the IEEE 802.11 MAC protocol stack [10]

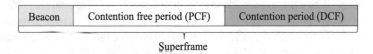

Fig. 4.14 Superframe structure for the co-existing of PCF and DCF modes [10]

called access point (AP) to coordinate medium sharing via a polling scheme. DCF and PCF can coexist simultaneously by using a superframe structure as illustrated in Fig. 4.14, which is divided into two segments: one for the PCF mode and the other for the DCF mode.

4.7.1.1 Inter-frame Space (IFS)

The IFS is defined by IEEE 802.11 to differentiate access priority for different MAC operations by forcing a node to wait an IFS at the end of a busy period before initiating a contention window, as illustrated in Fig. 4.15. Typically, four types of IFSs have been defined for the DCF mode as follows:

- Short IFS (SIFS) is the time interval between frame transmissions during an RTS-CTS-Data-ACK process.
- PCF IFS (PIFS) is the time interval to turn network operation in the PCF mode. That is, after a PIFS, if the medium is still idle, a beacon of the superframe is transmitted to prevent the network from operating in the DCF mode.
- DCF IFS (DIFS) is the time interval for the network operation to run in the DCF mode, i.e., after a DIFS, if the medium is still idle, the network enters the contention access period with a standardized backoff mechanism.

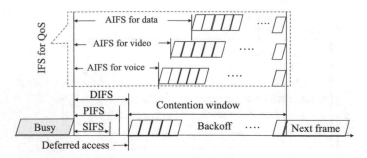

Fig. 4.15 IFS structure for the IEEE 802.11 MAC protocol [10]

- Arbitration IFS (AIFS) is used to enable QoS capability of the MAC protocol by providing different priorities to access a medium for applications with different QoS requirements.

To support the above prioritized medium access, it is defined that $SIPS < PIFS < DIFS < AIFS$. When a preceding frame transmission is in error, the node defers the frame retransmission by an extended IFS (EIFS), which is the sum of the transmission time of an ACK frame at lowest basic rate, an SIFS and a DIFS. To differentiate QoS support for voice, video and data, it is also defined that $AIFS_{voice} < AIFS_{video} < AIFS_{data}$.

4.7.1.2 Standardized CSMA/CA

The standardized operation of CSMA/CA for the DCF mode is described below. Assume nodes A and B are contending for medium access in the case of an ongoing transmission as illustrated in Fig. 4.16, following the steps below.

1. Both nodes A and B first listen to the medium through carrier sensing until the medium becomes idle. Then, they wait a time period defined by DIFS.
2. They further defer their transmissions by a randomly selected backoff period. Assume that the backoff period of node A is smaller than that of node B.
3. At the end of their backoff periods, they need to listen to the medium again. If it is still idle, the node wins the contention, and starts transmission; otherwise, it needs to repeat the contention process from step 1. Note that the backoff period of the new contention is set to the remaining part of the previous one, which is called rotation backoff.

For the example in Fig. 4.16, node A wins the contention since its backoff period is shorter and finds that the medium is still idle at the end of its backoff period. Node B fails since at the end of its backoff period, the medium becomes busy due to the transmission of node A. The backoff period of node B is set to 3 for the new contention.

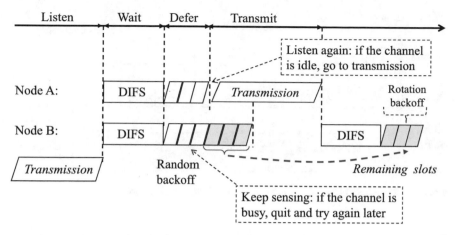

Fig. 4.16 CSMA/CA for the IEEE 802.11 DCF mode [10]

4.7.1.3 RTS/CTS Handshake

This protocol is situated on the top of CSMA/CA, and an optional part of the DCF mode. It is used by a sender-receiver pair before data frame transmission to handle the hidden terminal problem. The RTS and CTS are short and fixed-size signaling frames, and can be transmitted only after the sender wins medium access via CSMA/CA. The handshaking process is depicted in Fig. 4.17 and described below.

1. The sender sends an RTS frame to the receiver, indicating how long the sender is going to occupy the medium in a network allocation vector (NAV), which tells any node overhearing the RTS not to initiate any transmission in any case during this period [10].
2. Upon receiving the RTS, if the sender is ready for reception during the period indicated by the NAV, it returns a CTS frame to the RTS sender after waiting an SIFS. This CTS also indicates how long the CTS sender will not be available for other reception with another NAV.
3. Upon receiving the CTS, the RTS sender sends a data frame after waiting an SIFS.
4. Upon successfully receiving the data frame, the receiver (i.e., the CTS sender) returns an acknowledgement (ACK) to the frame sender (i.e., the RTS sender) after waiting an SIFS.
5. The other nodes overhearing either the RTS or the CTS have to defer transmissions at least by the period indicated by the NAVs plus a DIFS.

4.7.1.4 Comparison of RTS/CTS

A similarity between the RTS/CTS protocols proposed by MACA, FAMA and IEEE 802.11 is that both RTS and CTS contain some information useful for the receivers of

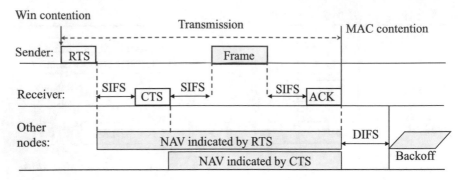

Fig. 4.17   RTS/CTS/DATA/ACK process in the IEEE 802.11 DCF mode [10]

RTS/CTS frames to schedule transmissions, such as the length of a scheduled transmission and reception. A difference between them is that the RTS frames of MACA are transmitted without carrier sensing, whereas this is not the case with FAMA and IEEE 802.11. Such setting of MACA does make sense in the following cases. (i) When hidden terminals exist, lack of carrier does not mean that the transmission can be received without error. (ii) When exposed terminals exist, the presence of a carrier does not always mean that the transmission is harmful to others. With FAMA, RTS frames can be transmitted only after the transmitter finds an idle medium via carrier sensing, and with IEEE 802.11, the more sophisticated CSMA/CA is used to ensure the transmitter to win medium access before transmitting RTS frames.

4.7.2   High Performance Local Area Network (HIPERLAN)

HIPERLAN [11] is a WLAN MAC protocol standardized by the European Union (EU). Similar to IEEE 802.11, this MAC protocol is also based on carrier sensing for collision avoidance. However, it uses an additional jamming mechanism to eliminate early low-priority nodes from a MAC contention so that is more efficient to support QoS.

As illustrated in Fig. 4.18, when a node wants to transmit a frame, it follows the steps below to contend for medium access. Assume that nodes A–D are contending for medium access in the case of an ongoing transmission.

- Each node needs to listen to the channel. Once the channel becomes idle, all nodes enter into the sensing period, whose length is determined by node's priority, with a lower priority node having a longer sensing period.
- At the end of this period, each node goes into the elimination period. That is, if a carrier is sensed by a node, it leaves the competition. In this example, node D leaves since its sensing period is the longest.

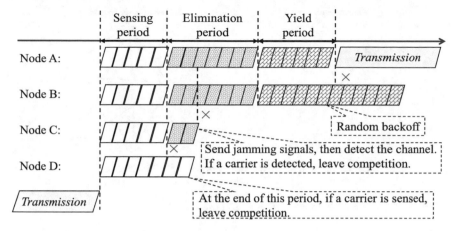

Fig. 4.18 Principle of the HIPERLAN MAC protocol [11]

- If a node has not sensed any carrier at the end of its sensing period, it sends a jamming signal to jam other nodes. The length of a jamming signal is also determined according to node's priority, with a lower priority node transmitting a shorter jamming signal.
- At the end of the jamming period, the remaining nodes enter the yield period. That is, each node detects the channel again. If a carrier is sensed by a node, it leaves the competition. Here, node C leaves since its jamming period is the shortest.
- If a node does not sense a carrier at the end of its jamming period, it needs to conduct a random backoff, similar to IEEE 802.11. At the end of the backoff period, if a carrier is sensed, it leaves the competition; otherwise, it wins the competition. In this example, node A wins the competition.

4.7.3 Wireless Personal Area Networks (WPANs)

There are several MAC protocol standards for WPANs, such as IEEE 802.15.1 for wireless ad hoc networks, IEEE 802.15.4 for data rate WPANs, and ECMA-368 for high data rate WPANs, each of which is described below.

4.7.3.1 IEEE 802.15.1

It is the standard of physical and MAC layer protocols for Bluetooth, which is one of the earliest ad hoc networks and developed for WPANs. As illustrated in Fig. 1.6, a Bluetooth WPAN consists of multiple piconets, each of which consists of a master unit and multiple salve units. A master unit is selected according to a pre-defined scheme, and can create a piconet. The MAC protocol is based on the

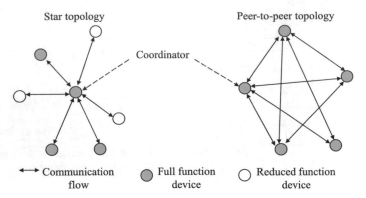

Fig. 4.19 Network topologies for IEEE 802.15.4 [12]

frequency hopping (See Fig. 3.9) and polling schemes described earlier. In a piconet, the master unit and the salve units run in the same hopping sequence, while hopping sequences are different for each piconet. The master unit polls each slave unit in the same piconet for transmission. The transmission is allowed only between a slave unit and the master unit but not between slave units.

4.7.3.2   IEEE 802.15.4

It defines a low data rate WPAN, which is typically applied for information collection and control systems at air data rates of 20, 40 and 250 kbps, respectively [12]. There are two types of devices in such network, namely, full function device (FFD) and reduced function device (RFD). An FFD can relay messages and function as a Personal Area Network (PAN) coordinator, while an RFD cannot be a PAN coordinator. Two network topologies are supported, i.e., star topology and peer-to-peer topology, as illustrated in Fig. 4.19. With a star topology, inter-device communication is coordinated by the coordinator, while in a peer-to-peer topology, devices can communicate directly. In both topologies, a PAN coordinator is used to create the network.

Superframe Structure

As illustrated in Fig. 4.20, the MAC protocol adopts a superframe structure, which starts with a beacon signal transmitted by the coordinator, and is followed by an active portion and an inactive portion. The PAN coordinator only interacts with its PAN members during the active portion, and enters a sleep mode during the inactive portion to reduce energy consumption. A simple active portion is composed of only a contention access period (CAP). Another type of action portion comprises a CAP and a contention free period (CFP), which is divided into guaranteed time slots (GTSs), which are allocated by the coordinator. A GTS can occupy more than one slot period.

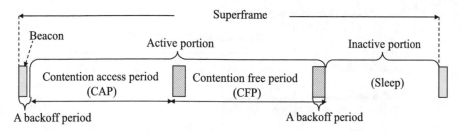

Fig. 4.20 Superframe structure for IEEE 802.15.4 [12]

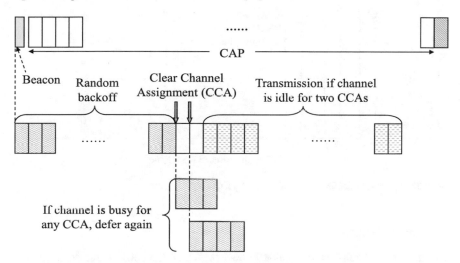

Fig. 4.21 Principle of slotted CSMA/CA in IEEE 802.15.4 [12]

Both CSMA/CA and slotted CSMA/CA are standardized for CAP medium access. With CSMA/CA, a node can sense carrier at anytime, and the backoff time can be measured in any time units. With the slotted CSMA/CA, sensing operation can be carried out only at the beginning of a slot, and the backoff period is measured in slots.

Slotted CSMA/CA

For a beacon-enabled network, the slotted CSMA/CA is applied for CAP medium access. As illustrated in Fig. 4.21, the slots are aligned with the beginning of a beacon transmission. A device needs to locate the boundary of the next backoff slot, and waits a random number of backoff slots. If the channel is busy at the end of the backoff period, the device shall wait another random number of backoff slots before trying again. If the channel is idle and the congestion window (CW) expires, i.e., $CW = 0$, then the device can transmit from the boundary of the next slot. The CW defines the number of backoff periods to be clear before the transmission, and is set to 2 initially. Both ACK and beacon are sent without processing CSMA/CA.

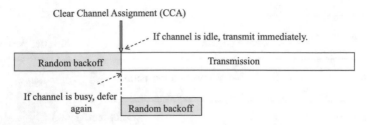

Fig. 4.22 Principle of non-slotted CSMA/CA in IEEE 802.15.4 [12]

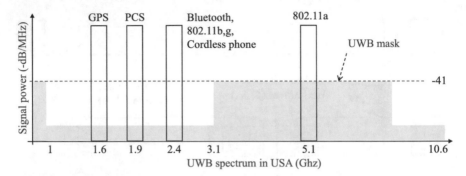

Fig. 4.23 UWB mask defined by FCC of USA [13]

Non-slotted CSMA/CA

It is applied in a non-beacon-enabled network, where the coordinator does not emit regular beacons. As illustrated in Fig. 4.22, each time a device wishing to transmit a data frame or a MAC command shall wait a random period. If the channel is sensed idle at the end of this period, it can start its transmission; otherwise, it needs to wait another random period before trying to access the medium again. The transmission of ACK frames just follows the reception of the frame without processing CSMA/CA.

4.7.3.3 ECMA-368

It defines a high data rate WPAN based on the Ultra-Wide Band (UWB), which is able to transmit data at rates from 54 Mpbs up to 480 Mbps within short ranges. The basic idea of UWB is to reuse spectrums already allocated to other communication systems as illustrated in Fig. 4.23. To this end, the transmission power is strictly constrained in order to limit the interference to those existing communication systems using the same spectrum, leading to short transmission ranges allowable for UWB nodes. Figure 4.23 depicts the mask for UWB transmission power defined by Federal Communications Commission (FCC) of USA, which also suggests that UWB should be mainly used in indoor environments. One typical application of UWB is wireless USB due to its very high transmission rate available in very short transmission ranges.

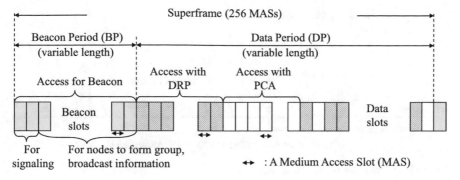

Fig. 4.24 Superframe structure defined by ECMA-368 [13]

The standard ECMA-368 [13] defines the physical layer and MAC layers of UWB-based high data rate WPANs for WiMedia [14]. The entire communication channel is divided into superframes, each of which consists of 256 medium access slots (MASs), as depicted in Fig. 4.24. A superframe consists of a beacon period (BP) and a data period (DP), each of which has a variable length. The BP is the most important and sophisticated component of the MAC protocol standard because it is used for signaling, to create a group and broadcast information as described below.

Superframe Structure

As illustrated in Fig. 4.24, a BP is located at the beginning of each superframe. The first two beacon slots are used for signaling, while the remaining ones are used by a node to join or create a logical group for communicating each other. A node wishing to communicate with other nodes has to join an existing logical group if any; otherwise, it has to create a new beacon group. For both, the node has to occupy a beacon slot in the BP. This beacon slot will be used to broadcast information, which includes (i) a list of neighboring nodes of the information sender used to solve the hidden terminal problem, (ii) the information on data slot reservation and (iii) the availability of data slots in the DP. Data slots in the DP can be accessed by a node through either the distributed reservation protocol (DRP) or the prioritized contention access (PCA). No central coordinator is required for the above MAC operations.

Logical Group

An active node shall transmit a beacon and listen for other beacons during each superframe period. A new node wishing to communicate with other needs to first scan the channel to find out existing logical groups for at least one superframe period before sending anything. If no beacon transmission is received, it just creates a new one by sending its own beacon in the first beacon slot following the signaling slots in the BP. If one or more invalid beacon signals are received, it needs to scan the channel for an additional superframe period. If one or more beacon signals are successfully received, it joins one of them by sending its beacon in an available beacon slot. In this case, the node acts differently according to the following conditions:

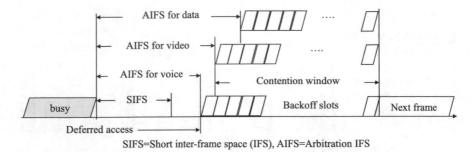

Fig. 4.25   IFS structure defined by ECMA-368 [13]

- If a beacon collision is detected, the node retransmits the same beacon in another available beacon slot.
- If two BPs overlap, i.e., the beginning of the related superframes are not aligned, the node may wait a random number of superframes before sending a beacon in a signaling slot.
- If the beacon slot chosen for beacon transmission is located beyond the BP length of any of its neighbors, the node should also transmit the same beacon in a randomly chosen signaling slot in the BP.

Distributed Reservation Protocol (DRP)

It enables nodes to reserve one or more data slots located at the beginning of the data period in each superframe for transmission. Reservation is made through negotiation, and reservation results are announced in the beacons of the reservation requesters. Reservation negotiation is always initialized by a node wishing to transmit. If the unused reserved data slots cannot be used by any nodes other than their reservation owner, such kind of reservation is called hard reservation. On the contrary, with soft reservation, the unused reserved data slots can be used by any nodes via the PCA.

There are two mechanisms used for negotiation: explicit and implicity. With an explicit mechanism, a node sends an explicit request to the receiver for data slot reservation. Upon receiving a request, the receiver should send back a response to the requester. With an implicit mechanism, a node announces its reservation request in its beacon, and the receiver also includes its response to this request in its beacon. For both, the reservation result is announced in the beacons of both the requester and the receiver. With an explicit negotiation, the reservation request frames and response frames are sent via the DRP if there are reserved slots or via the PCA.

Prioritized Contention Access (PCA)

The remaining part of a data period following the reservation made via the DRP in each superframe is shared following the PCA protocol. Basically, the PCA is similar to the MAC protocol defined by IEEE 802.11e: CSMA/CA with an arbitration IFS (AIFS) for prioritizing nodes as depicted in Fig. 4.15. Differently, neither PIFS nor DIFS used in IEEE 802.11 is adopted here, as illustrated in Fig. 4.25, because the PCF

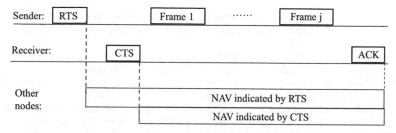

Fig. 4.26 RTS-CTS-frames-ACK handshake defined by ECMA-368 [13]

and DCF have been replaced by DRP and PCA, respectively. Similar to IEEE 802.11, the RTS/CTS and ACK mechanisms are also used here. Additionally, EMCA-368 defines a block acknowledgement (B-ACK) scheme, with which, one MAC ACK can acknowledge multiple received MAC frames, as illustrated in Fig. 4.26.

4.8 Summary

This chapter introduces some typical MAC protocols and standards for wireless local area networks (WLANs) and ad hoc networks. We can find that no MAC protocol is suitable for all types of wireless networks since each type of wireless network has some unique characteristics that affect MAC design. For example, for wireless sensor networks, energy efficiency is a critical issue. More discussion can be found in the literature such as [15, 16]. For satellite networks, the long propagation delay between the satellite and ground nodes raises new challenge to MAC design, which makes the above mentioned MAC protocols not suitable. This issue is similar to the long propagation delay caused by slow acoustic wave speed in UWANs, which will be discussed in Chaps. 9 and 10.

References

1. Walke, B.H.: Mobile Radio Networks: Networking, Protocols and Traffic Performance, 2nd edn. Wiley, New York (2002). ISBN 0-471-49902-1
2. Jiang, S.M.: State-of-the-art medium access control (MAC) protocols for underwater acoustic networks: a survey based on a MAC reference model. IEEE Commun. Surv. Tutor. **20**(1), s (2018)
3. Kleinrock, L., Tobagi, F.A.: Packet switching in radio channels: Part I-carrier sense multiple-access modes and their throughput-delay characteristics. IEEE Trans. Commun. **23**(12), 1400–1416 (1975)
4. Tanenbaum, A.S.: Computer Network, 4th edn. Pearson Education International (2003)
5. Kleinrock, L.: Queueing Systems, volume I: Theory. Wiley, New York (1975)

6. Tobagi, F.A., Kleinrock, L.: Packet switching in radio channels: Part II-the hidden terminal problem in carrier sense multiple access and the busy tone solution. IEEE Trans. Commun. **23**(12), 1417–1433 (1975)
7. Wu, C.S., Li, V.O.K.: Receiver-initiated busy tone multiple access in packet radio networks. ACM SIGCOMM Comp. Commun. Rev. (CCR) **17**(5), 336–342 (1987)
8. Karn, P.: MACA - a new channel access method for packet radio. In: Proceedings of the 9th Computer Networking Conference. London, Ontario Canada (1990)
9. Fullmer, C.L., Garcia-Luna-Aceves, J.J.: Floor acquisition multiple access (FAMA) for packet-radio networks. In: Proceedings of ACM SIGCOMM. Cambridge, USA (1995)
10. IEEE Std 802.11, Medium Access Control (MAC) sub layer and 3 Physical Layer Specifications (1997)
11. European Telecommunication System Inst (ETSI), Radio Equipment and Systems (RES): High Performance Radio Local Area Network (HIPERLAN) Type 1; Functional Specification ETS 350 652 ed (1996)
12. IEEE Std 802.15.4, Wireless Medium Access Control (MAC) and Physical Layer (PHY) Specifications for Low-Rate Wireless Personal Area Networks (WPANs) (2006)
13. ECMA International, Standard ECMA 368: High Rate Ultra Widehband PHY and MAC Standard (2005)
14. WiMedia Alliance, WiMedia Specifications. http://www.wimedia.org/en/index.asp
15. Naik, P., Sivalingam, K.M.: A survey of MAC protocols for sensor networks. In: Karl, H., Willig, A., Wolisz, A. (eds.) Wireless Sensor Networks, pp. 93–107. Kluwer Academic Publishers (2004)
16. Demirkol, I., Ersoy, C., Alagoz, F.: MAC protocols for wireless sensor networks: a survey. IEEE Commun. Mag. **44**(4), 115–121 (2006)

Chapter 5
Routing for RWNs

Abstract This chapter describes fundamental issue about routing protocol design, and typical routing strategies as well as de-facto routing protocol standards for Mobile Ad Hoc Networks (MANETs) reported in the literature.

5.1 Overview

A basic question that routing has to answer is how a router can relay every incoming packet to its destination as quickly as possible in a cost-effective way. The answer to this question is complex, mainly depending on network topologies, routing process, and the information used as well as networking units involved in routing.

5.1.1 When to Relay

Figure 5.1a depicts a wireless network, in which, node A tries to send packets to node D, which is within 1-hop range. In this case, packet delivery can be easily realized by allowing each packet to carry the address of node D because node A's transmissions can be received by all its neighbors due to the broadcast nature of wireless media. A question is how other nodes (e.g., nodes B, C and E) should handle their received packets destined for D. Of course, here these nodes can just drop all packets not destined for them to avoid unnecessary transmission.

However in Fig. 5.1b, packet dropping does harm because the distance between nodes A and D is larger than one hop. Dropping causes node D unable to receive the packet sent by node A. In this case, each packet from node A destined for node D should be relayed first by node B and then node C, while node E can still drop these packets, How can a node be smart enough to distinguish the above mentioned situations to take proper action accordingly? One way is to pre-establish a network path between each source-destination pair for packet delivery to minimize bandwidth waste. For this case, a possible path from A to D is A→B→C→D. However, if node B or C leaves or suddenly powers off, this path becomes invalid immediately.

© Springer Nature Singapore Pte Ltd. 2018
S. Jiang, *Wireless Networking Principles: From Terrestrial to Underwater Acoustic*,
https://doi.org/10.1007/978-981-10-7775-3_5

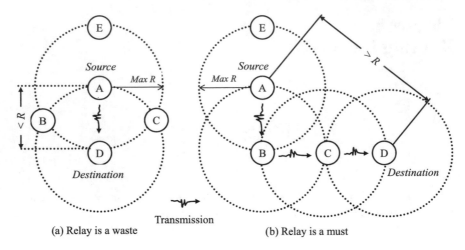

(a) Relay is a waste   Transmission   (b) Relay is a must

Fig. 5.1 Scenarios for packet forwarding in wireless networks

Actually, routing has been studied extensively in wired packet switched networks such as IP, in which, a router has a unicast physical link to each neighbor. Such a physical link is fully under the control of the connected two nodes. Usually a router has multiple input/output ports, each of which connects one upstream/downstream neighbor, respectively. Once a path is established, it is seldom changed due to relatively stable network topologies, but may frequently adjusted in order to adapt dynamic traffic loads.

In the following, we summarize some basic routing schemes developed for wired networks before discussing routing in mobile ad hoc networks.

5.1.2 Implicity and Explicit Routing

Once a packet arrives at a router, it can forward this packet according to a pre-established path (explicit routing) or without any pre-established path (implicit routing).

With implicit routing, only local information or even no information is used for packet forwarding, and no routing table is required by the routing protocol. The primary advantages of such routing include simplicity and small routing overhead as well as robustness. A packet can be forwarded quickly upon its arrival at a router. However, it wastes network resource and causes network congestion because the router does not have an overview on the global traffic load distribution to balance traffic load.

With explicit routing, the packet delivery process consists of two phrases: path establishment and packet relaying. Path establishment aims to establish at least one network path between a source-destination pair, while packet relaying forwards each packet along the established path according to the destination address carried by the packet. The path information can be either stored in a routing table located at the relevant router, or carried directly by each packet travelling along this path. The former is often called table-driven routing, while the latter is called source routing.

Compared to implicit routing, explicit routing is more complex due to path establishment and path maintenance. Particularly for table-driven routing, delay for routing table lookup affects routing performance. This delay primarily depends on search algorithms and table size, which are determined by address space and address structure. Explicit routing can balance traffic load in the network.

5.1.3 Unicast, Multicast and Broadcast

The minimum routing capability of a routing protocol is to support transmission between one source-destination pair, which is called unicast transmission. For multicast transmission, a packet is delivered to multiple destination nodes. With broadcast transmission, each packet has to be delivered to all nodes in a network domain. Although both multicast and broadcast transmissions can be realized through unicast transmission, i.e., the source node simply repeats the transmission of the same packet to all the related nodes, this method is not cost-effective because the same transmission will be repeated many times, which wastes bandwidth. Therefore, efficient routing protocols specifically designed for multicast and broadcast transmissions are necessary.

Actually, broadcast transmission can be realized through flooding without using pre-established paths, with which, a node just copies each received packet and broadcasts it to all its neighbors except the source of the packet (See more in Sect. 5.2.1.1). For multicast transmission, one issue is the expression of multiple destinations for each multicast packet. Obviously, carrying the addresses of multiple destination nodes in each packet is not cost-effective. In a connection-oriented network such as Asynchronous Transfer Mode (ATM), an one-to-many network connection can be set up for multicast transmission through a signaling protocol. In a connectionless network such as IP, a membership forwarding scheme is adopted. That is, all nodes tending to simultaneously receive messages from the same source node make up a membership group, which is stored in the routers of the network. The group identity (e.g., a multicast IP address) is carried in every multicast packet. Once such packet arrives at a router, it looks up the membership table stored locally according to the address carried by the packet for routing.

With the membership approach, it is easy for implicit routing to support multicast transmissions in the last-mile link directly connecting the member nodes of a multicast group, because it does not need a pre-established path for packet forwarding. If the last-mile link is a broadcast medium (e.g., wireless media), the last-mile router can simply broadcast each packet, and each receiving node can filter out those packets not destined for it according to the membership information stored locally. If the last-mile link is a unicast medium (e.g., wireline), the last-mile router needs to replace the membership identity carried by the packet with the destination address of the member. With explicit routing, a path that can reach every member of a multicast group needs to be established first to forward multicast packets. Routing for multicast and broadcast transmission is a complex issue, and more discussion can be found in [1].

5.1.4   Routing Metrics

Routing metrics are used by a routing protocol to make a local routing decision. With implicit routing, an outport is selected for packet forwarding, while with explicit routing, network paths need to be decided. This issue is relative simple with implicit routing because the routing protocol only needs to consider local situations, such as congestion status or link cost available in the router. With explicit routing, the routing protocol needs to consider more issues such as the cost and length of a path crossing the whole network rather than only a link.

In wired networks, a shortest path metric is often used to select paths, i.e., the selected path should be shortest in terms of either time delay or the number of hops that the path will pass. A sophisticate metric can take into account more factors such as link economic cost and even QoS capability. Since changes in network topologies seldom occur in wired networks, route reliability is not a big issue as compared with mobile networks, in which, terminal mobility causes frequent changes in network topologies. In this case, route reliability becomes critical, and should be considered by routing metrics. Furthermore, network security is important especially in mobile networks, and should be regarded as a routing metric too.

5.2   Basic Routing Protocols

Here routing protocols are roughly classified into implicit routing and explicit routing as depicted in Fig. 5.2. The protocols discussed below are initially investigated for wired networks, and constitute the fundamental of many other routing protocols for both wired and wireless networks.

5.2.1   Implicit Routing

Typical implicit routing protocols include flooding, random forwarding and local adaptive forwarding [2].

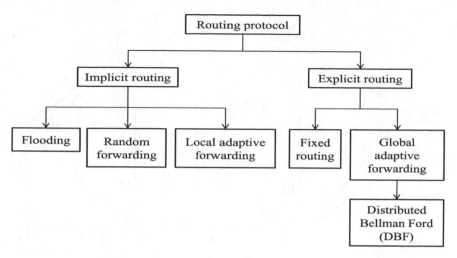

Fig. 5.2 A routing protocol classification

5.2.1.1 Flooding

Flooding simply lets each node retransmit every received packet that is not destined for itself to all its neighbors except the source node that just sent the packet. As illustrated in Fig. 5.3, after a packet from node 1 arrives at node 2, it copies this packet to its neighboring nodes 3, 4 and 5 except node 1. A similar operation will be taken by other nodes that receive this packet. However, one packet may still be retransmitted several times by the same node. For example, if node 3 receives a packet from node 2, node 3 will copy it to node 4, which then may copy it to node 2, and node 2 may copy it to node 3 again. To minimize such extra retransmission, an optimized scheme is to assign each broadcast packet an unique identity (ID), so that the same packet is retransmitted only once by the same node. However, such optimization cannot prevent a node from receiving the same packet several times.

Simplicity and robustness are the primary advantages of flooding because no network information is used, and a packet can always get through if at least one route exists between the source and destination nodes. Furthermore, the shortest path in terms of delay is automatically used. Therefore, flooding is often used by other routing protocols to search routes. However, flooding will cause too much extra traffic because the same packet is retransmitted by every node at least once except the destination node. Therefore, flooding is suitable for small networks with small numbers of nodes. Furthermore, its blind forwarding may lead to security problem if packets themselves are not well protected.

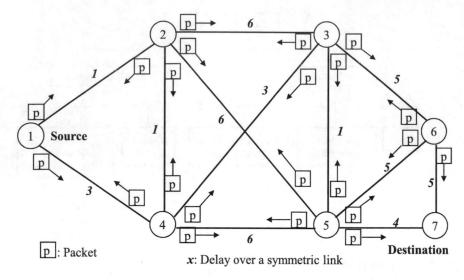

Fig. 5.3 An example of a flooding routing protocol

5.2.1.2 Random Forwarding

To reduce extra traffic caused by flooding, a random forwarding protocol suggests that an outgoing link is selected randomly to forward an incoming packet by excluding the link that the packet comes from. This protocol shares many similarities with flooding except less extra traffic caused by packet forwarding. However, the final route selected by this protocol might not be of least-cost nor minimum-hop.

The probability for a link, say link i, to be selected for packet forwarding, P_i, can be determined according to link capacity. For example, it can be determined according to link transmission rate as follows:

$$P_i = \frac{R_i}{\sum_{j=1}^{n} R_j},$$
(5.1)

where R_j is the transmission rate of link j and n is the total number of links of the router. In this case, a higher rate link will be selected more frequently, which may cause congestion in this node.

5.2.1.3 Local Adaptive Forwarding

To prevent congestion in high-rate links caused by random forwarding, a local adaptive routing protocol suggests a router to select an outgoing link based on some information available locally, such as the queue length corresponding to outgoing links. Let Q_i denote the queue length for outgoing link i and B_i the bias toward

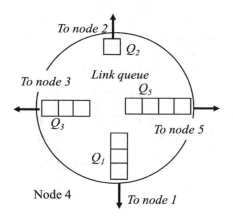

To	B_i	Q_i	B_i+Q_i
1	9	3	12
2	6	1	7
3	2	3	5
5	3	4	7

-Queue length for a link: Q_i
-Direction bias for a link: B_i
-Ideal link gives $min(B_i+Q_i)$

Fig. 5.4 Local adaptive routing in node 4

the link connecting node i. In reality, B_i can be determined according to link cost in a way that the higher B_i, the less frequently link i should be selected for packet forwarding. An ideal outgoing link should give a $min(B_i+Q_i)$.

As illustrated in Fig. 5.4, the bias toward the links connecting the neighbors of node 4, i.e., nodes 1, 2, 3 and 5, are set artificially to 9, 6, 2 and 0, respectively, corresponding to packet forwarding. Suppose that the current queue lengthes for the links to nodes 1, 2, 3 and 5 are 2, 3, 1 and 5, respectively. Then we can find that $(B_3+Q_3)=5$ is the minimum so that the outgoing link connecting node 3 should be selected.

5.2.2 Explicit Routing

The typical explicit routing protocols described here include fixed routing and global adaptive routing such as Distributed Bellman Ford (DBF) routing [2].

5.2.2.1 Fixed Routing

A route is selected for each source-destination pair in a network, and fixed unless there is any change in the network topology. The link cost for route selection cannot take into account dynamic variable such as traffic load. Such routing is neither adaptive to changing traffic load nor suitable for networks with dynamic topologies such as mobile networks. Albeit its simplicity, fixed routing protocols are seldom deployed in practice.

Figure 5.5 illustrates an implementation of such kind of routing protocol for the network depicted in Fig. 5.3. A routing table is used to store the information on routes

Local directory at node 1

	1	2	3	4	5	6	7
1		1	4	2	3	3	5
2	2		4	2	3	3	5
3	2	4		3	3	3	5
4	2	4	4		3	3	5
5	2	4	5	3		5	5
6	2	4	6	3	6		6
7	2	4	5	3	7	7	

Source nodes (columns) — *Destination nodes* (rows) — *Identity of next hop*

A path from nodes 1 to 7 is $1 \rightarrow 2 \rightarrow 4 \rightarrow 3 \rightarrow 5 \rightarrow 7$

Fig. 5.5 Example of a fixed routing implementation

between all node pairs. The first row stores the identities of all source nodes, and the first column stores all destination nodes. Such table is usually implemented in a central directory rather than in each individual node. For example, node 1 only maintains the routing information listed in the second column, where the identities of the neighboring nodes corresponding to the destination nodes are listed. From this table, we can find a route from node 1 to node 7, which is composed by nodes 2, 4, 3 and 5. Once a route is established, it is seldom changed.

5.2.2.2   Global Adaptive Routing

Similar to the local adaptive routing discussed earlier, with global adaptive routing, routing decisions change with network conditions such as traffic load. The difference is that, with global adaptive routing, routing decision is made based on the global information rather than the local information.

Overview

The major factors that may influence route selection include link failure and congestion status. With a link failure, a link can no longer be used as part of a route. Another route should be found, or at least, the broken links should be replaced by other working links. If a link is heavily congested, packets should be re-routed around rather than through this link, and the corresponding routing information has to be changed accordingly.

This kind of routing can improve routing performance and balance traffic load. However, the routing protocol becomes more complex because it requires frequent update of the information on the network status, which will consume extra bandwidth.

Particularly in a network with highly dynamic traffic load and topology, it is difficult and costly to maintain such information.

As mentioned above, a routing protocol adapting only according to local information is rarely used because it does not make a better use of available information such as delays to adjacent nodes. One way to make use of such information is to let each node report the link delay to its neighbors to a central unit, which then constructs routes based on this information, and broadcasts the selected routes to the other nodes. However, the central unit may become a bottleneck and even an one-point failure of the entire network. Another type of adaptive protocol shares the global information in a distributed manner, i.e., each node exchanges with its neighboring nodes the delay information, which is then used to estimate the delay throughout the whole network for route selection. A typical such protocol is the Distributed Bellman Ford (DBF) routing protocol discussed below.

Distributed Bellman Ford (DBF) Protocol

The data structure of DBF consists of delay vector (D_i) for node i and successor node vector (S_i) as defined below:

$$D_i = [d_{i,1}, d_{i,2}, ..., d_{i,n}], \text{ and } S_i = [s_{i,1}, s_{i,2}, ..., s_{i,n}], \tag{5.2}$$

where $d_{i,j}$ is the latest estimation of the minimum delay from node i to node j, $s_{i,j}$ the next hop (NH) corresponding to the route with this minimum delay from nodes i to j, and n is the total number of nodes in the network. Periodically, each node, say node i, exchanges its estimation of D_i with all its neighbors. Upon receiving a new estimation D_i, the receiving node, say node k, updates D_k and S_k to find out a neighboring node h such that

$$d_{k,h} = \min[l_{k,i} + d_{i,j}], i = 1, 2...., n, \tag{5.3}$$

where $l_{k,i}$ is the delay from node k to node i, which can be estimated locally according to the corresponding queue length. Once node h is found, set $s_{k,j} = h$.

Figure 5.6 depicted the routing table and its updating process following the DBF protocol for the network showing in Fig. 5.3. Node 1 has its current delay (D_1) and successor vector (S_1) as illustrated in the table on the left side of the figure. Now node 1 receives the updated distance vectors from nodes 2 and 4, i.e., D_2 and D_4 in the middle of the figure. It also finds new link delays to nodes 2 and 4, i.e., $l_{1,2} = 2$ and $l_{1,4} = 1$, respectively, as illustrated in the right side of the figure. Then it updates the route to destination node 5 as follows: re-calculating the delay from node 1 to node 5 along all possible paths to node 5, each of which will go through one neighboring node, i.e., $l_{1,i} + d_{i,5}$ for $i = 2$ and 4. Finally, node 4 is newly selected as the next hop because the delay via it is 5 shorter than that via node 2, as illustrated in the figure.

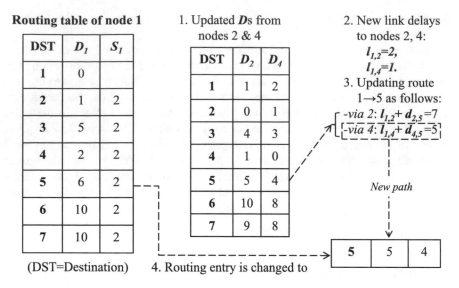

Fig. 5.6 Routing table and updating process in node 1 with DBF

5.2.3   Optimal Broadcast Routing

Broadcast routing is adopted almost by every routing protocol for path search, especially at the initial stage of routing, and it also be used for forwarding data to all nodes in a network such as TV broadcast. The simplest and robust broadcast routing is flooding, which however cause lots of extra transmissions as mentioned earlier, and is not suitable for bandwidth-limited mobile networks. Therefore, optimal broadcast routing is studied to minimize bandwidth consumption.

A reverse path forwarding protocol is investigated in [3]. It aims to help a node to decide whether or not to forward an incoming broadcast packet. An extended reverse path forwarding protocol decides which outgoing links that an incoming broadcast packet should be forwarded to. Their basic idea is that the forwarding of a broadcast packet initiated by a source node should follow the source tree rooted at itself. The generation of a source tree through flooding still yield redundant transmissions. With an established source tree, broadcast packet forwarding is optimal in terms of that a broadcast packet is received only once by each node. These two protocols are discussed below.

5.2.3.1   Shortest Reverse Path Tree

Here, an incoming broadcast packet is forwarded to other links by the receiving node if and only if this packet comes from a link that is a part of the shortest reverse path tree for the source node [3]. Such a tree for a source node can be derived from the shortest

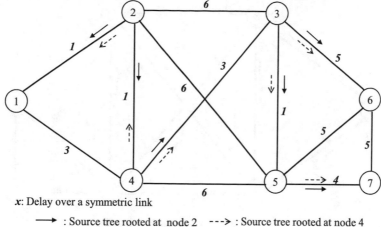

x: Delay over a symmetric link

⟶ : Source tree rooted at node 2   ---➤ : Source tree rooted at node 4

Fig. 5.7 Spanning trees rooted at nodes 2 (solid line) and 4 (dashed line) with the shortest paths to each node

path tree rooted at the source node. Actually, this tree is a spanning tree that covers all nodes in the network with a shortest path between the root node to each leaf node. If any branch is removed, the remaining subtrees of this branch are also disconnected from the tree. According to graph theory, a spanning tree is a tree-like subgraph including all of the vertices of the graph, and there is a unique path between any two vertices [3]. The shortest reverse path tree for a source node consists of the shortest paths from each other node to the source node. With symmetric links, the shortest path tree for a source node and the shortest reverse path tree are isomorphic [3].

A spanning tree with a shortest path to each other node can be generated through flooding initiated by the root node through broadcasting a packet. When this broadcast packet arrives at a node via an incoming link, it copies the packet on all links except the link that the packet comes from. For shortest path generation, only the earliest arriving packet is forwarded by the receiving node. Figure 5.7 demonstrates such process for the network depicted in Fig. 5.3, with two source trees rooted at nodes 2 and 4, respectively.

As to be discussed in Sect. 5.5.4, the above spanning tree is called source tree in Topology Dissemination Based on Reverse-Path Forwarding (TBRPF) protocol [4], which aims to address a challenging issue on how to maintain a source tree in a dynamic topology network such as mobile networks.

5.2.3.2 Extended Reverse Path Forwarding

Figure 5.8 further illustrates how a broadcast packet initiated by source node 2 is forwarded following the reverse path forwarding scheme mentioned above, where the sequence $Pi, j, ..., k$ defines the genealogy of the packet as used in [3]. However,

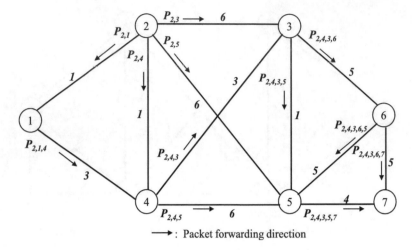

Fig. 5.8 Forwarding of a packet initiated by node 2 with reverse path forwarding

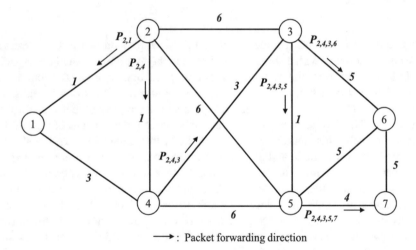

Fig. 5.9 Forwarding of a packet initiated by node 2 with extended reverse path forwarding

there are still some redundant transmissions, such as $P_{2,1,4}$, $P_{2,3}$ and $P_{2,5}$ etc., due to redundant copies of the packet. Thus, the extended reverse path forwarding scheme suggests that an incoming broadcast packet is only copied to an outgoing link that is a branch of the shortest reverse path tree, as illustrated in Fig. 5.9. Such kind of outgoing link can be identified by allowing each node to tell its neighbors from which link it receives the first arrival of the same broadcast packet.

5.3   Routing in Mobile Ad Hoc Network (MANET)

Mobile networks can be roughly classified into infrastructured networks and infrastructureless networks, each with different complexity of their routing protocols. In an infrastructured network, nodes are clustered, and the cluster head coordinates communications. A simple such network consists of only one cluster, which is often used in the last mile as access networks, e.g., mobile cellular networks and wireless local area networks (WLANs). In this case, a base station (BS) or an access point (AP) can be a coordinator. The links between BSs or between APs can be of wireline. A route can be split into wired segments and wireless part. Routing for a wireless part is relatively simple due to available coordinators and the broadcast nature of wireless media. The major routing issue is due to handoff support, which will be discussed in Chap. 7.

In an infrastructureless network, no node can be dedicated to coordinate communication, and each node is functionally identical. A typical infrastructureless mobile network is Mobile Ad Hoc Networks (MANETs). Routing in a small MANET is simple, especially when the distance between any nodes is one hop so that every node can hear each other. In this case, routing is carried out by packet filtering according to the destination address. However, if the distance is larger than one hop, routing becomes complex because relaying is required for successful packet delivery, and establishing a route before sending data packets is required to minimize resource consumption, as discussed earlier in this chapter. The following sections will focus on routing in MANETs.

5.3.1   Challenges

Routing protocols for MANETs have taken into account the following characteristics:

- Terminal mobility is the major factor that affects routing performance because it causes frequent changes in network topology to break established routes.
- Small channel capacity and limited power supply are major characteristics of most mobile networks. These features make it undesirable to use sophisticate routing protocols especially those are energy-hungry and bandwidth-greedy.
- The exposure of wireless media is vulnerable to security attacks, and unreliable links further cause unstable routing performance. These features complicate routing protocol design to achieve better routing performance.

Routing becomes more difficult when changes in network topology happen frequently due to irregular user behaviors, such as random motion of mobile terminals and unpredictable power-on/off operations. In this case, it is costly to maintain accurate global information for routing. Mobility also changes traffic load distribution in the network. These features require routing protocols able to adapt to those changes.

On the other hand, routing in wired networks is address-centric, i.e., each data packet is forwarded according to its destination address. However, in wireless sensor networks (WSNs), routing is data-centric, because no explicit destination is available for routing, which actually aims to find the location of the enquired data.

5.3.2 Loop Avoidance

When DBF is applied directly in MANETs, there may be routing loops because of the following factors. (i) A node determines the next hop of a route based on only delay vector (D) calculated locally, which may be stale and incorrect due to propagation delay and packet loss. (ii) Link delay (l) is a key parameter to calculate new routes; but it may change rapidly so that the local perception of a shortest path could change while a packet is still en route.

5.3.2.1 Example of Routing Loop

Figure 5.10 illustrates an example of loop generation for the network depicted in Fig. 5.3. Consider an existing path from node 1 to node 7, i.e., $1 \rightarrow 2 \rightarrow 4 \rightarrow 3 \rightarrow 5 \rightarrow 7$. The following actions taken by node 3 will cause a routing loop as discussed below.

- Suppose that node 3 observes the following changes in link delay: $l_{3,2}$ is reduced from 6 to 1, while $l_{3,5}$ is increased from 1 to 7. However, it does not receive any update from node 2, which means that $D_{2,5} = 5$ is still valid.
- Node 3 calculates a new shortest path to node 5, which goes through node 2 with a new cost equal to $l_{3,2}+D_{2,5}=1+5=6$, which is smaller than $l_{3,5}=7$.
- Now node 1 is sending a data packet destined for node 7 using the above-mentioned old path because no node has been updated of the new path modified by node 3.
- When the packet arrives at node 2, it still forward the packet to node 4, which then sends it to node 3. However, node 3 follows the new path to forward the packet to node 5 via node 2, generating a routing loop as illustrated in Fig. 5.11, i.e., $2 \rightarrow 4 \rightarrow 3 \rightarrow 2$.

Figure 5.11 illustrates how the routing loop is generated and explained below. A route can be divided into two segments by an intermediate node (e.g., node 3) located between the source and destination nodes, called upstream segment and downstream segment, respectively. The upstream segment starts from the intermediate node all the way up to the source node, while the downstream segment is the remaining part down to the destination node. The cause of the above-mentioned loop is that node 3 in the path fails in assuring its downstream segment free of loop. That is, in the original downstream segment $3 \rightarrow 5 \rightarrow 7$, there is a loop $3 \rightarrow 2 \rightarrow 4 \rightarrow 3$ after node 3 changes its routing to node 5 via node 2.

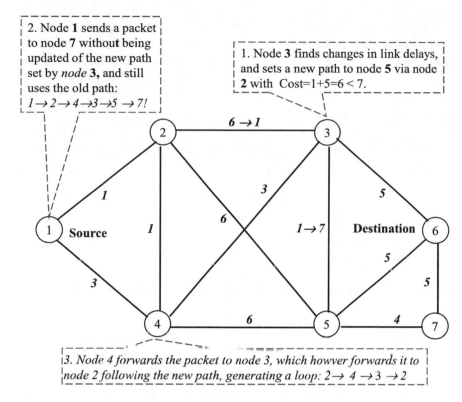

Fig. 5.10 An example of routing loop generation with DBF

5.3.2.2 Destination Sequenced Distance Vector (DSDV)

The DSDV protocol [5] modifies DBF to prevent routing loop by stipulating that the process of updating an existing path can only be initiated by the destination node of the path, and the other nodes can only follow up the process to update the path as explained below.

As illustrated in Fig. 5.11, since the upstream segment is a part of the original path, it is loop-free. If the downstream segment is also free of loop after a change in the path, the entire path will be still loop-free. One way to do this is to let the destination node of a path initiate route updating process. This process is continued hop by hop in the reverse direction of the path such that each intermediate node keeps its downstream segment free of loop. It can be easily proofed that the whole updated path is still free of loop.

DSDV also proposes the following modifications to DBF: (i) using a sequence number (SN) to indicate a new route for loop avoidance, and (ii) delaying the advertisement of unstable routes to reduce fluctuations of routing tables and the number

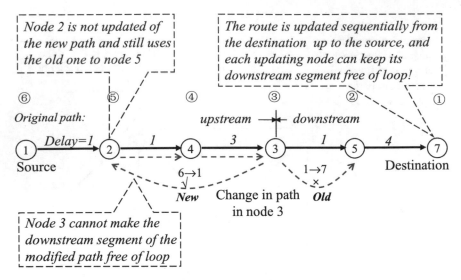

Fig. 5.11   Course of loop generation with DBF and principle of loop avoidance with DSDV

of rebroadcasts of the same route entries. The route updating process is summarized below.

- A path to be updated is assigned a new SN larger than the old one by the destination node, i.e., $1 + SN \rightarrow SN$. The SN is broadcast through a routing update message in a form of [DST,Cost,NH,SN]. The destination node sets DST (destination) to its address, Cost (e.g., link delay) to 0 and NH (next hop) to null, and then broadcasts it.
- Upon receiving an updating message, the receiver adds the cost (c) of its link (e.g., link delay) to Cost, i.e., $c + Cost \rightarrow Cost$, NH is set to the address of the message sender, and then rebroadcasts the updated message.
- A path with the largest SN is always selected first. If there are multiple paths with the same SN, the one with the lowest cost is selected.

The above route updating process may cause a broadcast storm of messages, which will consume too much bandwidth and energy. Even worse is that all expensive effort will end up with null for inactive routes. Furthermore, every node needs to maintain a routing table for the whole network, and update frequently the paths to itself in order to maintain shortest path settings.

5.3.3   Routing Strategies

Many routing protocols have been proposed for MANETs in the literature, which are briefly summarized in Fig. 5.12. There are also several surveys available such

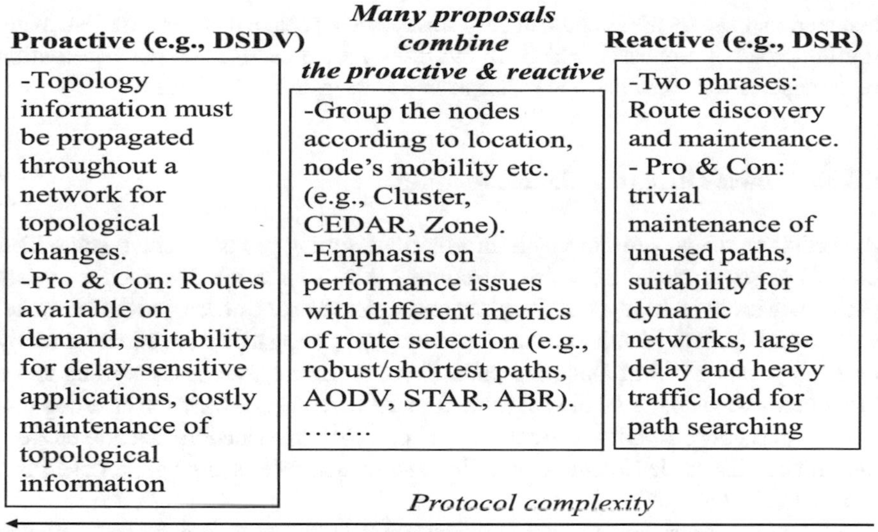

Proactive (e.g., DSDV)	***Many proposals combine the proactive & reactive***	**Reactive (e.g., DSR)**
-Topology information must be propagated throughout a network for topological changes. -Pro & Con: Routes available on demand, suitability for delay-sensitive applications, costly maintenance of topological information	-Group the nodes according to location, node's mobility etc. (e.g., Cluster, CEDAR, Zone). -Emphasis on performance issues with different metrics of route selection (e.g., robust/shortest paths, AODV, STAR, ABR). …….	-Two phrases: Route discovery and maintenance. - Pro & Con: trivial maintenance of unused paths, suitability for dynamic networks, large delay and heavy traffic load for path searching

Protocol complexity

Fig. 5.12 Summary on routing protocols for MANETs

as [6–9]. The majority of routing protocols are explicit routing, and can be further classified from various aspects as discussed below. So far four routing protocols have been rationalized for different types of MANETs by Internet Engineering Task Force (IETF), and one for opportunistic networks by Internet Research Task Force (IRTF), in the form of request for comments (RFC). Please refer to Sect. 5.5 for more detail.

5.3.3.1 Proactive and Reactive Routing

According to the availability of a required path for an arriving packet, a routing protocol can be either proactive or reactive. With a proactive routing, a path is available whenever it is required because each node maintains the routes to each reachable destination node, no matter whether there is a need for the paths. In contrast, with reactive routing, a route is searched on demand. Proactive routing can make an arriving packet not to wait for path search, and is suitable for delay-sensitive applications. However, it requires network topology information to be propagated throughout the whole network periodically or immediately when any change in network topology occurs. This operation may cause lots of traffic overhead, especially in mobile networks. Reactive routing can minimize maintenance cost for less used or unused paths. However, an arriving packet needs to wait for route searching, and heavy traffic load may be generated for instant route searching operations.

As summarized in Fig. 5.12, a typical proactive routing protocol is Ad Hoc On-Demand Distance Vector (AODV) protocol [10], and a typical reactive routing protocol is Dynamic Source Routing (DSR) protocol [11]. Both AODV and DSR have been rationalized by IETF, and will be described below. Note that there are also many routing protocol proposals, which are hybrids of the proactive and reactive routing with emphasis on some particular design issues for specific MANETs.

5.3.3.2   Table-Driven and Source Routing

According to the way to store path information, routing protocols can be classified into source routing and table-driven routing. With source routing, each sent data packet carries in its header the complete and ordered list of nodes which the packet will go through. The identities of the next hops composing a path are carried directly by every packet traveling along this path. These nodes are arranged according to the order in which the packet will visit them, and the first one is always the next hop that the packet is going to visit. Once a node has been visited, its identity is removed from the next hop list carried by the packet. However, it generates a large packet overhead especially for long paths.

With table-driven routing, path information is stored in each node that a packet will visit. That is, a routing table is used by a router to store path information, which typically includes next hop and delay. Routing entries in the table are indexed according to destination addresses. Once a path has been established, the path information is stored in the table, and will be updated when necessarily. To forward an incoming packet, the router looks up the routing table according to the destination address of the packet to find the next hop toward the destination node. The maintenance of routing table is a complex issue.

5.3.3.3   Distance Vector and Link State Routing

According to the way that topology information is used, a routing protocol can be of either distance vector or link state. With a distance vector protocol, a node does not possess information on the entire network topology, and advertises or receives distance information only to or from its neighboring nodes. Differently, a link state protocol computes a shortest path according to the topology information on the entire network. A node can possess the information on the entire network topology and connectivity map because each node sends the information on its neighbours to the entire network. Based on this information, a node can independently calculate the best next hop for every possible destination node in the network.

5.4   Opportunistic Routing

All the above routing protocols assume that at least one route exists between any source-destination pair, and the routing protocol just tries to find it out. However, in opportunistic networks, this assumption does not hold because sometime no route

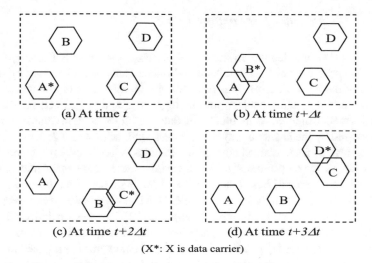

(a) At time t (b) At time $t+\Delta t$

(c) At time $t+2\Delta t$ (d) At time $t+3\Delta t$

(X*: X is data carrier)

Fig. 5.13 Example of opportunistic routing [12]

exists between a source-destination pair when there are data packets to forward. In the following, some related issues are discussed.

5.4.1 Delay and Disruption Tolerant Network (DTN)

A DTN was originally proposed for Interplanetary Internet as depicted in Fig. 1.12, in which a network connection is not always-on for message transfer. Now a DTN refers to any kind of wireless networks with intermittent connectivity leading to excessive delays. Such network typically includes WSNs with intermittent connectivity between sleeping state and waking-up state, satellite networks with periodic connectivity caused by periodic rotation around the earth, and underwater wireless acoustic networks (UWANs) with frequent interruptions due to environmental factors [12].

Connection intermittence is a consequence of changes in network connectivity, which is mainly contributed by either node mobility and dynamics of link quality caused by interference or power exhaustion. Actually, node mobility may cause some nodes engaging in a network connection to create opportunistic routing by occasionally bringing new relaying nodes to cure a disrupted path, or even carrying data to its destination directly, as illustrated in Fig. 5.13. A similar consequence can be brought by upgrading link quality or joining-in of newly active nodes (e.g., turning on power).

5.4.2 DTN on Top of Transport Layer

An overlay DTN architecture on the top of the transport layer is formally defined
in [12]. The protocol unit is called "bundle" rather than the packet used by routing
protocols of the network layer. This protocol unit bundles data of an entire session,
including both data and metadata, into a single message [13]. Each bundle has a
defined format, which contains blocks of data. A block contains the information
usually carried by the header or payload of other types of protocol data units [12]. A
bundle can also be fragmentized into smaller bundles called bundle fragments during
transmission, which may be reassembled again into a new bundle in the middle way to
the destination. The identifiers of the source and destination of a bundle are expressed
syntactically as a Uniform Resource Identifier (URI) [14], which can express various
types of names or addresses with variable lengthes [12]. Bundles are delivered toward
destinations over occasional paths, which are generated by collections of contacts
between meeting nodes [15]. A Postal-Style Delivery scheme is proposed to deliver
bundles, using a similar priority class of the postal system, which consists of Bulk,
Normal and Expedited, to differentiate delivery service. An essential element of
the proposed bundle forwarding structure is that every forwarding node should have
sufficient place to queue bundles so that they can wait for communication opportunity.

A contact is primarily defined by its start time, duration, endpoints, forwarding
capacity, and latency of a link in the topology graph [13]. The following types of
contacts are defined in [12]:

- A persistent contact is an always-on connection, which does not require connection
  initiation action.
- An on-demand contact is a connection that requires some action to set it up, which
  is similar to the connection-oriented network discussed earlier.
- A scheduled contact will be available at the scheduled time points lasting for a
  particular duration, such as a link with a non-GEO satellite.
- A predicted contact can be predictable statistically following certain patterns based
  on historic information.
- An opportunistic contact is an unexpected meeting without a prior information
  about its presence.

The last three types of contacts show intermittent connectivity, and only the opportunistic contact has random intermittence.

5.4.3 Challenges

Actually, an intermittent connection can also be found in other mobile networks,
such as the mobile cellular network when a mobile node is moving out the coverage
of the currently serving base station or moving into a signal hole. A major difference
between these networks and DTNs is the time scale that the network can have to

handle the link intermittence, which largely depends on the application running over the network. In a mobile cellular network, the link intermittence is handled by handoff mechanisms (See Chap. 7), which must be completed in very short time for voice application such that the user will not feel any disruption for the communication in progress; otherwise, the voice call will be dropped. To this end, mobile cellular networks are designed carefully and deployed accordingly for good service coverage. On the other hand, a DTN cannot be designed artificially or controlled completely, and the applications running over it have to be able to tolerate long delay caused by intermittent network connections. For example, the transfer of large volume of non-delay-sensitive data (e.g., product statistic data) can allow long processing time, whereas the inter-planet communication cannot find a faster way due to the inherent distances between the source and the destination.

In traditional wired and wireless networks, where there is always at least one network path between any source-destination pair, the routing protocol is used to find the best path between them. Then, the source node forwards all packets along the established path. If the path is disrupted, re-routing is invoked, and the intermediate nodes will drop all packets that cannot be forwarded along the disrupted path. With opportunistic routing, once a node receives data (Here "data" refers to either packets or bundles) that cannot be delivered to the destination directly, it stores them, and looks for an opportunity to forward them to other nodes that might enable the data to reach or be closer to the destination. The primary issue for opportunistic routing is to maximize successful data delivery to its destination. To this end, flooding is still the simplest and the most robust routing method in this case, but at cost of consuming more precious wireless bandwidth and energy. Therefore, an optimal tradeoff should be found between the maximal successful delivery and the minimum network resource consumption in bandwidth and energy as well as buffer space.

To maximize successful delivery, nodes in DTNs should exploit the information on mobility pattern to schedule transmission and make routing decisions because in many practical cases, mobility is not completely random; instead it can be predicated statistically, especially for large mobile nodes. For example, a satellite always periodically rotates around the earth in a fixed earth orbit. Air planes, trains and various types of vehicles (e.g., cars and ships etc.) normally move along the pre-established roads. Only small mobile nodes, such as the human and small boats, show random mobility pattern. However, the capacity of small mobile nodes in terms of communication and energy capacity is usually much smaller than that of large mobile nodes. Thus, such non-randomness of mobility pattern should be exploited to maximize data delivery.

There are many opportunistic routing proposals for DTNs in the literature, such as, Epidemic routing [16], Probabilistic Routing Protocol using History of Encounters and Transitivity (PRoPHET) [13], Spray and Wait [17] and MaxProp [18]. Several surveys can also be found in the literature such as [19, 20]. A comprehensive survey can be found in [21]. Since PRoPHET is a RFC of Internet Research Task Force (IRTF) [15], we give more description on it in Sect. 5.5.5.

5.5   De-Facto Standards for Routing Protocols

Four routing protocols for MANETs are approved by IETF between 2003 and 2007, namely, Ad hoc On-Demand Distance Vector (AODV, RFC 3561) [10], Optimized Link State Routing Protocol (OLSR, RFC 3626) [22], Topology Dissemination Based on Reverse-Path Forwarding (TBRPF, RFC 3684) [23] and Dynamic Source Routing (DSR, RFC 4728) [11]. Recently, a routing protocol for DTNs, called Probabilistic Routing Protocol using History of Encounters and Transitivity (PRoPHET, RFC 6693) [15], is also rationalized by Internet Research Task Force (IRTF). Their principles are introduced below.

5.5.1   Dynamic Source Routing (DSR)

The DSR protocol [11] uses explicit routing to allow a sender to select and control loop-free routes for packet forwarding. It consists of two mechanisms: route discovery (RD) and route maintenance (RM).

5.5.1.1   Route Discovery (RD)

It is a mechanism by which a source node wishing to send data packets obtains a source route to the destination node. It is used only when a source node attempts to send packets but does not know a route to the destination node [11].

A source node broadcasts a route request packet (RRP), which contains the addresses of the source and destination nodes. Each node that receives this RRP first adds its own address to the route carried in this packet, and then rebroadcasts it with the extended route. The route carried in the first RRP received by the destination node is selected and returned to the source node via the reverse route. To minimize redundant broadcast, the receiver will not rebroadcast the request in the following cases. (i) The receiver is just the destination node of the received RRP. (ii) The same request (i.e., carrying the same destination-source pair) has just been rebroadcast. (iii) The requested route already exists in the receiver.

As illustrated in Fig. 5.14, node 1 as the source node tries to find a route to node 7. It first broadcasts an initial RRP carrying only 1 and 7. Upon node 3 receives this RRP, it just adds its address to this RRP and rebroadcasts it, and the same for nodes 2 and 4. However, the RRPs sent by nodes 2 and 4 for this RD are ignored because the request for the same route has already been sent. Upon node 5 receives the RRP sent by node 3, it adds its own address into this RRP and rebroadcasts it. The destination node 7 will receive two RRPs, one from node 5 and the other from node 6. Because the one from node 5 arrives earlier, the route carried in this RRP, i.e., $1 \to 3 \to 5 \to 7$, is selected and returned to node 1 via the reverse path, i.e., $7 \to 5 \to 3 \to 1$.

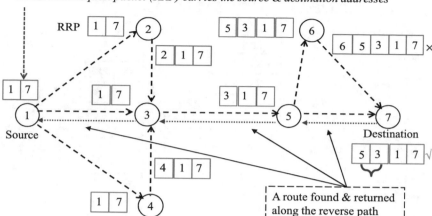

Fig. 5.14 Route discovery procedure of DSR

5.5.1.2 Route Maintenance (RM)

It is a mechanism by which a node is able to detect whether a route to a destination cannot be used because a link along the route no longer works [11]. In order to detect link availability, after a node forwards a packet, it needs to check whether the packet has been received or not by the next hop. This can be realized by using an acknowledgement scheme. That is, a receiver must acknowledge the sender if it has received the packet sent by the sender. If a node detects a link failure of a route, it must inform the other nodes using this route by sending a notice to them.

5.5.2 Ad Hoc On-Demand Distance Vector (AODV)

Like DSR, the AODV protocol [10] is also a reactive routing protocol but is a table-driven protocol. It also consists of RD and RM mechanisms.

5.5.2.1 Route Discovery

It consists of the following operations: generation and dissemination of route request (RREQ) and route reply (RREP) as well as reverse route setup.

As illustrated in Fig. 5.15, when a route to a new destination is needed and it is not available, the node broadcasts an RREQ to find a route to this destination. When a node receives the PREQ, it caches the reverse route to the originator of the RREQ. A route can be determined when the RREQ reaches either the destination or an

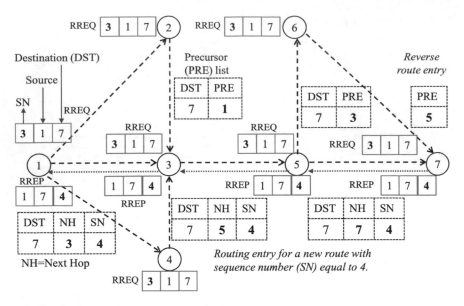

Fig. 5.15 Route discovery procedure of AODV

intermediate node that has an active route to the destination. The route is announced by either the destination node or any intermediate node that is able to satisfy the request through unicasting a RREP back to the originator of the RREQ along the reverse path.

When an RREQ is generated for a destination, the destination sequence number (DSN) in the RREQ is set to the last known DSN for this destination. If no such number is known, the unknown DSN is set. If the RREQ times out without receiving an RREP, the originator broadcasts the RREQ again with the time-to-live (TTL) incremented by a unit until the TTL reaches a threshold.

To reduce congestion in a network, repeated attempts by a source node during route discovery for a single destination must utilize a binary exponential backoff. To prevent unnecessary network-wide dissemination of RREQs, the originating node should use an expanding ring search scheme, with which the originating node initially sets the TTL in the IP header of the RREQ packet and a timeout to receive an RREP.

Once the RREQ arrives at a node, an RREP is generated if the node is the desti-nation or has an active route to the destination node whose DSN is valid and not less than the DSN carried by the RREQ. The RREP is unicast to the next hop toward the originator of the RREQ, and the hop count is incremented by one at each hop. Note that in DSR, the number of hops of a path is explicitly indicated by the path informa-tion. Similar to DSDV discussed earlier, the sequence number (SN) is also used here to indicate a new path. When a destination node generates an RREP, it increments its own SN by one if the SN in the RREQ packet is equal to that incremented value; otherwise, it does not change its SN before generating the RREP. When an immediate node generates the RREP, it copies its known SN corresponding to the destination into the DSN field of the RREP.

The route is only updated if the new SN is either higher than the DSN in the route table, or the SNs are equal, but the hop count for the new path plus one is smaller than the existing hop count in the routing table. This tries to assure that the final route is the shortest in terms of hop count.

5.5.2.2   Route Maintenance

Each node monitors the status of the link to the next hops in the active routes. When a link break is detected, a route error (RERR) message is broadcast to announce that the indicated destination is no longer reachable through the broken link. In order to enable this reporting mechanism, each node keeps a precursor list, which contains the IP address of each its neighbor that is likely to use it as a next hop towards the destination. The information in the precursor list is acquired during the generation of an RREP message, which has to be sent to a node in the precursor list.

5.5.3   Optimized Link State Routing (OLSR)

In both DSR and AODV, routing messages are broadcast over the entire network for path search, leading to bandwidth wastage. Differently, the OLSR protocol [22], which is a table-driven and proactive link state protocol, adopts a multipoint relay (MPR) scheme to coordinate the flooding of control messages for path search in the network. An MPR is a node selected by another node, which is called MPR selector. It is responsible for broadcast message forwarding during path search to avoid redundant retransmissions. To reduce control message transmission in the network, link state information is generated only by an MPR node, which may choose to report links only between itself and its MPR selectors so that only partial link state information is distributed.

5.5.3.1   Multipoint Relay Selection

Each node selects a set of nodes in its symmetric (i.e., bi-directional) 1-hop neighborhood to form an MPR set in order to cover all symmetric strict 2-hop nodes. The MPR set of a node is an arbitrary subset of its symmetric 1-hop neighborhood which satisfies the following condition: every node in the symmetric strict 2-hop neighborhood must have a symmetric link to the MPR of the node for full coverage. Any node in the MPR may retransmit the messages sent by the selector. The neighbors of a node that are not in its MPR set can receive and process broadcast messages, but do not retransmit broadcast messages received from the MPR selector.

Fig. 5.16 An example for
MPR [22]

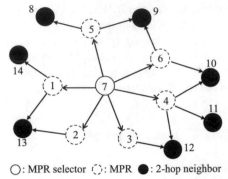

○: MPR selector ◌: MPR ●: 2-hop neighbor

MPR={1,5,4} are used to construct routes to all destinations

   The MPR set of a node is computed such that for each interface, it satisfies that
a broadcast message retransmitted by the selected neighbors will be received by all
nodes 2-hop away via the selected MPR nodes. In other words, a node should select
an MPR set such that any strict 2-hop neighbor is covered by at least one MPR
node. Keeping the MPR set small ensures that the protocol overhead is kept at the
minimum. The information required to perform this calculation is acquired through
the periodic exchange of HELLO messages. The recalculation of an MPR set should
occur when changes are detected in the symmetric neighborhood or in the symmetric
strict 2-hop neighborhood.

   Figure 5.16 depicts an example of an MPR set. Nodes in set MPR = {1,4,5}
retransmits control messages from node 7, while nodes in set {2,3,6} do not. The link
state information is generated only by nodes in set {1,5,4}, which also broadcast this
information and topology control (TC) messages. Nodes in set {1,5,4} are sufficient
to construct the routes from node 7 to all destinations, and it is not necessary for
nodes 2, 3 and 6 to broadcast every message.

5.5.3.2 Control Message Dissemination

An MPR node declares link-state information to its MPR selector. A node which has
been selected as an MPR by one or more neighbor nodes announces this information
(i.e., being selected) periodically in its control messages, indicating that it has a
reachability to the nodes which have selected it as an MPR. In route calculation,
the MPRs are used to create the route from a given node to any destination in the
network.

   Each node maintains the information on the set of neighbors that have selected it as
an MPR, which is called multipoint relay selector set. A node obtains this information
from periodic HELLO messages sent by its neighbors. A broadcast message from
any MPR selector of a node will be retransmitted by this node if it has not received

this message yet. This set can change over time, e.g., when a node selects another MPR set, and is indicated by the selector nodes in their HELLO messages [24].

A TC message is diffused to provide each node in the network with sufficient link-state information for route calculation. Link sensing and neighbor detection operations aim to offer to each node a list of neighbors with which it can communicate directly. A node must at least disseminate links between itself and the nodes in its MPR-selector set to provide sufficient information to enable routing via TC message. An advertised link set must include at least the links to all nodes of its MPR selector set, i.e., the neighbors which have selected the sender as an MPR. Then routes are constructed through advertised links and links with neighbors.

5.5.3.3 Route Computation

Given the link state information acquired through periodic message exchange and the interface configuration of the nodes, the routing table in each node can be computed. Routing entries are recorded in the routing table for each destination in the network for which a route is known. The routing table is based on the information contained in the local link information base and the topology set. If any of these sets are changed, the routing table is recalculated to update the route information on each destination in the network.

5.5.4 Topology Dissemination Based on Reverse-Path Forwarding (TBRPF)

Similar to OLSR, the TBRPF protocol [23] is also a proactive, table-driven link-state routing protocol. It aims to further reduce control messages dissemination for path search. It is a hop-by-hop routing protocol suitable for highly mobile networks, by which each hop decide the optimal outgoing link for each incoming packet.

5.5.4.1 Overview of TBRPF

The following schemes are adopted to minimize control message overhead.

- The extended reverse path forwarding scheme discussed in Sect. 5.2.3.2 is used to control the redundant transmission of the same broadcast packet.
- A node disseminates only partial topology information to its neighbors.
- Differential update reports only changes in link status, and allows more frequent updates for faster detection of topological changes in mobile networks.
- A node will receive only one copy of the update for the same link.

Actually, TBRPF is abbreviated for Topology Broadcast Based on Reverse-Path Forwarding originally proposed for broadcast packet dissimilation in [25]. It is a full-topology routing protocol [23], and provides each node with the information on all links in a network [4]. The current TBRPF of this RFC mainly follows the Partial Tree-Sharing Protocol (PTSP) [4] discussed below, which provides partial topology information to neighbors of a node. The TBRPF of this RFC consists of two modules: neighbor discovery module and routing module, each of which is described below after the introduction of PTSP.

5.5.4.2 Partial Tree-Sharing Protocol (PTSP)

As illustrated in Fig. 5.7, although broadcasting along each source tree is optimal as discussed earlier, a node may still receive multiple updates for the same link. For example, node 1 receives two copies of update for link (2,1), and similarly for node 3, it also gets two for link (4,3). The basic idea of the PTSP [4] is that only the neighbors selected by a node will report the link status for each node in the source tree of the selected neighbors, as described below.

Each node i computes its source tree (T_i) to provide a shortest path to each reachable node. It selects neighbors from T_i as the next hops to each source node u. The selected node called parent (p) is then responsible for updating to node i the states of links in T_p. Consider a source node u in T_i that is indicated by node i to parent p. The links to be updated to node i by p include all links (u,v) in T_p. In the case of several parents selected by node i, the sets of source nodes indicated by node i to each parent should be different to prevent a node from receiving multiple identical updates for the same link.

Consider Fig. 5.7, in which node 1 is selecting its parents. It is easy to show that T_1 goes through node 2, which is the only neighbor of node 1 in T_1. So node 1 selects node 2 as the parent to update the links for all source nodes except itself in T_2. The links to be updated by node 2 to node 1 include (2,1), (2,4), (4,3), (3,5), (3,6) and (5,7). Once a link status changes such that the affected node needs to change its parent selection, the nodes send messages to both the old and new parents. Again in Fig. 5.7, if the delay between nodes 1 and 4 decreases to 1, T_1 can also go through node 4. In this case, T_4 also changes as follows: the original branch (2,1) is replaced by the new branch (4,1). Then node 1 selects node 4 as another parent to update source nodes 4 and 5, while lets node 2 (old parent) update source nodes 2 and 3, for example, which means links (4,1), (4,2) and (5,7) to be updated by node 4, and links (2,1), (2,4), (3,5) and (3,6) by node 2, respectively.

However, the current TBRPF adopts a different method to achieve the above objective, mainly with the following operations. First, each update message is only propagated to the 1-hop neighbors of the sender. Once a node receives update messages, it calculates the source tree rooted at itself to find any changes in the related link status. Only the changed part will be updated to its 1-hop neighbors in the next round of update so that the identical update for the same link will not be sent again. The remaining issue is how to select the updated part of the source tree, which is handled by the routing module discussed below.

5.5.4.3 Neighbor Discovery Module

This module aims to discover the 1-hop neighbors of a node, and the discovery of 2-hop neighbors is handled by the routing module. The discovery process allows each node to quickly detect its neighbor nodes of existing bidirectional links between them. Similarly, it can also allow a node to know when a bidirectional link breaks or becomes unidirectional. These discovery operations are based on HELLO messages sent by each node periodically.

Each node needs to maintain a neighbor table to store the link status (i.e., 1-WAY, 2-WAY, or LOST) for each neighbor indicated from HELLO messages. This table determines the contents of HELLO messages, which will report only changes in neighbor status to reduce protocol overhead. This kind of message is called differential HELLO message.

5.5.4.4 Routing Module

The computation of a source tree is based on the topology information stored in the topology table of each node, which stores the information for each known node and link in the network. A combination of periodic and differential updates is used to keep all neighbors of a node updated of changes in topology.

Topology Update

With the routing module, each node maintains a source tree rooted at itself in order to provide a shortest path to each reachable node. A node computes and updates its source tree based on the stored topology information, using a modified Dijkstra algorithm [23]. To minimize traffic overhead, only a part of a source tree is reported to its neighbors, which is called reported subtree (RT) to be defined below. RTs are reported in periodic topology updates, while changes in an RT are reported in more frequent differential updates. After a node receives a topology update, it does not forward this update. Instead, it recalculates its RT based on the received update, which may result in changes in its RT. These changes will be reported in the next round of differential or periodic update. The number of control packets sent for topology update is expected to be minimized through (i) 1-hop propagation of each update, (ii) differential update reporting, and (iii) using the same packet to carry multiple updates as much as possible.

Reported Node Set (RN)

An RN is a set of nodes to be reported in the topology update sent by a node, and is determined as follows:

- Step 1: A node, say node i, first determines which neighbors should be selected to its RN. It first includes its neighbor j in its RN if and only if node i may be selected by another neighbor as a next hop on the shortest path to j. To this end, node i computes the inter-neighbor shortest paths (up to 2 hops) that uses only its

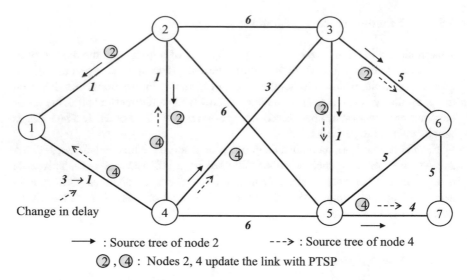

⟶ : Source tree of node 2   ---> : Source tree of node 4

②, ④ : Nodes 2, 4 update the link with PTSP

Fig. 5.17 Principle of PTSP and reported node set (RN) computation of TBRPF

neighbors or node i as an intermediate node. So we can obtain RN = {1,2,3} for node 4 in Fig. 5.17.

- Step 2: Each node u (u ≠ i) in the topology table of node i is included in the RN if and only if the next hop on the shortest path to node u is also in the RN. Equivalently, node u is included in the RN if and only if it is in the subtree rooted at neighbor j of the source tree. A topology table stores the information on every known node and link in the network [23]. Then, RN becomes {1,2,3,5,6,7} for node 4 in Fig. 5.17.
- Step 3: Node i also includes itself in the RN, hence RN={1,2,3,4,5,6,7} for node 4 in Fig. 5.17.

Each node updates its RN immediately before generating periodic or differential topology updates.

Reported Subtree (RT)

The RT of a node consists of links (u,v), of which node u is in its RN calculated above, and called the predecessor of node v. A leaf node of a source tree is not the predecessor of any other node of this tree. In other words, the RT consists of subtrees of the source tree rooted at neighbors in the RN. Since node i includes itself in its RN, the RT also includes each local link (i,j) in the RN, where j is in the RN. For node 4 in Fig. 5.17, its current RT actually is just its source tree. Each node should report its RT, but may also report additional links to increase robustness in highly mobile networks [23].

5.5.5 Probabilistic Routing Protocol Using History of Encounters and Transitivity (PRoPHET)

The PRoPHET protocol [15] is a variant of the Epidemic routing protocol (ERP) [16] proposed for DTNs. ERP assumes a fully random mobility pattern, and a node will forward its carried data to each node that it encounters, like flooding. So it can ensure high data delivery but is wasteful of scarce channel resource, energy and buffer space. The enhancement proposed by PRoPHET is based on the fact that node mobility patterns in some cases are not completely random. Instead, it may follow certain statistical patterns determined by their mobility history, e.g., daily or weekly periodic activities determine human mobility patterns.

5.5.5.1 Delivery Predictability

PRoPHET defines a metric called delivery predictability to measure delivery oppor tunity, i.e., $0 \leq P(A, B) \leq 1$, which indicates a degree at which node A can successfully deliver data to node B. Here, node A applies a strategy to deliver data, and node B is the destination of the data to be delivered by node A. For example, $P(A, B) > P(C, B)$ means that the data destined for node B is preferably forwarded to node A rather than node C. Note that, for an asymmetric connection, it is most likely that $P(A, B) \neq P(B, A)$. The calculation of this metric conducted by node A consists of two parts: (i) increasing $P(A, B)$ for every new encounter with node B, following

$$P(A, B) = P(A, B)_{old} + [1 - \delta - P(A, B)_{old}] \times P_{encounter}, \qquad (5.4)$$

and (ii) decreasing $P(A, B)$ for each pre-defined time unit during which node A does not meet node B because the value of this metric evolves over time, following

$$P(A, B) = P(A, B)_{old} \times \gamma^K, \qquad (5.5)$$

where $P(A, B)_{old}$ is the value of $P(A, B)$ that stored before a new encounter. If no delivery predictability value is stored for node B, $P(A, B) = 0$. As a special case, the metric value for a node itself is always set to 1, i.e., $P(A, A) = 1. 0 \leq P_{encounter} \leq 1$ is a scaling rate at which the predictability increases on encounters after the first encounter. Parameter δ is a small positive number that effectively makes each predictability less than 1. $0 \leq \gamma \leq 1$ is the aging constant, and K is the number of time units that have elapsed since the last time when the metric was aged [15].

A distortion of delivery predictability may be caused by multiple encounters due to frequent up and down of the same physical link during a short time period, which actually can only stand for one encounter. To minimize such effect on $P(A, B)$ estimation, $P_{encounter}$ is set as a function of the time interval since the last encounter that invoked an update of $P(A, B)$, as shown in Fig. 5.18, which is formulated by (5.6).

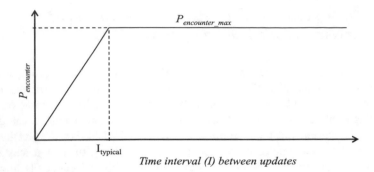

Fig. 5.18 $P_{encounter}$ as function of time interval (I) between updates [15]

Thus, both the increment and decay of $P(A, B)$ are related to the update interval for short encounter intervals. For encounter intervals longer than the typical time ($I_{typical}$), $P_{encounter}$ is set to the fixed $P_{encounter\_max}$, which is chosen such that the increment of $P(A, B)$ significantly exceeds the decay of $P(A, B)$ over an I_{typ} interval.

$$P_{encounter}(I) = \begin{cases} P_{encounter\_max} \times \frac{I}{I_{typical}} & \text{for } 0 \leq I \leq I_{typical}, \\ P_{encounter\_max}, & \text{for } I > I_{typical} \end{cases} \tag{5.6}$$

With a symmetric connection, delivery predictability also has a transitive property as discussed below. If node A can frequently encounter node B, and likewise for node B to encounter node C, then node C can relay data destined for node A. This transitive property is used to determine delivery predictability as follows:

$$P(A, C) = \max[P(A, B)_{old}, P(A, B) \times P(B, C)_{recv} \times \beta], \tag{5.7}$$

where $0 \leq \beta \leq 1$ is a scaling constant to control the impact of the transitivity on the delivery predictability, and $P(B, C)_{recv}$ is a metric that node A has received from the encountered node B.

5.5.5.2 Forwarding Strategies

Given a delivery predictability $P(A, B)$, it is not straightforward to make a forwarding decision based on only this metric. For example, when node A encounters a node with a low delivery predictability, if node A does not forward the carried data to this node, it might not be able to encounter another node with a higher metric within a reasonable time interval. Thus in some cases, other issues are necessarily considered jointly to make a forwarding decision. Therefore, several forwarding strategies have been defined in [15]. Some of them are listed below for the case that the encountered node is not the destination of the carried data. Here the destination is set to node D. When nodes A and B meet, node A can forward the carried data to node B by applying one of the following strategies.

1. Only if $P(B, D) > P(A, D)$: Node A sends its carried data to node B if the delivery predictability of the destination of the carried data is higher at node B. Node A still stores the data after sending if it has sufficient buffer space because it might encounter a better node or even the destination in the future.
2. Only if $P(B, D) > P(A, D)$ AND $NF < NF_{max}$: This strategy is similar to the first one, but with an additional condition that the number of other nodes that node A has forwarded the carried data (NF) should be limited by an upper bound NF_{max}.
3. Only if $P(B, D) > P(A, D)$ OR $P(B, D) > FORW_{thres}$: This strategy is identical to the first one if $P(B, D) > P(A, D)$ holds. In any cases where $P(B, D)$ is larger than a threshold value $FORW_{thres}$, the carried data is spread epidemically.

When two nodes have a communication opportunity, they will first exchange the delivery predictability for all destinations that they know. This information is used by each node to update the local delivery predictability following the algorithms mentioned above. Then, the nodes further exchange information about the data carried by each node, including the destination and size of the carried data. This information is used by a node in conjunction with the updated delivery predictability to make forwarding decisions on the carried data following the defined forwarding strategies.

5.5.6 Discussion

Table 5.1 summarizes the five routing protocol RFCs discussed above, and is explained below.

Table 5.1 Summary of the RFCs for routing protocols

RFC for (Network)	Routing action	Used Info	Route Expr	Routing metric	Routing manner	Protocol stack	Protocol unit
DSR (MANET)	Reactive	Dist vector	Source routing	Earliest arrival	End-to-end	Network layer	Packet
AODV (MANET)	Reactive	Dist. vector	Table-driven	Shortest path	End-to-end	Network layer	Packet
OLSR (MANET)	Proactive	Link state	Table-driven	Shortest path	End-to-end	Network layer	Packet
TBRPF (MANET)	Proactive	Link state	Table-driven	Shortest path	Hop-by-hop	Network layer	Packet
PRoPHET (DTN)	Reactive	Local Info	No path Expr	Maximum MDP	Hop-by-hop	Above transport	Message

Info = Information, Dist = Distance, Expr = Expression, MDP = Message delivery predictability

With DSR, intermediate nodes do not need to maintain any routing information for packet forwarding because the path of each packet is carried by the packet itself. Any nodes receiving these packets can cache the path information for future use. It operates entirely on demand without periodic broadcast for route maintenance. DSR is suitable for a high mobility network of up to about two hundred nodes, and all protocol state is soft state so that the loss of any state will not impact the correctness of protocol operation [11]. However, since DSR is a reactive routing protocol, it takes time to find a path, which may affect the QoS of delay-sensitive applications. In addition, path information carried in every packet is an overhead proportional to the number of hops that a route will pass through.

AODV is also a reactive protocol, and avoids the large packet overhead of DSR. However, a node has to maintain a list for active forwarding and reverse routes. When a link breaks, the affected nodes will be notified so that they can respond timely to the link breakage and related changes in network topology. Routing loop is avoided by using a sequence number created by the destination node for each route entry, which is similar to DSDV. The final route is selected in terms of the smallest hop count indicated by the RREP among the ever recorded routes, which probably results in a sub-optimal shortest path in comparison with link-state routing protocols [10].

OLSR is a proactive routing protocol, and can provide optimal routes in terms of the number of hops. It generates much less control messages flooded in the network for path search by using a multipoint relay (MPR) scheme. It does not rely on any central entity, and is suitable for large and dense networks. Control messages periodically sent by each node can make the protocol to sustain a reasonable loss of such messages without using a reliable transmission scheme. However, a large protocol overhead will be generated for the selection and maintenance of MPR especially in high mobility network, and the protocol is more complex than DSR and AODV.

In comparison with OLSR, TBRPF adopts an extended reverse-path forwarding scheme along with a partial and differential reporting for control message dissemination to further reduce overhead for path search. No sequence number is used for topology updates to avoid the wraparound problem. It does not require reliable or sequenced delivery of messages so that no ACK or NACK schemes are used. It adopts a hop-by-hop routing scheme to adapt the protocol for high mobility networks. However, the protocol operation requires more computation than OLSR.

PRoPHET for DTNs is also a reactive routing protocol sitting above the transport layer. Since probably no path is available in DTNs when routing, path expression is not necessary, and only local information is mainly used to make routing decisions.

5.6 Summary

This chapter mainly focuses on routing protocols for mobile ad hoc networks (MANETs). Mobility in MANETs may cause frequent topological change, which becomes a challenge for routing protocol design. Routing for Delay and Disruption Tolerant Networks (DTNs) is also discussed. Four routing protocols for MANETs have been ratified by IETF and one for DTNs has been approved by IRTF. Routing for all the above-mentioned networks is address-centric routing, where every node is identified by a network address like the IP address, and the routing protocol aims to find paths between a node pair according to their network addresses. However, in wireless sensor networks, the routing protocol is data-centric routing, aiming to find particular data distributed in different nodes. Such routing has different issues to be addressed in comparison with the address-centric routing, and more discussions can be found in [26, 27]. In both networks, it is assumed that a network path can be found between any pair of nodes when there is data to transmitted between them. However, this may not be the case in DTNs [19, 20].

References

1. Paul, S.: Multicasting on The Internet and Its Applications. Kluwer Academic Publishers, Alphen aan den Rijn (1998)
2. Stallings, W.: Data and Computer Communications, 5th edn. Prentice-Hall, Upper Saddle River (1997)
3. Dalal, Y.K., Metcalfe, R.M.: Reverse path forwarding of broadcast packets. Commun. ACM 21(12), 1040 1048 (1978), Dec
4. Ogier, R.G.: Efficient routing protocols for packet-radio networks based on tree sharing. In: Proceedings of Sixth IEEE International Workshop on Mobile Multimedia Communications (MOMUC), pp. 104–113. New York, USA (1999)
5. Perkins, C.E., Bhagwat, P.: Highly dynamic destination-sequenced distance-vector routing (DSDV) for mobile computers. In: Proceedings of ACM SIGCOMM, pp. 234–244. London England, UK (1994)
6. Royer, E.M., Toh, C.K.: A review of current routing protocols for ad hoc mobile wireless networks. IEEE Personal Commun. Mag. 46–55 (1999)
7. Tarique, M., Tepeb, K.E., Adibi, S., Erfani, S.: Survey of multipath routing protocols for mobile ad hoc networks. J. Net. Comp. App. 32(6), 1125–1143 (2009)
8. Boukerchea, A., Turgutb, B., Aydinc, N., Ahmadd, M.Z., Bölönid, L., Turgutd, D.: Routing protocols in ad hoc networks: a survey. Comput. Net. 2011(55), 3032–3080 (2011)
9. Mahmood, B.A., Manivannan, D.: Position based and hybrid routing protocols for mobile ad hoc networks: a survey. Wireless Pers. Commun. 83(2), 1009–1033 (2015)
10. Perkins, C.E., Belding-Royer, E.M., Das, S.R.: Ad hoc on-demand distance vector (AODV) routing. IETF RFC3561 (2003)
11. Johnson, D.B., Maltz, D.A., Hu, Y.C.: The dynamic source routing protocol (DSR) for mobile ad hoc networks for IPv4. IETF RFC 4728 (2007)
12. Cerf, V., Burleigh, S., Hooke, A., Torgerson, L., Durst, R., Scott, K., Fall, K., Weiss, H.: Delay-Tolerant Networking Architecture. IETF RFC 4838 (2007)
13. Lindgren, A., Doria, A., Schéln, O.: Probabilistic routing in intermittently connected networks. In: Proceedings of ACM MobiCom. San Diego, CA, USA (2003)
14. Berners-Lee, T., Fielding, R., Masinter, L.: Uniform Resource Identifier (URI): Generic Syntax. IETF RFC 3986 (2005)
15. Lindgren, A., Doria, A., Davies, E., Grasic, S.: Probabilistic Routing Protocol for Intermittently Connected Networks. Internet Research Task Force (IRTF) (2012)
16. Vahdat, A., Becker, D.: Epidemic routing for partially connected ad hoc networks, Technical Report CS-2006-067. Duke University, Department of Computer Science (2000)
17. Spyropoulos, T., Psounis, K., Raghavendra, C.S.: Spray and wait: an efficient routing scheme for intermittently connected mobile networks. In: Proceedings of ACM SIGCOMM Workshop on Delay-Tolerant Networking, pp. 252–259. Philadelphia, USA (2005)

18. Burgess, J., Gallagher, B., Jensen, D., Levine, B.N.: MaxProp: routing for vehicle-based disruption-tolerant networks. In: Proceedings of IEEE INFOCOM. Barcelona, Spain (2006)
19. Cao, Y., Sun, Z.L.: Routing in delay/disruption tolerant networks: a taxonomy, survey and challenges. IEEE Commun. Surv. Tutor. 15(2), 654–677 (2013). Second Quarter
20. Wei, K.M., Liang, X., Xu, K.: A survey of social-aware routing protocols in delay tolerant networks: applications, taxonomy and design-related issues. IEEE Commun. Surv. Tutor. 16(1), 556–578 (2014). First Quarter
21. Chakchouk, N.: A survey on opportunistic routing in wireless communication networks. IEEE Commun. Surv. Tutor. 17(4), 2014–2041 (2015). Fourth Quarter
22. Clausen, T., Jacquet, P.: Optimized link state routing protocol (OLSR). IETF RFC 3626 (2003)
23. Ogier, R., Templin, F., Lewis, M.: Topology Dissemination Based on Reverse-Path Forwarding (TBRPF). IETF RFC 3684 (2004)
24. Qayyum, A., Viennot, L., Laouiti, A.: Multipoint relaying: An efficient technique for flooding in mobile wireless networks. In: Proceedings of Hawaii International Conference on System Sciences (HICSS). Hawaii, USA (2001)
25. Bellur, B., Ogier, R.G.: A reliable, efficient topology broadcast protocol for dynamic networks. In: Proceedings of IEEE INFOCOM, vol. 1, pp. 178–186. New York City, NY, USA (1999)
26. Al-Karaki, J.N., Kamal, A.E.: Routing techniques in wireless sensor networks: a survey. IEEE Wirel. Commun. Mag. 6–28 (2004)
27. Sara, G.S., Sridharan, D.: Routing in mobile wireless sensor network: a survey. J. Telecom. Syst. 57(1), 51–79 (2014)

Chapter 6
End-to-End Transmission Control in RWNs

Abstract End-to-end transmission control is the major function of the transport layer, which is a bridge between networks and applications, providing reliable and unreliable end-to-end transmissions. An end point here refers to a user terminal such as a desktop computer or a notebook. Reliable end-to-end transmissions guarantee successful data delivery between source-destination pairs, whereas there is no such a guarantee with unreliable end-to-end transmissions. The functions used for those transmissions are installed in user terminals, and unnecessarily in networking units like routers or switches. Applications on top of this layer will not be aware of the particulars of underlayer networking operations.

The primary transport layer protocols include User Datagram Protocol (UDP), Transmission Control Protocol (TCP), Datagram Congestion Control Protocol (DCCP) and Stream Control Transmission Protocol (SCTP). The most popular protocols are UDP and TCP. UDP provides unreliable end-to-end transmissions, while TCP provides reliable ones. As illustrated in Fig. 6.1, most popular Internet applications available today are based on them, including email (SMTP, port 25), HTTP (port 80) for web page transmission, File Transfer Protocol (FTP, port 21), remote login (Telnet, port 23), Secure Shell (SSH, port 22), and some streaming media applications, such as audio, video and playback etc. The following sections focus on UDP and TCP.

6.1 User Datagram Protocol (UDP)

UDP provides unreliable end-to-end transmissions in a connectionless mode, by which a source node does not need to inform the destination node before sending it any data segments (called datagram). A source node can send a datagram whenever it is available without traffic control. The destination node does not reserve any buffer space for datagram reception. Thus, congestion may happen in either routers or the destination node, which leads to datagram loss. However, the source node is not concerned about whether the sent datagrams have been received successfully or not, and no retransmission is triggered for any lost datagram.

© Springer Nature Singapore Pte Ltd. 2018
S. Jiang, *Wireless Networking Principles: From Terrestrial to Underwater Acoustic*,
https://doi.org/10.1007/978-981-10-7775-3_6

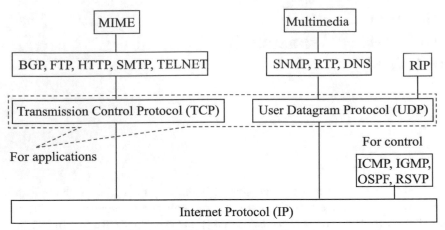

BGP = Border Gateway Protocol, DNS = Domain Name Server,
FTP = File Transfer Protocol, HTTP = Hypertext Transfer Protocol,
ICMP = Internet Control Message Protocol, IGMP = Internet Group Management Protocol,
MIME = Multi-Purpose Internet Mail Extension, OSPF = Open Shortest Path First,
RIP = Routing Information Protocol, RTP = Real-Time Transport Protocol
SMTP = Simple Mail Transfer Protocol, SNMP = Simple Network Management Protocol

Fig. 6.1 Internet protocol stack and typical applications

6.2   Transmission Control Protocol (TCP)

Different from UDP, TCP provides reliable end-to-end transmissions in a connection-oriented mode, by which a source node needs a permission from the destination node before sending it TCP data segments (or simply segments henceforth). Transmission reliability is realized by retransmitting the sent segments that have not been positively acknowledged by the destination node. To this end, the source node first sends a request to the destination node to set up a TCP connection. If the destination node is ready to receive, it replies positively with a buffer reservation for segment reception. During transmission, the source node needs to perform flow control and congestion control as well as transmission reliability control, each of which will be described below.

6.2.1   Transmission Reliability Control

This part consists of acknowledgement and retransmission timeout (RTO) schemes.

6.2.1.1 Acknowledgement (ACK)

Segment loss detection can be realized through acknowledgement schemes, which can be either positive (ACK) or (NACK). With ACK, if the destination node has successfully received a data segment, it returns an ACK segment to the source node. With NACK, if the destination node has not received the expected data segment, it returns an NACK segment to the source node. A loss may also happen to an acknowledgement segment. Particularly for NACK, the destination node should know a prior the transmission plan of data segments in order to issue an NACK. In this sense, the positive acknowledgement scheme plus a timer is better than the negative one. With the positive one, if a source node fails in receiving the excepted ACK after timeout, it may infer the reception failure of the related data segment at the destination node.

TCP adopts an accumulative and positive acknowledgement scheme, which works as follows. If segments 1 to n have been successfully received, sequence number $n+1$ is carried by an ACK to indicate: (i) the successful reception of segments up to n, and (ii) Segment $n+1$ is expected by the destination node. As illustrated in Fig. 6.2, segments 1–3 have been received, and an ACK carrying sequence number 4 is sent to the source node. If a segment is lost, the source node keeps retransmitting it until being acknowledged of its successful reception. To this end, once a data segment is sent out, the source node sets a retransmission timer. If the ACK for this segment is received before timeout, the timer is stopped; otherwise, a retransmission timeout (RTO) happens, and the segment is retransmitted. An RTO event is also treated by the source node as an indicator of a congestion event in the network. Therefore, how to set RTO affects TCP performance, which discussed below.

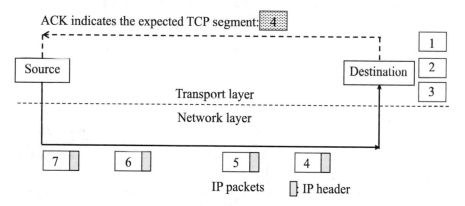

Fig. 6.2 TCP acknowledgement scheme

6.2.1.2 Retransmission Timeout (RTO)

The setting of RTO is calculated based on round-trip time (RTT), which is the time interval between when a segment is sent and when the ACK is received by the source node. An adaptive RTO estimation scheme consists of the following elements:

- Smoothed RTT estimation (RTT_s): It is estimated according to its historic record RTT (RTT') and its currently measured RTT (RTT_m) as follows:

$$RTT = \alpha RTT' + (1 - \alpha)RTT_m, \tag{6.1}$$

where $\alpha \leq 1$ is a positive weight, and usually set to $\frac{7}{8}$ as suggested by RFC 2988 [1]. Initially, RTT' is set to RTT_m.
- RTO estimation: According to RFC 2988, RTO is calculated as follows:

$$RTO = RTT_s + 4 \times RTT_D, \tag{6.2}$$

where RTT_D is calculated by

$$RTT_D = \beta RTT'_D + (1 - \beta)|RTT_s - RTT_m|, \tag{6.3}$$

where RTT'_D is the historic value, and $\beta \leq 1$ is a positive weight, and usually set to $\frac{3}{4}$ as suggested by RFC 2988 [1]. Initially, RTT_D is set to $0.5RTT_m$.

The above estimation algorithm can work well in the case of no retransmission. When a retransmission occurs, in some cases, it is difficult to determine RTT_m, which further affects the accuracy of the RTT estimation (6.1). Figure 6.3 depicts such a situation, in which the source node sends a segment but has not received the corresponding ACK after the timeout, so it retransmits this segment. When the destination node has successfully received the segment, which may be either the

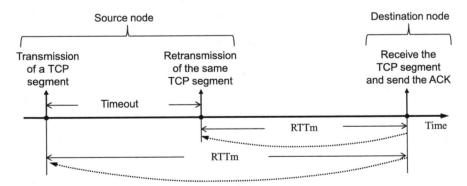

The source cannot know which transmission the ACK is corresponding to.

Fig. 6.3 Difficult situation for accurate RTT measurement [2]

first sent or the retransmitted, it sends an ACK. When the source node receives this ACK, it cannot distinguish which transmission, the first sent one or the retransmitted one, the received ACK acknowledges. Note that the ACK segments to these two transmissions correspond to different RTT_m settings.

Thus, Karn algorithm [3] suggests that in the case of retransmission, RTT_m is not used to estimate RTT, while RTO is estimated as follows:

$$RTO = \gamma \times RTO', \tag{6.4}$$

where RTO' is the historic value, and γ is set to 2 as suggested by RFC 2988.

6.2.2 Flow Control

It is used by a source node to prevent buffer overflow at the destination node, which uses a receiver window ($rwin$) to limit the number of segments to be received. The source node regulates the number of segments to be sent with a sliding window as illustrated in Fig. 6.4. The window slides from the left to the right according to ACKs received by the source node. Here, the maximum size of the window is 9 segments, which is the minimum of rwin and the congestion window ($cwnd$). There are total 15 segments indexed from 0 to 14, of which, segments 0–3 have been sent and acknowledged positively so they are removed from the buffer, and the window slides forwards by 4 segments. Only segments 4–12 are under the control, of which, segments 4–8 have been sent but still waiting for the ACKs, and the number of such segments is called FlightSize. These segments must stay in the window because if any of them is lost, the corresponding retransmission needs to be carried out. Segments 9–12 can be sent anytime, which depends on the availability of underlayer service. Segments 13–14 are still out of the window and cannot be sent until the window slides forwards.

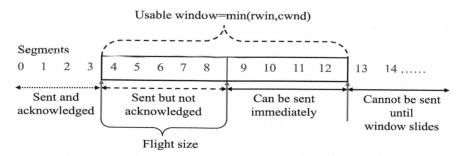

Fig. 6.4 Slide window for flow control in TCP

6.2.3 Congestion Control

Since TCP is a transport layer protocol, it is unable to intervene with the buffer management on the network layer to control congestion in routers. Furthermore, TCP cannot know the exact congestion situation on the network layer, and can only infer the situation according to the status of segment reception with the following assumption: when a congestion happens in a network, the arriving segments are dropped so that segment loss indicates congestion events. In other words, no segment loss means no congestion in the network. A source node justifies the loss of a sent segment through the corresponding ACK reception. That is, if the ACK of the sent segment is received within the RTO, there is no congestion along the TCP connection; otherwise, a congestion is conjectured. Note that the loss of a segment can also be indicated by ACKs for other segments with a selective acknowledgement scheme, which is an optional implementation.

Congestion control in TCP is manipulated through $cwnd$ mentioned above, which defines the number of bytes that can be sent at one time. Its size is adjusted according to the congestion situation in the network. TCP throttles the number of data segments to be sent according to cwnd, Consider the flow control mentioned earlier, the maximum amount of data that can be sent is limited by $\min(rwin, cwnd) - FlightSize$. The manipulation of cwnd varies with TCP variants, and some standards are discussed below.

6.3 TCP Standards

Several TCP standards are established during its evolution, typically includes TCP Tahoe, TCP Reno and TCP NewReno. These standards are identical in terms of their reactions to a congestion state indicated by an RTO event as illustrated in Fig. 6.5 (i.e., the solid line), where the source node sets $ssthresh$ to a half of the FlightSize of the current cwnd, and resets cwnd to 1 to force the window to run a slow start. Their differences are related to a congestion state indicated by 3 duplicate ACKs as discussed below.

6.3.1 TCP Tahoe

It is one of the earliest TCP standards that performs congestion control, using an Additive Increase and Multiplicative Decrease (AIMD) scheme. It consists of three parts: slow start (SS), congestion avoidance (CA) and fast retransmission.

The additive increase operation is carried out during slow start and congestion avoidance processes. During slow start, $cwnd$ is increased by one maximum segment size (MSS) for every received ACK ($cwnd = cwnd + 1$) until $ssthresh$ is reached.

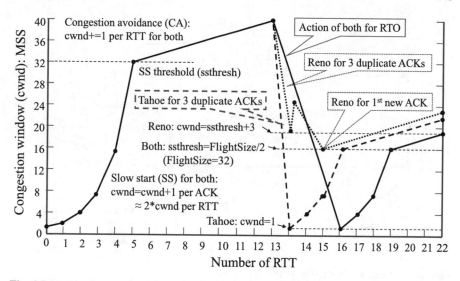

Fig. 6.5 Congestion window manipulation with TCP Tahoe and TCP Reno

Since there are up to $cwnd$ MSSs to be sent during one RTT, roughly $cwnd$ can be doubled every RTT ($cwnd \backsim 2 \times cwnd$), resulting in an almost exponential increase as follows: $cwnd \approx 2^n$ with initially $cwnd = 1$, where n is the number of RTTs, as illustrated in Fig. 6.5. Once $cwnd > ssthresh$, the window manipulation enters the congestion avoidance period, during which $cwnd$ is increased by one MSS for every RTT to avoid congestion, approximately $cwnd = cwnd + 1/cwnd$ per received ACK.

The multiplicative decrease is conducted once a congestion is inferred by the source node. If the congestion is indicated by an RTO event, the window manipulation is the same as those described above. However, in some cases, a congestion indication is very obvious to the destination node, but it may take a long time for the source node to learn a congestion with an RTO event. Therefore, a fast retransmission scheme using duplicate ACKs is proposed to allow the source node to react earlier.

When a segment arrives at the destination node out of the order, it re-sends the last ACK for the previously received segment, i.e., a duplicate ACK, to the source node immediately. As illustrated in Fig. 6.6, segment 3 is a such segment acknowledged by an ACK carrying 4 sent by the destination node. However, segment 5 rather than segment 4 is received, which triggers an immediately sending of a duplicate ACK, and so do for each subsequential reception. As illustrated in Fig. 6.6, 3 duplicate ACKs have been sent out following the reception of segments 5, 6 and 7.

The absence of segment 4 may be due to mis-order reception rather than loss, which means that it might be received later. Therefore, the source node does not need to react to the first two duplicate ACKs but better to the third one. Once upon the arrival of the third one, the source node retransmits the segment indicated by this ACK immediately as depicted in Fig. 6.6, and then runs slow start. Usually the

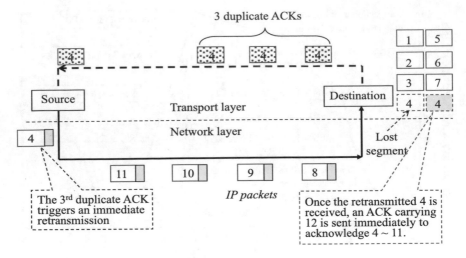

Fig. 6.6 Fast retransmission with TCP Tahoe

time for receiving 3 duplicate ACKs is shorter than the RTO setting, which leads to a faster reaction to a congestion as indicated by the dashed line in Fig. 6.5.

6.3.2 TCP Reno

TCP Tahoe treats every congestion as a serious congestion, which may lead to over reaction to a mild congestion, resulting in an unnecessary throttling of segment transmission. As illustrated in Fig. 6.6, segments 1, 2 and 3 are successfully received but not segment 4, which indicates a congestion in the network. However, segments 5, 6 and 7 have gone through the network, which means that the congestion situation is not so severe. Thus, it is unnecessary for the source node to react so aggressively as done by TCP Tahoe. Therefore, TCP Reno [4] is proposed to overcome this problem, jointly using fast retransmission and fast recovery schemes as described below.

As depicted in Fig. 6.5, the fast recovery state follows the fast retransmission with the following different manipulations:

1. Set *ssthresh* to a half of the current FlightSize of *cwnd*, i.e., $ssthresh = \frac{1}{2} \times FlightSize$.
2. Set *cwnd* to *ssthresh* plus 3 MSSs, i.e., $cwnd = ssthresh + 3$, instead of resetting cwnd to 1 as done in TCP Tahoe. An addition of 3 to cwnd corresponds to the three duplicate ACKs for the three successfully received segments.
3. Increase *cwnd* by one *MSS* each time another duplicate ACK is received, i.e., $cwnd = cwnd + 1$.
4. Set *cwnd* to *ssthresh* (similar to Tahoe) upon the arrival of the first ACK that acknowledges the reception of a new segment.

The remaining operations just follow the congestion avoidance. During the fast recovery process, if *cwnd* permits, the source node can send other segments if any.

6.3.3 TCP NewReno

TCP Reno can work well only if there is only one segment loss during one congestion window. In this case, the first ACK for new segments sent by the destination node can acknowledge all the segments that have been sent before the fast retransmission operation and the retransmitted segment itself. As pointed in Fig. 6.6, an ACK carrying 12 is sent back immediately to acknowledge the reception of segments 4 and 8 ~ 11 once the retransmitted segment 4 has been received.

When there are multiple losses during a congestion window, the above mentioned ACK may only acknowledge some of the segments that have been successfully received. This is called partial acknowledgement, which may cause the source node to shrink *cwnd* repeatedly, making TCP Reno to perform similarly to TCP Tahoe. As illustrated in Fig. 6.7, where both segments 4 and 6 are lost, once the retransmitted segment 4 is received, the destination node sends immediately an ACK carrying the expected 6. Actually, the destination node has already received segments 8 ~ 11. If this node continues to receive segments 12 and onwards, duplicate ACKs carrying 6 will also be sent repeatedly to the source node. Once receiving the third duplicate ACK, the source node halves *cwnd* again following steps (1)–(4) of TCP Reno, which makes *cwnd* much smaller. If there are several losses, *cwnd* will approach to 1 quickly, which is similar to the initial setting of TCP Tahoe.

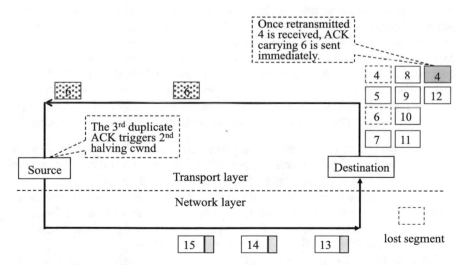

Fig. 6.7 TCP Reno in the case of multiple losses in one cwnd

TCP NewReno [5] is proposed to overcome the above weakness through simply delaying the exit of the fast recovery state described in step (4) of TCP Reno. That is, upon receiving the first ACK for new segments, the source node checks whether this ACK has acknowledged the last segment sent just before the retransmission, e.g., segment 11 in Fig. 6.7 corresponding to the retransmission of segment 4. If this is the case, e.g., an ACK carrying 12 in Fig. 6.6, the source node just follows the steps defined by TCP Reno to manipulate $cwnd$; otherwise, e.g., an ACK carrying 6 in Fig. 6.7, the source node only retransmits the segment indicated by this ACK, e.g., segment 6 in Fig. 6.7, and continues staying in the fast recovery state as defined by TCP Reno without halving $cwnd$, until the last transmitted segment has been acknowledged.

6.4   TCP for Multi-hop Mobile Networks

It is proofed by several experiments that in multi-hop mobile networks, TCP cannot perform as good as in wired networks. This section briefly summarizes the primary challenges facing TCP in mobile networks with a discussion on some typical approaches proposed to enhance TCP.

6.4.1   Challenging Issues

Since the transport layer cannot know what happens on the network layer, a source node infers congestion status according to the ACK reception for the sent segments, jointly using retransmission timer. Inevitably, such conjecture may lead to misjudgment on the real congestion status. For example, a failed ACK reception in time may be due to too a long queueing delay rather than a congestion. Even in case of a congestion, it may not occur in the direction from the source node to the destination node but in the opposite direction. Similar misjudgement happens more frequently in multi-hop mobile networks because many failed segment receptions are not caused by congestions as illustrated in Fig. 6.8, which severely impacts TCP performance.

- Typical BER of radio channels ranges from 10^{-1} to 10^{-3} in comparison with $10^{-6} \sim 10^{-9}$ of wired channels. Such unreliable channels cause more erroneous receptions, which eventually leads to more failed segment receptions.
- A wireless medium is usually shared by multiple terminals, and concurrent transmissions cause collision. Both bursty segment arrival and ACK traffic from the reverse direction intensifies such contention over the shared medium. These features cause collisions leading to failed segment receptions.
- Hidden terminals further increase collisions, and the more hops of a network path, the more hidden terminals may present, which also impacts successful segment reception in time.

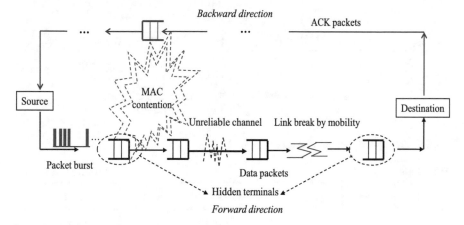

Fig. 6.8 Factors leading to misjudgement on congestion status in multi-hop mobile networks

Table 6.1 Major causes of reception failure in multi-hop mobile networks

Cause of failed reception	Consequence	TCP judgement
Congestion	Packet loss	Correct
Unreliable channel	Frame corruption	Wrong
Terminal mobility	Overdue packet	Wrong
Hidden terminals	Frame collision	Wrong
Medium Access Control (MAC) contention	Frame collision	Wrong

- Mobile terminals may move during communication in progress, resulting in link breakage, which further causes communication disruption and consequent rerouting. In this case, more time is required to forward segments so that some of them cannot be received before the RTO.
- In multi-hop mobile networks, link intermittence due to mobility makes long RTT, which causes packets to take more time than expected in their journey to the destinations, eventually causing more RTO events.

In the above cases, a source node will make wrong judgements on congestion situation as listed in Table 6.1, leading to unnecessary decrease of $cwnd$, which determines the number of segments to be sent to the network. Furthermore, the increment of $cwnd$ largely depends on RTT. The longer RTT, the slower increment of $cwnd$, which reduces network throughput.

Many TCP enhancement strategies are investigated to tackle the above problems [6, 7], typically including splitting of TCP connection, accurate congestion judgement, MAC contention relief and ATM-like control as well as decoupling of TCP function. Among them, the strategy of accurate congestion judgement consists of the following methods: precise congestion inference, route failure notification and

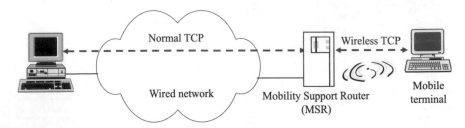

Fig. 6.9 Principle of indirect TCP (ITCP) [10]

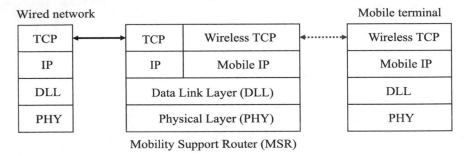

Fig. 6.10 Protocol stack for the end-to-end TCP connection with ITCP [10]

congestion state probe. Many traffic control approaches originally investigated for Asynchronous Transfer Mode (ATM) network [8, 9] are also adopted here.

6.4.2   Splitting of TCP Connection

When a TCP connection spans both wired networks and mobile networks, TCP performs poor due to the problems in mobile networks as mentioned earlier. One solution is to split the end-to-end TCP connection into two segments: one for the wired network and the other for the mobile network, such as Indirect TCP (ITCP) [10]. In this case, a TCP connection for the wired network runs the normal TCP, and a specific TCP is designed for the mobile network as illustrated in Fig. 6.9. The TCP for the wired network keeps unchanged.

This design violates the fundamental Internet design principle: end-to-end arguments, which basically suggests that the network functions should be placed as much as possible at the end points, rather than inside the network in order to retain a simple, robust and high-performance network [11, 12]. With ITCP, a networking unit located between wired networks and mobile networks, called mobility support router (MSR), have to run TCP as illustrated in Fig. 6.10, which originally should be installed only at end points as mentioned earlier. Such splitting complicates these networking units, and also introduces extra delay caused by such as encapsulation and decapsulation between TCP segments and IP packets.

6.4.3 Precise Congestion Inference

Many efforts have been made to allow a source node to make more accurate judgement on the real congestion status in the network without modifying the protocol stack. For example, the Fixed RTO [13] tries to find the cause of packet loss through heuristic analysis on received ACKs. TCP Veno [14] tries to distinguish congestive losses and non-congestive loss. Jitter-based TCP (JTCP) [15] estimates the congestion level according to a jitter ratio.

Moreover, some of them try to modify the protocol stack slightly. For example, TCP Detection of Out-of-Order and Response (TCP-DOOR) [16] allows TCP ACKs to carry more information. A Delayed Congestion Response TCP (TCP-DCR) [17] delays the response to duplicate ACKs by a certain time to allow the data link layer to recover frames received in error as much as possible. Only if the link layer fails in doing so, the fast retransmission and fast recovery processes of TCP are invoked. This modification can be realized using an inter-layer interaction scheme in the end-point.

The following sections provide more descriptions on congestion inferring methods proposed by TCP Veno, JTCP and TCP-DOOR.

6.4.3.1 TCP Veno

It jointly exploits TCP Vegas [18] and TCP Reno by adopting a simple mechanism proposed by Vegas to distinguish congestive losses and non-congestive loss [14]. Vegas uses the following RTT settings to determine the network condition: the minimum of measure RTT ($BaseRTT$) and the smoothed RTT measured (RTT) calculated by (6.5). When $RTT > BaseRTT$, there should be a bottleneck node, in which the IP packets of the TCP connection accumulate. Consider the current source rate $\frac{cwnd}{RTT}$ over an RTT with the current $cwnd$, we can have an estimation as follows:

$$RTT = BaseRTT + N \times \frac{RTT}{cwnd}, \tag{6.5}$$

from which, we can have an estimated backlog (N) in the queue of the bottleneck node as follows:

$$N = (RTT - BaseRTT) \times \frac{cwnd}{RTT}. \tag{6.6}$$

The major objective of Vegas is to keep N to a small range so that a proactive adjustment of $cwnd$ can be used [14].

However, the above adjustment puts Vegas in an unfavorable position when Reno flows coexist [19] because well-behaving Vegas is less aggressive than Reno in increasing $cwnd$. Furthermore, in an asymmetric network where the bottleneck node on the reverse is path caused by ACK, Vegas would not continue to increase $cwnd$ to transmit more segments on the forward path, whereas Reno continues increasing $cwnd$ [14].

The main idea of Veno is to use the measurement of N as a congestion indication for the TCP connection as follows. If $N < \beta$, when a segment loss is detected, Veno infers this loss as a non-congestive loss. In this case, its window adjustment is the same as Reno's, i.e., $cwnd = cwnd + 1$ for every received new ACK. If $N \geq \beta$, a detected segment loss is treated as a congestive loss, and its window adjustment is different from Reno's as follows: $cwnd = cwnd + 1$ for every other new ACK, which actually corresponds to two RTTs so that the congestion window can stay longer in this operating region. The experimental result in [14] shows that $\beta = 3$ should be a good setting.

6.4.3.2   Jitter-Based TCP (JTCP)

JTCP tries to use an interarrival jitter ratio to reflect the queue state of the bottleneck node. If any packet is queued, the queue length will grow, which may lead to congestion. A jitter ratio estimation is proposed to detect congestion. The interarrival jitter between two consecutive segments, say $D(i, i - 1)$ between segments i-1 and i, is the difference between the segment spacings at the source node and the destination node [20], i.e.,

$$D(i, i - 1) = [t_R(i) - t_R(i - 1)] - [t_S(i) - t_S(i - 1)]$$
$$= [t_R(i) - t_S(i)] - [t_R(i - 1) - t_S(i - 1)], \qquad (6.7)$$

where S and R denote the sending time from the source node and receiving time at the destination node for segments i-1 and i, respectively.

Consider the queueing system at a bottleneck node of a TCP connection, and a TCP flow arriving at the system with a fixed interarrival spacing t_A. Let t_D be the packet delay in the system. The service rate of the system can be approximated by $\frac{1}{t_D}$. Then the ratio of the packets queued to those arriving at the system is approximated as follows [21]:

$$1 - \frac{\frac{1}{t_D}}{\frac{1}{t_A}} = \frac{t_D - t_A}{t_D}$$
$$\approx \frac{[t_R(i) - t_R(i - 1)] - [t_S(i) - t_S(i - 1)]}{t_R(i) - t_R(i - 1)}$$
$$= \frac{D(i, i - 1)}{t_R(i) - t_R(i - 1)}. \qquad (6.8)$$

Based on (6.8), an average jitter ratio (Jr) is calculated based on the samples for one RTT interval, and another ratio $\frac{w}{k}$ is defined for congestion detection, where w denotes the current $cwnd$, and $k < 1$ is a control parameter. That is, if $Jr > \frac{w}{k}$, a congestion state is inferred.

6.4.3.3 TCP Detection of Out-of-Order and Response (TCP-DOOR)

TCP-DOOR [16] tries to find out the following two events in segment receptions that might mislead congestion control in the source node. The first event is out-of-sequence due to a retransmission of lost segments, which causes a previous sequence number to be repeated. The second event is out-of-order, with which a segment sent earlier arrives later than a subsequent one. This results because the routing protocol dynamically changes network paths according to realtime traffic load in the network.

The first event is already handled by TCP, but not for the second one, which may happen to both data segments received by the destination node and ACKs arriving at the source node. Particularly, an out-of-order ACK leads to halving both $cwnd$ and $ssthresh$, or even invoking the slow start process if an RTO happens. All these operations may cause an unnecessary decrease of TCP throughput.

A scheme is proposed to detect these events. An out-of-order ACK event can be detected through sequence number comparison except duplicate ACKs. For the out-of-order data segment event, the time-stamp [22] sealed by the source node to each data segment can be used by the destination node to distinguish the transmission order provided that a network-level time synchronization is available. Once detected, the destination node notifies the source node by setting some bit in the ACK. When the source node receives this notification, it temporarily disables congestion control operations by keeping its state variables constant (e.g., RTO). If some window control operations (e.g., halving $cwnd$) were taken due to misjudgement from either an RTO or three duplicate ACKs before the explicit information is received, an instant recovery is invoked to allow the node to operate immediately in the previous state before the action was invoked [16].

6.4.4 Route Failure Notification

In a multi-hop mobile network, route failure may happen frequently due to either link breakage caused by mobility or node failure because of battery exhaustion, for example. Once such a failure happens, packets traveling along the affected route will be delayed due to rerouting. If such delay is longer than what the packets can tolerate, they are dropped. Apparently, these packet losses are not caused by congestion, and the network layer should notify the source node to avoid unnecessary actions.

Typical schemes in the literature include TCP with Feedback (TCP-F) [23], TCP-BuS [24] and the Explicit Link Failure Notification (ELFN) [25]. Since these schemes require cooperation from the network layer to detect route failure, they lose the

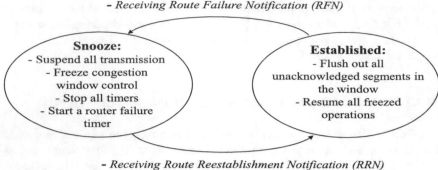

- Receiving Route Failure Notification (RFN)

- Receiving Route Reestablishment Notification (RRN)
- Or router failure timeout

Fig. 6.11   The TCP-F state machine [23]

end-to-end semantics of TCP. Furthermore, both TCP-BuS and ELFN follows
TCP-F, and are mainly designed for particular routing protocols, i.e., Associativity-
Based Routing (ABR) [26] and Dynamic Source Routing (DSR) [27], respectively.
Since DSR is ratified by IETF as mentioned earlier, more descriptions on TCP-F and
ELFN are provided below.

6.4.4.1   TCP with Feedback (TCP-F)

TCP-F [23] allows a source node to enter a snooze state instead of the window
operation when being notified of route failure, and restore to the normal operation
when a route is established, using two messages: route failure notification (RFN)
and route re-establishment notification (RRN). Once an intermediate node detects a
route disruption, it sends back an RFN message to the source node. Once receiving an
RFN, the source node enters the snooze state with a timer setting, and all operations
are suspended until either the arrival of an RRN or the timeout. When this happens,
the node enters the active state and restores all normal operations as illustrated in
Fig. 6.11.

6.4.4.2   Explicit Link Failure Notification (ELFN)

ELFN [25] aims to provide the source node with information on link and route failures, jointly using the "destination unreachable" ICMP message. Alternatively, some
routing protocols can detect link breakage and route failure, and this information can
be piggy-backed on a message to be sent to the source node. In [25], the DSR's
route failure message is modified to carry such kind of information, which can also
identify a particular TCP connection through the source-destination address pair.

Once a source node receives such information, it suspends congestion control operation until the route has been restored. That is, when a source node receives an ELFN, it disables its retransmission timers and enters a standby state (similar to the persist state of TCP-Peach), in which, a probe segment is sent periodically to check if a route has been established. If an ACK is received, the node leaves the standby state and restores its retransmission timers with normal operation [25].

6.4.5   Congestion State Probing

Since a source node cannot know congestion status in the network, probing schemes are proposed to allow a source node to act properly, such as TCP-Peach [28], TCP-Peach+ [29] and TCP-Probing [30]. TCP-Probing is similar to TCP-Peach and TCP-Peach+ in terms of sending segments to probe the network state. The major differences include (i) the time when probing operation is initiated, (ii) responses to the probing results and (iii) probing segment structure, as discussed below.

6.4.5.1   TCP-Peach

It is proposed for satellite networks, in which bandwidth delay product is large, and TCP cannot perform well due to its slow start process, which largely depends on RTT [28]. Satellite networks are similar to underwater wireless acoustic networks (UWANs) in terms of long propagation delay, so research results discussed here can provide a reference to UWANs.

According to [31], if a TCP ACK is transmitted for every received segment, the time for the slow start process to reach a bit rate B is given by

$$t_{SlowStar} = RTT \times (1 + \log_2 \frac{B \times RTT}{l})$$
$$= RTT \times (1 + \log_2 RTT + \log_2 B - \log_2 l), \quad (6.9)$$

where l is the average segment length in bits. Obviously, (6.9) shows that $t_{SlowStar}$ increases quickly with RTT. As calculated in [28], for $B = 155 Mbps$, $t_{SlowStar}$ for LEO, MEO and GEO is respectively equal to 0.55s, 3.31s and 7.91s, each corresponding to RTTs of 50ms, 250ms and 550ms. In this case, the available bandwidth is under utilized.

To tackle the above problem, TCP-Peach proposes two algorithms: sudden start and rapid recovery, in addition to the original fast retransmission and collision avoidance. To probe available network resource in terms of bandwidth and buffer space, a source node sends low priority dummy segments, which actually is the copy of the last transmitted segment. When a dummy segment arrives at an intermediate node, it is dropped in the case of a congestion; otherwise, it passes through the node. The destination node recognizes dummy segments, and acknowledges the source node with corresponding ACKs. Therefore, the source node can estimate the available bandwidth based on the number of acknowledged dummy segments, and increases $cwnd$ accordingly. To save bandwidth, the dummy segment is replaced by low priority data segments in TCP-Peach+ [29].

With the sudden start, at the beginning of a TCP connection, the source node sends many dummy segments following the first data segment so that $cwnd$ increases quickly. Let $rwnd$ denote the maximum size of the congestion window defined by the destination node. A number of $(rwns - 1)$ dummy segments will be sent with a transmission interval equal to $\frac{RTT}{rwnd}$ between two consecutive dummy segments.

The rapid recovery aims to tackle the throughput degradation problem due to the misjudgement of a loss actually caused by link error as a congestive loss. It is invoked following the original fast retransmission for a loss indicated by duplicate ACKs, and lasts for one RTT until the ACK for the retransmitted segment is received. During this period, $cwnd$ is increased by 1 every ACK reception for either a data segment or a dummy segment. Upon receiving an ACK for a data segment, the node sends two dummy segments if the allowed dummy segment number is positive, which enables a rapid recovery.

6.4.5.2   TCP-Probing

With this TCP, a source node initiates a probing operation whenever it observes either an RTO event or the arrival of three duplicate ACKs, but no window control operation defined by TCP corresponding to these events is performed [30]. Instead, the source node initiates a probing cycle by sending out some specific segments to probe the network status, and acts accordingly according to probing results. For random loss, the probe cycle will finish much more quickly, and neither $cwnd$ nor $ssthresh$ is adjusted downwards, which is called "immediate recovery". Particularly for duplicate ACKs, $cwnd$ is increased when more duplicate ACKs are received. However, when a congestion happens, the duration of the probe cycle will be longer. In this case, the normal congestion control operations defined by TCP are carried out, i.e., the slow start for RTO and halving $cwnd$ and $ssthresh$ for three duplicate ACKs.

6.4.6   MAC Contention Relief

MAC contention severely impacts TCP performance in multi-hop mobile networks because it may cause many non-congestive losses. This contention will be further intensified by hidden terminals and bursty packet arrivals as well as TCP ACKs in the reverse direction. Some schemes proposed to reduce MAC contention in the literature are briefly described below.

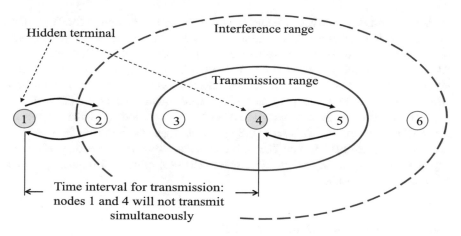

Fig. 6.12 Principle of TCP-AP

6.4.6.1 Hidden Terminal Avoidance

The hidden terminal problem in multi-hop mobile networks cannot be resolved completely by the RTS/CTS protocol of IEEE 802.11. A transport layer scheme, called TCP with Adaptive-Pace (TCP-AP) [32], is investigated to address this issue. That is, a source node controls the time interval of sending segments close to a 4-hop propagation delay, and jointly uses congestion window control to reduce the number of collisions caused by hidden terminals.

As illustrated in Fig. 6.12 for an IEEE 802.11 network, the above time interval can be adjusted corresponding to the distance between two mutually hidden terminals so that they will not transmit concurrently, such as nodes 1 and 4. It has been demonstrated that TCP-AP can perform better than TCP-NewReno [33] in multi-hop mobile networks. Obviously, its efficiency largely depends on the estimation accuracy of the 4-hop propagation delay, which however is dynamic by nature in multi-hop mobile networks due to time-varying traffic loads and mobility.

6.4.6.2 ACK Transmission Control

An ACK is much smaller than a data segment in sizes. However, to send an ACK, the same MAC contention process is invoked with the same amount of protocol overhead as for sending a data segment in wireless networks. This process consists of CSMA/CA and the RTS/CTS handshake as well as MAC ACK with IEEE 802.11. Therefore, reducing the number of ACK transmissions in the reverse direction can reduce the MAC contention in the forward direction because a wireless channel is often shared by transmissions in both directions.

Actually, an delayed ACK scheme has been defined as an option for TCP ACK in [34], stipulating that an ACK is generated for every d segments within a certain time interval (t). That is, if no more data segment arrives beyond the time interval, an ACK is sent anyway. By default, $d = 2$ and $t = 100$ ms. The study in [35] shows that letting $d > 2$ and $t > 100$ ms can further improve TCP performance in multi-hop wireless networks. Furthermore, Adaptive Delayed Acknowledgement (TCP-ADA) in [36] delays sending the accumulative ACKs by a time period equal to $\alpha \times \tau$, where α is a waiting factor, which is set to $0.8 \sim 1.2$ as suggested by the authors. And τ is the inter-arrival time between TCP data segments, which is averaged by using the exponential averaging algorithm. With these schemes, more data segments can be acknowledged per ACK.

6.4.6.3 Disjunction of Data and ACK Routes

A wireless link is often shared by transmissions in forward and reverse directions of TCP connections, e.g., data segments are sent along the forward path, while ACKs are sent in the reverse path. These transmissions may become synchronized when the two paths go through the same link, resulting in high MAC contention over this link. In this case, ACKs may be either delayed or even dropped, which affects TCP window operation. Thus, the Contention-based Path Selection (COPAS) protocol [37] tries to disjoint the routes for data segments and ACKs, using a routing protocol as described below.

Basically, COPAS is an on-demand routing protocol, jointly using a metric that can reflect medium contention degree in path selection. This metric is calculated according to the number of MAC backoff operations conducted by a node during a pre-defined time interval, and averaged with the exponential averaging algorithm. The larger this value, the higher MAC contention degree.

The above calculation is performed by each node. Once receiving an RREQ, a node will let this RREQ carry the newly averaged value of this metric, which is used by the destination node to determine the least contented node-disjoint paths in both directions for symmetric links. The destination node collects RREQs arriving during a predefined time interval, and selects node-disjoint paths first. From the selected paths, it computes least contended paths through evaluating the sum of the contention metric carried by each RREQ along the selected node-disjoint path. The path with the minimum sum is selected. Finally, the destination node replies to the source node with two RREPs, one for the forward path and the other for the reverse path.

6.4.7 ATM-like Control

The schemes discussed here include explicit congestion notification (ECN), traffic rate based control and *cwnd* based on available bandwidth estimation.

6.4.7.1 Explicit Congestion Notification (ECN)

Since many packet losses are not due to congestion, instead of notifying each non-congestive packet loss, it should be simple to only notify congestive packet loss to the source node so that it does not need to infer congestion status in the network. Thus, upon receiving an ECN message, the source node acts accordingly, such as Ad Hoc TCP (ATCP) [38].

ATCP takes advantage of an ECN scheme of the network layer to detect congestion status in the network. Three states are defined for a source node to take actions, namely, persist state, congestion control state and retransmit state. In the persist state when the network is partitioned, the node neither transmits nor retransmits without congestion control. When a non-congestive loss is detected, the node is in the retransmit state, where only retransmission is conducted for the failed packets without congestion control. When a congestive loss is detected, the node is in a congestion state, where the normal TCP congestion control is performed.

In order to render the original TCP implementation intact, an ATCP sub-layer is inserted between the network layer and the transport layer. This sub-layer listens to the network to detect congestion via ECN messages or the connectivity to the destination node via ICMP messages, and turns the source node into the appropriate state. For example, if the destination node is not reachable indicated by an ICMP message, the node enters the persist state.

6.4.7.2 Rate-Based Control

The window-based flow control of TCP causes segments to arrive at the lower layers in a burst, which intensifies MAC contention degree. A rate-based flow control can smooth packet arrival burstiness. Actually, the above-mentioned TCP-AP is a rate-based protocol, and the same for the Ad Hoc Transport Protocol (ATP) discussed below.

With ATP, congestion control is based on congestion information feedback from the network layer rather than TCP ACKs, which is originally coupled with transmission reliability control in TCP. However, here a selective ACK (SACK) scheme is adopted for reliability control, and with this decoupling, the following congestion estimation is proposed for end-to-end rate-based congestion control.

Each intermediate node in the network for a TCP connection maintains the sum of the average queueing delay (Q_t) and the average transmission delay (T_t) experienced by each packet passing it, i.e., $Q_t + T_t$. The node conducts an exponential averaging of Q_t and T_t, respectively, i.e.,

$$Q_t = \alpha Q_t + (1 - \alpha) Q_{sample} \tag{6.10}$$

and

$$T_t = \alpha T_t + (1 - \alpha) T_{sample}, \tag{6.11}$$

where Q_{sample} and T_{sample} are the queueing delay and transmission delay experienced by the outgoing packet, respectively. The node overwrites a by-passing packet with the updated $(Q_t + T_t)$ if it is larger than that carried by the packet. The destination node conducts an exponential averaging of the sums (D) carried by the received segments as follows:

$$D_{avg} = \beta D_{avg} + (1 - \beta)D. \tag{6.12}$$

The destination node performs flow control by observing the rate (R_{app}) at which the application is processing in-sequence data from the receive buffer. If R_{app} is smaller than $\frac{1}{D_{avg}}$, it records R_{app} to the D_{avg} field of the feedback segment to be sent to the source node; otherwise, the currently calculated D_{avg} is used instead. The source node determines its transmission rate (i.e., $\frac{1}{D_{avg}}$) according to the feedback.

6.4.7.3 Available Bandwidth Estimation

With TCP, whenever a source node perceives a segment loss either through the reception of three duplicate ACKs or an RTO event, it halves $cwnd$ and $ssthresh$, or even invokes the slow start process, regardless of the bandwidth available in the network. Thus, some enhancements are proposed to estimate the available bandwidth to properly set $cwnd$ and $ssthresh$. Examples include TCP Westwood [39], TCP Westwood+ [40] and TCP-Jersey [41].

With TCP Westwood, a source node monitors the incoming ACK flow to estimate bandwidth available to the corresponding TCP connection. As discussed in [42], a congestion occurs whenever the low-frequency input traffic rate exceeds the link capacity, so a lowpass filter is used to average sampled measurements for low-frequency components of the available bandwidth. Actually, TCP-Jersey takes the same idea as used by TCP Westwood to estimate the available bandwidth but with a simpler estimator [41] as discussed below.

With TCP-Jersey, a source node employs an available bandwidth estimator derived from the time sliding window estimator [43] to estimate the bandwidth available for each TCP connection through monitoring the rate of received ACKs as follows:

$$R_n = \frac{RTT \times R_n + L_n}{(t_n - t_{n-1}) + RTT}, \tag{6.13}$$

where R_n is the available bandwidth estimated when the n_{th} ACK arrives at time t_n, L_n is the size of the segment acknowledged by this ACK, and RTT is the actual round-trip propagation delay estimated at time t_n.

Then the source node calculates for every RTT the optimum congestion window size $(ownd_n)$ in units of TCP segments based on the estimated available bandwidth as follows:

$$ownd_n = \frac{RTT \times R_n}{seg_{size}}, \tag{6.14}$$

which is used to set the congestion window once a congestion is inferred, where seg_{size} is the segment size.

6.4.8 Decoupling of TCP Functions

Basically, TCP has two major functions: transmission reliability control and congestion control. They are coupled in a way that the congestion status in the network is judged through ACK reception, which is also used for transmission reliability control. The frequency of ACKs to be sent by a destination node affects the $cwnd$ setting at the source node. The more frequently ACKs are sent, the larger $cwnd$ can increase, resulting in more segments to be sent into the network. However, more ACKs sent by a destination node intensifies MAC contention in the reverse direction, while any ACK loss will be misjudged by the source node as a congestion event in the forward direction, leading to unnecessary reduction of $cwnd$ and eventually degrading packet forwarding performance in the forward direction.

One solution to this problem is to decouple these two functions, and use other methods to infer congestion status. Indeed, several proposals investigated the above decoupling approach, typically including two types. The first ones, which may be different in particular designs, are common in terms of that the decoupled two controls are still performed by TCP, such as ATP [44] mentioned above. The second type goes further by moving congestion control function down to lower layers such as the data link layer. In this case, only transmission reliability control is retained in TCP for end-to-end reliable transmission. A typical example is Semi-TCP [45], which is introduced below.

6.4.8.1 Constraint of Congestion Window

The setting of $cwnd$ in TCP largely depends on the RTT between the source node and the destination node. The RTT mainly consists of the packet queueing delay, transmission time and propagation delay. The former two both increase with the number of hops in the path and transmission rate of each hop, while the third one increases with path length.

According to [46], in a wired network, the maximum sending rate for a TCP connection over a single cycle in the steady state is given by

$$\lambda_m \leq \frac{1.5l}{R} \sqrt{\frac{2}{3p}}, \tag{6.15}$$

where R is the minimum RTT, p is link loss probability in the steady-state, and l is the maximum segment size.

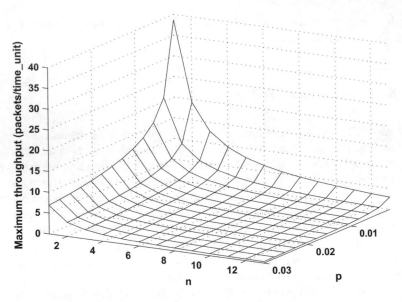

Fig. 6.13 Maximum throughput of TCP ($\frac{l}{\tau} = 1$) versus p and n [47]

Consider a TCP connection consisting of n links each with a link RTT of τ and a link loss probability p. In this case, R can be estimated by $n\tau$, and the goodput at which packets can reach the destination node is bounded by [47]

$$\mu_m \leq \frac{l}{n\tau} \sqrt{\frac{3}{2p}} (1 - p)^n. \qquad (6.16)$$

The numerical results are depicted in Fig. 6.13, which shows that the goodput decreases rapidly with n especially with small p, hence that n is the dominant factor of TCP performance. This phenomenon has been observed empirically in the literature such as [48].

6.4.8.2   Semi-TCP

It can avoid all the misjudgments on congestion status in multi-hop mobile networks mentioned above because congestion control is performed completely by the data link layer. Therefore, congestion control window is no longer used by a source node to arbitrate the number of segments to be sent to the network. Instead, the buffer space available on the data link layer can be used in traffic control. Similarly, the amount of traffic load from one node to the other node also depends on the availability of the buffer space at each node along the network path. In this case, the above constraint of the RTT-determined $cwnd$ on TCP performance improvement no longer exists.

It can also leverage hop-by-hop congestion control to speed up the reaction to both a congested state and a released congestion. This feature is especially useful to highly mobile networks, in which congestion status may change rapidly due to mobility. More discussion is given below.

6.4.8.3   Hop-by-Hop Congestion Control

With Semi-TCP, congestion control is carried out by the data link layer hop-by-hop. With such control, the upstream node of a link controls its traffic load to the downstream node according to the congestion status therein. If the downstream node is congested, the upstream node should reduce or even stop transmission to it. In the case of no congestion, the amount of traffic to be sent should be also controlled to avoid possible congestion. Such kind of congestion control is much more efficient than end-to-end congestion control adopted by TCP due to its faster and more accurate reaction to the congestion status in the network.

With TCP, a source node cannot react quickly to an ongoing congestion especially with a large RTT, because once a congestion happens, the minimum reaction time for the source's reaction to arrive at the congested node is one RTT. It will also take at least one RTT for the source node to learn whether a congestion state is relieved. More time is needed by the source node to restore its $cwnd$ to the normal level due to the slow start and congestion avoidance processes. With Semi-TCP, the upstream node can take an immediate action to the congestion at its downstream node by a link RTT, which is much shorter than an end-to-end RTT. Similarly, the upstream node can also learn quickly a relieved congestion in its downstream node. Furthermore, with Semi-TCP [45], duplicate ACKs are not necessary, while a delayed ACK can be adopted to accommodate more received data segments. Both features can relive the effect of ACK traffic in the reverse direction on data segment forwarding in the forward direction.

A major concern for hop-by-hop congestion control is its implementation complexity due to per-node's participation. In shared-media wireless networks, each node may need to detect activities of other nodes and even interact with them for error control and MAC. Thus, there are some mechanisms used to capture and exchange information between neighbors. In this case, it is relatively easy to implement a hop-by-hop congestion control through piggy-backing it with these mechanisms without increasing implementation complexity. In the IEEE 802.11 DCF [49], the RTS/CTS handshake is used to solve the hidden terminal problem. It can be slightly modified by including the information on congestion status for hop-by-hop congestion control [50]. Actually, IEEE 802.11s also consolidates such function into the MAC sub-layer for wireless mesh networks [51, 52].

There are also some other hop-by-hop congestion control schemes at the data link layer such as those discussed in [53, 54], while cross-layered hop-by-hop congestion control schemes have also been investigated [55–57]. Note that all these schemes do not decouple congestion control from TCP.

6.4.8.4   Connection with TCP

When Semi-TCP and TCP compose an end-to-end connection, the end-to-end semantics can be maintained because Semi-TCP does not change the formate of both the data segment and ACK segment, while the behavior of both destination nodes and source nodes to ACK segments can also be kept intact. The primary problem is how to achieve good TCP performance over the end-to-end connection, which actually is an issue that all TCP proposals enhanced for multi-hop mobile networks have to address. Two approaches discussed above can be used here to solve this problem. One is to split the end-to-end connection into two parts, similar to I-TCP, as illustrated in Fig. 6.14. In this case, all advantages of Semi-TCP can be exploited to maximize TCP performance over multi-hop mobile networks. However, the gateway between TCP and Semi-TCP needs to impersonate the destination node of each TCP connection to relay segments towards both sides.

An alternative implementation able to preserve the semantics is to add a sub-layer above the data link layer like ATP. The sub-layer of the gateway generates TCP ACKs according to the its available buffer space to prevent TCP source nodes from unnecessary *cwnd* shrinking for non-congestive loss, and buffers the data frame sent over the wireless network of a TCP data segment because impersonated ACK sent by the gateway may cause the source node to flush out TCP segments still waiting for acknowledgement.

A hybrid implementation of Semi-TCP can combine the above two approaches. The sub-layer can delay TCP ACKs sent to a TCP source node in the case of non-congestive losses to create a longer RTT, which is used to calculate a larger RTO to

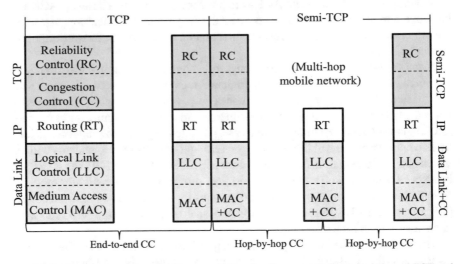

Fig. 6.14 Displacement of congestion control function for a connection consisting of TCP and Semi-TCP [58]

eventually reduce RTO occurrence. This sub-layer needs to exploit routing information jointly to check whether the destination node of a TCP ACK is a TCP source node.

6.5   Discussion

Since the actual TCP originally proposed for wired networks have been widely implemented in practice, a major difficulty in tackling the problems of TCP in multi-hop mobile networks is to allow the enhanced TCP to work with the actual TCP without changing the end-to-end semantic, while improving the overall performance of the entire TCP connection. However, current research mainly focuses on improving TCP over multi-hop mobile networks with less attention on seamless end-to-end connectivity between enhanced TCP and the actual TCP.

The above discussion show that some modifications to the actual TCP or other network functions are necessary to improve TCP performance over multi-hop mobile networks. The major issue is how a modification may affect end-to-end semantics and the compatibility with the implemented TCP, and complicate the network nodes especially battery-operated wireless terminals. Some related issues and possible development are discussed below.

6.5.1   End-to-End Semantics

Here the end-to-end semantics of TCP are discussed for two scenarios: (i) only wireless networks are under consideration, and (ii) wireless networks connect wired networks. For the first scenario, if a modification to networking functions are conducted or a new sub-layer are added both below the transport layer, the original TCP implementation can keep unchanged, such as ATCP and COPAS. Almost all early Internet applications were developed based on the actual TCP designed for wired networks, and it is impossible to conduct modification to them to improve TCP performance over wireless networks. Thus, the enhanced TCP except I-TCP and Semi-TCP cannot work well in the second scenario as discussed above. More discussion on TCP for last-mile access networks can be found in [7].

6.5.2   Compatibility with TCP

The modified TCP proposals discussed in Sect. 6.4 have different compatibilities with the actual TCP. Consider the second scenario. A TCP compatible modification can work smoothly with the implemented TCP without protocol conversion between them nor severe side-effect on the normal window operation of the implemented

TCP. Examples include TCP Veno, JTCP, TCP-AP, COPAS, TCP Westwood and TCP Jersey as well as Semi-TCP. These modifications do not change the formats of the data segment and ACK segment without any special segments that the actual TCP cannot recognize. However, such kind of modifications cannot completely get rid off the problems of TCP because congestion inference and congestion control are still performed at the transport layer except Semi-TCP.

6.5.3   Implementation Complexity

Once a modification is carried on the transport layer or layers below, it affects every node in MANETs because every node in this network is functionally identical. A node is a terminal that implements all protocol stacks including TCP, while it is also a router relaying packets. Therefore, the proposed modifications increase the complexity of wireless nodes. Many proposals modify only one layer, such as I-TCP, TCP-DOOR, TCP Veno, JTCP and TCP-AP. They require sender-side changes in TCP. For those relying on explicit notification, they need cooperation from the network layer or even the data link layer to detect congestion status. In this case, more layers are involved. For example, TCP-F, ELFN, TCP-Peach and ATCP modify the actual implementation or require assistance from the network layer. Semi-TCP is an exception although it affects the transport layer and data link layer simultaneously. It removes the congestion control from TCP, which is the most complex operation of TCP, while this function can be piggy-backed to the existing functions already implemented at the data link layer.

6.5.4   Evolution of TCP Research

When wireless networks are used as last-mile access network to wired networks, mainly addressed issues are to handle transmission failures due to channel error and mobility, such as I-TCP. In multi-hop mobile networks, research mainly focuses on improving TCP performance over wireless networks, especially being inspired by several reports of TCP's poor performance in such networks (e.g., IEEE 802.11) [48, 59, 60]. However, less research is conducted e to handle issues on whole connections crossing both wired and wireless networks as discussed below.

One objective of many proposals is to obtain the information on the congestion status more accurately for more efficient reaction. Thus, more sophisticated congestion inferring methods are devised, such as monitoring ACK rates or sending probing segments as proposed by TCP-DOOR, TCP Veno and JTCP. Eplicit state information on either route failure or congestion state is further adopted so that more precise decision can be made. These proposals include TCP-BuS, ELFN, TCP-Peach and ATCP. Some problems of wireless networks that affect TCP performance are also tackled, such as the hidden terminal problem with TCP-AP, and MAC contention relief with

COPAS and TCP-Jersey, for example. Rate-based congestion control approach is also investigated to replace the window-based congestion control of TCP, such as TCP-AP and TCP-Jersey. A "more aggressive" modification is to decouple transmission reliability control and congestion control of TCP (e.g., ATP), both of which rely on ACK reception. Such decoupling can provide more space to improve congestion control performance without impact on reliability control. However, ATP still keeps these two functions in TCP, so congestion control still face the same problem as TCP does. Semi-TCP moves congestion control function to the data link layer, which can know the congestion status in the network precisely and immediately, and take corresponding actions quickly. Note that IEEE 802.11s has ratified a MAC-level congestion control framework [52].

6.6 Summary

TCP over wireless networks has been studied for more than two decades, and there is not yet a TCP standard that considers the specific problems that TCP faces for multi-hop mobile networks as mentioned earlier. Although there are many proposals reported in several surveys available [6, 7, 61–65], there are still some key points that have not been addressed. It seems that most proposals treat a multi-hop mobile network as an isolated island, and do not investigate its interconnection with the actual TCP implementation, which however cannot be modified following the enhanced TCP proposals for multi-hop mobile networks. Different from the TCP for wired networks, which is 1-layer issue of the transport layer, TCP for multi-hop mobile networks may involve other layers such as the network layer and the data link layer. There is not yet a standard framework for such cross-layer design.

References

1. Paxson, V., Allman, M.: Computing TCP's Retransmission Timer. IETF RFC 2988 (2000)
2. Xie, X.L.: Computer Networks (in Chinese), 6th edn. Electronic Industry Press, Beijing (2013). ISBN 978-7-121-13072-4
3. Karn, P., Partridge, C.: Estimating round-trip times in reliable transport protocols. In: Proceedings of ACM SIGCOMM. Stowe, VT, USA (1987)
4. Allman, M., Paxson, V., Stevens, W.: TCP Congestion Control. IETF RFC 2581 (1999)
5. Floyd, S., Henderson, T., Gurtov, A.: The NewReno Modification to TCP's Fast Recovery Algorithm. IETF RFC 3782. (2004)
6. Hanbali, A.A., Altman, E., Nain, P.: A survey of TCP over ad hoc networks. IEEE Commun. Surv. Tutor. **7**(3), 22–36 (2005). 3rd Quarter
7. Sardar, B., Saha, D.: Survey of TCP enhancements for last-hop wireless networks. IEEE Commun. Surv. Tutor. **8**(3), 20–34 (2006)
8. Jain, R., Ramakrishnan, K.K.: Congestion avoidance in computer networks with a connectionless network layer: concepts, goals and methodology. In: Proceedings of Computer Networking Symposium, pp. 134–143 (1988)

9. Ramarkrishnan, K.K., Jain, R.: A binary feedback scheme for congestion avoidance in computer networks. ACM Trans. Comput. Syst. **8**(2), 158–181 (1990)
10. Solomon, J.D.: Mobile IP. The Internet Unplugged. Prentice-Hall, Upper Saddle River (1998)
11. Saltzer, J.H., Reed, D.P., Clark, D.D.: End-to-end arguments in system design. ACM Trans. Comput. Syst. **2**(4), 277–88 (1984)
12. Clark, D.: The design philosophy of the DARPA internet protocols. In: SIGCOMM '88: Symposium Proceedings on Communications Architectures and Protocols, pp. 106–114. New York, USA (1988)
13. Dyer, T., Boppana, R.: A Comparison of TCP performance over three routing protocols for mobile ad hoc networks. In: Proceedings of ACM International Symposium Mobile Ad Hoc Networking & Computing (MobiHoc), pp. 56–66. Long Beach, CA, USA (2001)
14. Fu, C.P., Liew, S.C.: TCP veno: TCP enhancement for transmission over wireless access networks. IEEE J. Sel. Areas Commun. **21**(2), 216–228 (2003)
15. Wu, E.H.K., Chen, M.Z.: TJTCP: jitter-based TCP for heterogeneous wireless networks. IEEE J. Sel. Areas Commun. **22**(4), 757–766 (2004)
16. Wang, F., Zhang, Y.: Improving TCP performance over mobile ad hoc networks with out-of-order detection and response. In: Proceedings of ACM International Symposium Mobile Ad Hoc Networking & Computing (MobiHoc), pp. 217–225. Lausanne, Switzerland (2002)
17. Bhandarkar, S., Sadry, N.E., Reddy, A.L.N., Vaidya, N.H.: TCP-DCR: a novel protocol for tolerating wireless channel errors. IEEE Trans. Mob. Comput. **4**(5), 517–529 (2005)
18. Brakmo, L.S., Oar?Malley, S.W., Peterson, L.L.: TCP vegas: new techniques for congestion detection and avoidance. In: Proceedings of ACM SIGCOMM, pp. 24–35. London England, UK (1994)
19. Mo, J., La, R.J., Anantharam, V., Walrand, J.J.: Analysis and comparison of TCP reno and vegas. In: Proceedings of IEEE INFOCOM, vol. 3, pp. 1556–1563. New York City, NY, USA (1999)
20. Schulzrinne, H., Casner, S., Frederick, R., Jacobson, V.: RTP: A transport protocol for real-time application. IETF RFC 1889 (1996)
21. Chen, S.Y., Wu, E.H.K., Chen, M.Z.: A new approach using timebased model for TCP-Friendly rate estimation. In: Proceedings of IEEE International Conference on Communication (ICC), vol. 1, pp. 679–683. Anchorage, Alaska, USA (2003)
22. Jacobson, V., Braden, R., Borman, D.: TCP extensions for high performance. IETF RFC 1323 (1992)
23. Chandran, K., Raghunathan, S., Venkatesan, S., Prakash, R.: A feedback-based scheme for improving TCP performance in ad hoc wireless networks. IEEE Pers. Commun. Mag. **8**(1), 34–39 (2001)
24. Kim, D., Toh, C., Choi, Y.: TCP-BuS: improving TCP performance in wireless ad hoc networks. J. Commun. Net. **3**(2), 175–186 (2001)
25. Holland, S.G., Vaidya, N.: Analysis of TCP performance on mobile ad hoc network on wireless. ACM Wirel. Netw. (WINET) **8**(2–3), 275–288 (2002)
26. Toh, C.K.: Associativity-based routing for ad hoc networks. Wireless Pers. Commun. **4**(2), 103–139 (1997)
27. Johnson, D.B., Maltz, D.A., Hu, Y.C.: The dynamic source routing protocol (DSR) for mobile ad hoc networks for IPv4. IETF RFC 4728. (2007)
28. Akyildiz, I.F., Morabito, G., Palazzo, S.: TCP-peach: a new congestion control scheme for satellite IP networks. ACM/IEEE Trans. Netw. **9**(3), 307–321 (2001)
29. Akyildiz, I.F., Zhang, X., Fang, J.: TCP-Peach+: enhancement of TCP-peach for satellite IP networks. IEEE Commun. Lett. **6**(7), 303–305 (2002)
30. Lahanas, A., Tsaoussidis, V.: Improving TCP performance over networks with wireless components using probing devices. Int. J. Commun. Syst. **15**(6), 495–511 (2002)
31. Partridge, C., Shepard, T.J.: TCP/IP performance over satellite links. IEEE Netw. Mag. **11**(5), 44–49 (1997)
32. ElRakabawy, S.M., Alexander, K., Christoph, L.: TCP with adaptive pacing for multihop wireless networks. In: Proceedings of ACM International Symposium Mobile Ad Hoc Networking & Computing (MobiHoc), pp. 288–299. New York, NY, USA (2005)

33. Floyd, S., Henderson, T.: The new-reno modification to TCP's fast recovery algorithm. IETF RFC 2582 (1999)
34. Braden, R.: Requirements for Internet Hosts - Comunication Layers. IETF RFC 1122 (1989)
35. Altman, E., Jimenez, T.: Novel delayed ACK techniques for improving TCP performance in multihop wireless networks. In: Proceedings of IEEE International Conference on Personal Wireless Communications, pp. 237–242. Venice, Itlay (2003)
36. Singh, A.K.: Kankipati, K.: TCP-ADA: TCP with adaptive delayed acknowledgement for mobile ad hoc networks. In: Proceedings of IEEE Wireless Communication & Networking Conference (WCNC), vol. 3, pp. 1685–1690. Atlanta, Georgia, USA (2004)
37. Cordeiro, C., Das, S., Agrawal, D.: COPAS: Dynamic contention-balancing to enhance the performance of TCP over multi-hop wireless networks. In: Proceedings of IEEE Conference on Computer Communication and Network (ICCCN), pp. 382–387. Miami, USA (2003)
38. Liu, J., Singh, S.: ATCP: TCP for Mob. Ad Hoc Netw. IEEE J. Sel. Areas Commun. 19(7), 1300–1315 (2001)
39. Casetti, C., Gerla, M., Mascolo, S., Sanadidi, M.Y., Wang, R.: TCP westwood: end-to-end congestion control for wired/wireless networks. ACM Wirel. Netw. (WINET) 8(5), 467–479 (2002)
40. Mascolo, S., Grieco, L.A., Ferorelli, R., Camarda, P., Piscitelli, G.: Performance evaluation of westwood+ tcp congestion control. Perform. Eval. 55(1–2), 93–111 (2004)
41. Xu, K., Tian, Y., Ansari, N.: TCP-jersey for wireless IP communications. IEEE J. Sel. Areas Commun. 22(4), 747–756 (2004)
42. Li, S.Q., Hwang, C.: Link capacity allocation and network control by filtered input rate in high speed networks. ACM/IEEE Trans. Netw. 3(1), 10–25 (1995)
43. Clark, D.D., Fang, W.J.: Explicit allocation of best-effort packet delivery service. ACM/IEEE Trans. Netw. 6(4), 362–373 (1998)
44. Sundaresan, K., Anantharaman, V., Hsieh, H.Y., Sivakumar, R.: ATP: a rellable transport pro tocol for ad hoc networks. IEEE Trans. Mob. Comput. 4(6), 588–603 (2005)
45. Jiang, S.M., Zuo, Q., Wei, G.: Decoupling congestion control from TCP for multi-hop wireless networks: semi-TCP. In: Proceedings of ACM MobiCom Workshop on Challenged Networks (CHANTS). Beijing, China (2009)
46. Floyd, S., Fall, K.: Promoting the use of end-to-end congestion control. ACM/IEEE Trans. Netw. 7(4), 458–472 (1999)
47. Cai, Y.G., Jiang, S.M., Guan, Q.S., Yu, R.: Decoupling congestion control from TCP (semi-TCP) for multi-hop wireless networks. EURASIP J. Wirel. Commun. Netw. 149, 2013 (2013)
48. Xu, S., Saadawi, T.: Does the IEEE 802.11 MAC protocol work well in multihop wireless ad hoc networks? IEEE Commun. Mag. 39(4), 130–137 (2001)
49. IEEE Std 802.11, Medium Access Control (MAC) sub layer and 3 Physical Layer Specifications (1997)
50. Zhai, H.Q., Wang, J.F., Fang, Y.G.: Distributed packet scheduling for multihop flows in ad hoc networks. In: Proceedings of IEEE Wireless Communication & Networking Conference (WCNC), vol. 2, pp. 1081–1086. Atlanta, Georgia, USA (2004)
51. Camp, J.D., Knightly, E.W.: The IEEE 802.11s Extended Service Set Mesh Networking Standard (2007)
52. IEEE Std 802.11s, Part 11: Wireless LAN Medium Access Control (MAC) and Physical Layer (PHY) specifications, Amendment 10: Mesh Networking (2011)
53. Sadeghi, B., Yamdad, A., Fujiwara, A., Yang, L.: A simple and efficient hop-by-hop congestion control protocol for wireless mesh networks. In: Proceedings of Annual International Wireless Internet Conference (WICON). Boston, USA (2006)
54. Scheuermann, B., Locherta, C., Mauve, M.: Implicit hop-by-hop congestion control in wireless multihop networks. Ad Hoc Netw. 6, 260–288 (2008)
55. Chen, K., Nahrstedt, K., Vaidya, N.: The utility of explicit rate-based flow control in mobile ad hoc networks. In: Proceedings of IEEE Wireless Communication & Networking Conference (WCNC), vol. 3, pp. 1921–1926. Atlanta, Georgia, USA (2004)

56. Yi, Y., Shakkottai, S.: Hop-by-hop congestion control over a wireless multi-hop network. ACM/IEEE Trans. Netw. **15**(1), 133–144 (2007)
57. Wang, X.Y., Perkins, D.: Cross-layer hop-by-hop congestion control in mobile ad hoc networks. In: Proceedings of IEEE Wireless Communication & Networking Conference (WCNC). Las Vegas, USA (2008)
58. Jiang, S.M.: Future Wireless and Optical Networks: Networking Modes and Cross-Layer Design. Springer, London, UK (2012)
59. Gerla, M., Tang, K., Bagrodia, R.: TCP performance in wireless multi-hop networks. In: Proceedings of IEEE Workshop on Mobile Computing Systems & Applications (WMCSA), pp. 41–50. New Orleans, LA, USA (1999)
60. Kawadia, V., Kumar, P.R.: Experimental investigations into TCP performance over wireless multihop networks. In: Proceedings of 2005 ACM SIGCOMM Workshop on Experimental Approaches to Wireless Network Design and Analysis, pp. 29–34. New York, USA (2005)
61. Barakat, C., Altman, E., Dabbous, W.: On TCP performance in a heterogeneous network: a survey. IEEE Commun. Mag. **38**(1), 40–46 (2002)
62. Chlamtac, I., Conti, M., Liu, J.: Mobile ad hoc networking: imperatives and challenges. Ad Hoc Netw. J. (Elsevier) **1**(1), 13–64 (2003)
63. Leung, K.C., Li, V.O.K.: Transmission control protocol (TCP) in wireless networks: issues, approaches, and challenges. IEEE Commun. Surv. Tutor. **8**(4), 64–79 (2006)
64. Lochert, C., Scheuermann, B., Mauve, M.: A survey on congestion control for mobile ad hoc networks. Wiley Wirel. Commun. Mob. Comput. **7**(5), 655–676 (2007)
65. Xu, C.Q., Zhao, J., Muntean, G.-M.: Congestion control design for multipath transport protocols: a survey. IEEE Commun. Surv. Tutor. **18**(4), 2948–2969 (2016)

Chapter 7
Mobility in RWNs

Abstract The major advantage of wireless networks over wired networks is mobility support for terminal mobility and service mobility. Terminal mobility allows a user along its terminals for instance mobile phones to be able to move anytime even during a communication in progress. Service mobility permits a user to use the registered services anywhere and anytime with different terminals. Now the term "mobility" may also refer to other capabilities such as session mobility, with which an ongoing session can be switched from one terminal to another without session interruption. Here we focus on the issues due to location change and user motion illustrated in Fig. 1.30, i.e.,

- Roaming service aims to maintain connectivity and services when a user changes the location on a large scale, such as moving from one city or even one country to another, so that such changes will not lead to loss of connectivity and service.
- Handoff denotes the process of changing the channel associated with a communication in progress, and handoff support aims to maintain the continuation of such a communication when the user moves in the course of communication.

Handoff usually occurs between the same type of radio systems, and such handoff is called horizontal handoff. Due to the rapid development of wireless networks, heterogeneous wireless networks such as IEEE 802.11 WLANs and 3G/4G mobile cellular networks co-exist in the same area. It is possible for a handoff to occur between a WiFi AP and a base station of mobile cellular networks, and such handoff is often called vertical handoff [1–3]. The IEEE 802.21 standard [4] defines functions and protocols to support vertical handoff between heterogeneous IEEE 802 networks and networking between IEEE 802 networks and cellular networks [5]. The major references cited for this chapter include [6–8].

7.1 Horizontal Handoff

Handoff support allows a user to move freely to anywhere and anytime during a communication in progress without suffering from communication interruption. Most handoff events are caused by user motion across boundaries between different

© Springer Nature Singapore Pte Ltd. 2018
S. Jiang, *Wireless Networking Principles: From Terrestrial to Underwater Acoustic*,
https://doi.org/10.1007/978-981-10-7775-3_7

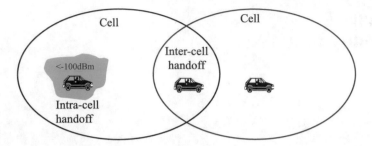

Fig. 7.1 Inter-cell handoff versus intra-cell handoff

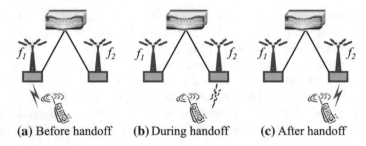

(a) Before handoff   **(b)** During handoff   **(c)** After handoff

Fig. 7.2 Hard handoff

service areas covered by different base stations in mobile cellular networks as illustrated in Fig. 7.1, and such handoff is called inter-cell handoff. A handoff may also be invoked to adjust the current channel assignment in order to optimize channel utilization, and this handoff is called intra-cell handoff.

Consider a mobile cellular network as depicted in Fig. 1.2. Handoff support consists of handoff, policies handoff initiation and handoff execution.

7.1.1 Handoff Policies

When a handoff is necessary, several questions arise, such as when the connection with the new base station should be setup and when the old one should be released. Accordingly, there are three handoff polices, namely, hard handoff, seamless handoff and soft handoff.

- Hard handoff: A connection with the new base station is set up only after the old one has been released as illustrated in Fig. 7.2. It is simple for implementation but the user suffers communication interruption. It is suitable for data applications but not for delay sensitive applications.

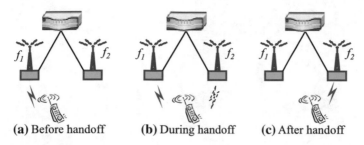

(a) Before handoff   **(b)** During handoff   **(c)** After handoff

Fig. 7.3 Seamless handoff

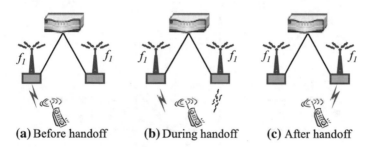

(a) Before handoff   **(b)** During handoff   **(c)** After handoff

Fig. 7.4 Soft handoff

- Seamless handoff: A connection with a new base station is set up in parallel with the old one to avoid the above communication interruption, as illustrated in Fig. 7.3. It is more complex than hard handoff for implementation.
- Soft handoff: Similar to seamless handoff, a connection with the new base station is also set up in parallel with the old one; but there is no channel switching as illustrated in Fig. 7.4. Now it is only supported by CDMA systems.

7.1.2 Handoff Initiation

Two issues are related to handoff initiation: initiation criteria and initiation process.

7.1.2.1 Initiation Criteria

Received signal strength (RSS) is often used as a criterion to trigger a handoff as illustrated in Fig. 7.5. RSS is the sum of the carrier signal power (C) and interference (I) of the channel, i.e., $RSS = C + I$. If RSS is less than a threshold, then a handoff is triggered [6]. Since it is easy to measure RSS, an initiation scheme based on RSS is simple. However, RSS may give wrong judgement on channel quality, leading to unnecessary handoff as discussed below.

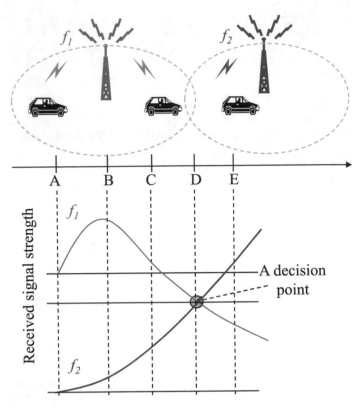

Fig. 7.5 A handoff trigger process based on RSS

Communication quality depends on ratio $\frac{C}{I}$ rather than C. If I is very strong, resulting in a very low $\frac{C}{I}$, a handoff should be triggered but actually not because $(C+I)$ may be still larger than the threshold. On the other hand, if I is very low so that RSS is low too, then a handoff is triggered according to the criterion. However, if $\frac{C}{I}$ is still above the threshold for the minimum communication quality, there is no need to trigger such handoff. Thus, a carrier-to-inference ratio (CIR) of C to I, i.e., CIR $= \frac{C}{I}$, is proposed to trigger handoff: only if CIR is less than a threshold, a handoff is triggered. Apparently, CIR decreases as I increases and as C decreases.

It is difficult to measure an accurate CIR because it is impossible to measure a realtime I and C exactly during a communication in progress since a receiver can measure only RSS in this case. Due to channel fading and multipath propagation effect, the signal strength upon its arrival at the receiver is much different from the original one from the transmitter. One way to handle this issue is to measure I before a call setup, and C is approximated by RSS $- I$. Then, CIR $= \frac{RSS}{I} - 1$. When C is very large in comparison with I, CIR can be approximated by $\frac{RSS}{I}$.

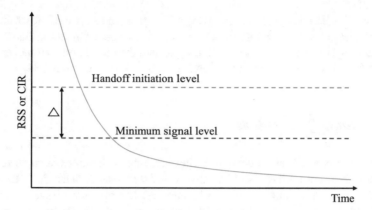

Fig. 7.6 Handoff initiation without delay [7]

7.1.2.2 Initiation Process

Since it takes time to process a handoff request, a time margin ($\triangle$) is set such that a handoff can be completed before communication quality degrades below a threshold to avoid communication interruption, as illustrated in Fig. 7.6. However, $\triangle$ should be calculated carefully because if it is too big, unnecessary handoff may happen, whereas if it is too small to complete the handoff, the call will be dropped.

In some cases, even if the RSS is the initiation criteria are satisfied, it may be still unnecessary to trigger a handoff. For example, when a user moves in a signal hole such as behind a high building, the user may suffer a sudden signal deterioration. This phenomenon may disappear immediately after the user moves away from the building. In this case, the handoff initiation should be delayed to avoid unnecessary handoff. However, if the communication quality deterioration continues, the handoff must be invoked to avoid communication interruption as illustrated in Fig. 7.7. Thus,

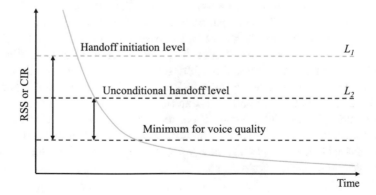

Fig. 7.7 Handoff initiation with delay [7]

an approach called delaying handoff using two threshold levels works as follows. If the RSS/CIR is less than an initiation threshold L_1, a handoff is not initiated unless the RSS/CIR of a new base station is stronger than the current one. However, when the RSS/CIR is lower than L_2, a handoff has to be initiated immediately.

7.1.3 Handoff Execution

During this process, besides channel switching, other operations such as resource allocation and/or re-routing should be also considered to smooth handoff. Therefore, when the handoff initiation condition is satisfied, the following issues have to be further addressed: who initiate a handoff process, who control the process and who are involved in the process. These issues will be discussed below by taking Global Systems for Mobile Communication (GSM) and IEEE 802.11 WLANs as examples.

7.1.3.1 Handoff Initiators

An ideal handoff initiator should have the information on the channel qualities in both uplink and downlink directions in order to make a proper handoff decision. Actually, both base stations and mobile nodes can only have such information in one direction for an asymmetric channel. In this case, message exchange between base stations and mobile nodes needs to be carried out for each to have the information on the channel quality in both directions.

A possible handoff initiator is the base station because it is much more powerful than the mobile node in terms of both computing capability and energy capacity with control of the network resource. However, the base station needs to monitor all terminals under its coverage, and will consume lots of resource to monitor channel quality for handoff initiation in the case of a large number of mobile nodes. Meanwhile, each mobile node needs to update the base station the information on the downlink channel quality, generating signaling traffic to the uplink channel.

Alternatively, the mobile node can be the handoff initiator to reduce signaling traffic in the uplink channel. In this case, the base station can simply broadcast a beacon for the mobile node to measure the downlink channel quality. When this quality is deteriorated to a certain threshold, the mobile node selects a base station with the best quality to initiate the handoff. However, whether a handoff request can be approved or not by the base station depends on the availability of network resource and the quality of the uplink channel. If the request is denied, the mobile node has to resubmit the request to another base station, which affects handoff performance.

As discussed below, an optimal strategy for selecting handoff initiators is to combine the base station and mobile nodes.

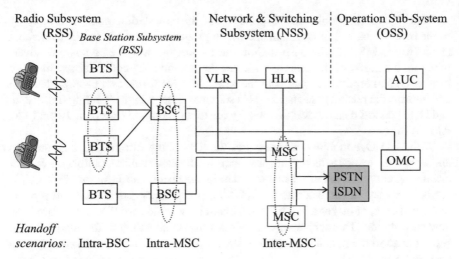

Fig. 7.8 System structure for mobility support in GSM [6]

7.1.3.2 Handoff in GSM

A typical mobility support architecture used by GSM is adopted here to discuss the above-mentioned issues.

System Structure

As illustrated in Fig. 7.8, this structure consists of a radio subsystem (RSS), a network and switching subsystem (NSS) and an operation subsystem (OSS). The RSS consists of mobile nodes and a base station subsystem (BSS). A BTS further consists of the base transceiver station (BTS) and base station controller (BSC). An NSS handles call switching between external networks and a BSC, and controls external access to customer databases. The mobile switching center (MSC) is the core unit of the RSS, consisting of a home location register (HLR), a visitor location register (VLR) and an authentication center (AUC). They are the key elements for roaming service, and will be discussed later.

As illustrated in Fig. 7.8, an intra-BSC handoff occurs between two BTSs controlled by the same BSCs. An intra-MSC handoff takes place in a service area covered by the same MSC. An inter-MSC handoff, which is also called inter-system handoff, occurs between two cells controlled by different MSCs. Such handoff process is more complex with longer handoff delay, such as a handoff between two systems managed by different countries, while the compatibility between two different MSCs is an important issue.

Handoff Process

According to who initiates, controls and is involved in a handoff, handoff schemes are classified into network controlled handoff (NCHO), mobile assisted handoff (MAHO) and mobile controlled handoff (MCHO).

With NCHO, the mobile node is passive, and the base station monitors the quality of the reverse channel of communications in progress, and sends the measurement results to the MSC. In the base station, the locator receiver (i.e., a spare receiver) monitors the RSSs of the mobile nodes in the neighboring cells that are likely to hand over, and reports the results to the MSC. The MSC makes a handoff decision and informs the related base station. This scheme causes a high signaling traffic load and large handoff delay. It was adopted by the first generation cellular network (e.g., AMPS) [6].

With MAHO, a mobile node measures the RSS of the surrounding base stations and reports the results to its serving base station. The base station monitors the current channel quality, and either (i) makes a handoff decision and informs MSC, or (ii) sends the measurement results to the MSC. The MSC supervises the handoff process in the first case, or makes a handoff decision in the second case (ii), then informs the new base station. This scheme also generates higher signaling traffic load but with shorter handoff delay, and is adopted by the second generation TDMA based system such as GSM.

With MCHO, the mobile node monitors the channel quality, selects the new base station, makes a handoff decision and informs its severing base station. The base station monitors the quality of the reverse channels, sends the results to the mobile node and passes the handoff decision to the MSC. The MSC supervises the handoff process and informs the new base station. Thus, the mobile node needs to be much more powerful. This scheme generates less signaling traffic with much shorter handoff delay.

Compared with NCHO, both MAHO and MCHO can distribute control burden between mobile nodes and base stations, and drastically reduce signaling traffic and handoff delay because the MSC is seldom or even not involved in making handoff decision. This feature is especially desirable in the personal communication system (PCS) consisting of micro-cells due to highly frequent handoff to occur therein.

7.1.3.3   Handoff in 802.11 WLANs

The early IEEE 802.11 WLAN was mainly for the use in small indoor environments, and handoff is not an issue. Now it becomes very popular as access networks, and is an important element to enable pervasive computing. In this case, handoff should be supported efficiently to provide mobile users with seamless network services.

Peculiar Features of WLANs

Similar to GSM cellular networks, in IEEE 802.11 WLANs, a handoff occurs between service areas covered by APs associated with different WLANs. The differences between them listed in Table 7.1 suggest that different handoff schemes should be designed following the specific characteristics of 802.11 WLANs.

The spectrum bands of mobile cellular networks are licensed to particular operators, who have to invest lots for the spectrum band and network deployment with deliberate network plans, design and management to maximize service and revenue.

Table 7.1 IEEE 802.11 WLANs versus GSM cellular networks

Items	GSM networks	802.11 WLANs
Spectrum type	Licensed	Free of license
Network operation	Chartered	Unlimited
Network deployment	Planned	Arbitrary
Coverage size per AP/BS	Larger	Smaller
Resource allocation	Circuit-switched[a]	Packet-switched
Networking mode	Connection-oriented	Connectionless
Data rate	Slower	Faster
Major applications	Delay-sensitive	Loss-sensitive

[a]Fully packet-switched networks appear in new generation cellular systems

802.11 WLANs exploit free unlicensed ISM spectrum bands (e.g., 2.4 and 5.1 GHz), which can be used by anyone. Although network operators may deploy their WLANs with a good plan, they cannot stop other users from deploying WLANs in the vicinities of their networks. Thus, many WLANs of the same series (i.e., 802.11a, 802.11b and 802.11g) may operate at the same frequencies in the same area. Although IEEE 802.11 WLANs divide the spectrum band into orthogonal subchannels (e.g., 14 subchannels in 802.11b and 32 subchannels in 802.11a), the same subchannel may be used simultaneously by multiple users. It takes time for a mobile node to find a proper AP for handoff because many subchannels have to be scanned to lock on a suitable AP with necessary authentication operation. Furthermore, co-channel interference may trigger unnecessary handoff even when a mobile node is in stationary state [9]. On the other hand, small coverage per AP (e.g., 250 m-radius for IEEE 802.11 WLANs) will cause more frequent handoff.

Handoff Process

The IEEE 802.11 WLAN can operate in two modes: DCF and PCF. In the DCF mode, handoff refers to switching the destination of the transmitted frames during the 4-way-handshake process: RTS-CTS-DATA-ACK. This handoff can be handled by retransmitting the frames to the nodes selected for the handoff. With PCF, a handoff denotes the process for a mobile node to switch APs during a communication in progress, which is similar to that of mobile cellular networks.

The handoff process in the PCF mode usually consists of the following steps:

• Detection is used to discover the necessity of a handoff through measuring RSS. If the RSS is lower than a threshold, the handoff process is triggered.
• Discovery is used to find a new AP for handoff through scanning all possible subchannels, i.e., 11 of the 14 subchannels in 802.11b and 32 subchannels in 802.11a. Two types of scanning modes have been defined: passive and active. In the passive mode, the mobile node listens to each subchannel to detect beacon sent by the AP. In the active mode, the mobile node sends a request first through a subchannel, and then keeps listening to this channel. Upon receiving the request,

the AP may accept it by sending a confirmation via the same subchannel. For both modes, the mobile node will select the best AP to hand over.

- Re-authentication is used by the AP to check the legitimacy of the requesting mobile node through the credentials from the old AP. Once the mobile node passes the authentication, it sends a reassociation request to the new AP, which then replies with a response to indicate an acceptance or a rejection.

The above handoff delay consists of the following parts: delays for new AP probing, authentication and reassociation. The probing delay for 802.11b is about 1100 ms and around 3200 ms for 802.11a. Such a long delay makes the entire handoff processing delay unsuitable for delay-sensitive applications such as voice over IP (VoIP). Thus, fast handoff is studied to reduce handoff processing delay. Please refer to [10] for more details.

7.2 Vertical Handoff

Today a mobile user can be served by heterogenous wireless networks simultaneously with one mobile terminal. The PCS is among the earliest developed and most popular mobile networks motivated by the Third Generation Partnership Project (3GPP) and 3GPP2. Now a PCS can also provide both data and video communication services. Another popular wireless network is WLANs motivated by IEEE mainly for data communication, and underdevelopment toward providing both voice and video communication in addition. Typical wireless LANs and MANs include IEEE 802.11 based WiFi and IEEE 802.16 based WiMAX.

Actually, wireless wide area networks (WWANs) based on satellite systems (e.g., Iridium and GlobeStar) were also developed very early. However, due to low cost-efficiency, they have not become as popular as expected. Today, the most of terrestrial wireless network services are provided by PCSs and WLANs, while satellite services are dominant in harsh environments such as oceans and deserts. Now 5G mobile technologies [11] is under development to unify all terrestrial wireless networks at higher speeds. For different networks mentioned above, their cell sizes are also different, namely, macro-cell, micro-cell, pico-cell and femto-cell. In the following, we mainly focus on issues on vertical handoff related to PCSs and WLANs.

7.2.1 Handoff Scenarios

When different wireless networks co-exist in the same area, they may overlap in coverage, e.g., cellular networks overlap with WLANs and WMANs, creating complex handoff scenarios as illustrated in Fig. 7.9, which are described below.

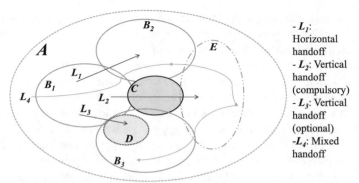

A – E each indicates the coverage of different wireless networks

Fig. 7.9 Vertical handoff scenarios

- Horizontal handoff: Following path L_1, the mobile user moves from B_1 to B_2, which belong to the same type of network. Such handoff has been discussed earlier.
- Vertical handoff (compulsory): Following path L_2, the mobile user moves from B_1 to C, which are two types of networks, while there is no other network available to support this handoff.
- Vertical handoff (optional): Following path L_3, the mobile user moves from B_1 to D, which are two types of networks. However, D overlaps with B_3. In this case, the handoff can be either horizontal between B_1 and B_3 or vertical between B_1 and D, which depends on other issues such as cost or QoS.
- Mixed handoff: Following path L_4, the mobile node moves from A into subareas of B_1, B_2, E and B_3. Since all motions occur in the area A, the mobile node can go without invoking any handoff. However, if the cost or QoS provided by A is not as good as those provided by the subareas, handoff is triggered, which include horizontal handoff (i.e., between B_1 and B_2) and vertical ones (i.e., between A and B_1, B_2 and E as well as E and B_3).

7.2.2 Handoff Operation

The policies, initiation and process of horizontal handoff can also be applied for vertical handoff but with more issues necessarily to be addressed as discussed below.

For handoff policies, hard handoff is suitable for non-delay sensitive applications such as email and ftp. It allows more time for handoff process. Since various wireless networks usually adopt different communication technologies at different frequencies, a handoff will require a change in channel so that soft handoff does not work here. For delay-sensitive applications such as voice, seamless handoff is preferable, and will take more time than horizontal handoff because it has to consider some high-layer issues, which typically include QoS, cost and security. If the networks are operated by different operators, it is also necessary to check user legitimacy, similar to processing an inter-MSC horizontal handoff.

Regarding handoff initiation, similar to horizontal handoff, RSS or SNR is still a fundamental indicator to trigger a handoff. For a horizontal handoff, the indicators of the current and possible candidate channels have to be considered in making a handoff decision. Since a vertical handoff has to consider some high-layer issues, some vertical handoff scenarios will not appear for horizontal handoff. For example, with cost-driven handoff, if a mobile node finds that the service cost of another network is much lower, it may trigger a vertical handoff, e.g., from a satellite network to a terrestrial network, and from cellular WiFi to home WiFi, without session interruption.

For handoff process, due to high-layer issues to be considered in order to make handoff decision, a mobile node cannot complete it alone, and the base station or the access point should be involved from the beginning. If networks operated by different operators, an agent that can communicate with each involved part is necessary to coordinate handoff process and re-route packets for handoff.

7.2.3   IEEE 802.21

As described in [4], the IEEE 802.21 standard defines a media independent handover[1] (MIH) framework to (i) facilitate handoff between IEEE 802 networks and cellular networks, and (ii) enable the optimization of handoff between heterogeneous IEEE 802 networks, for both mobile and stationary users. It also supports an optimized handoff-link adaptation, with which a node can choose a link with a higher data rate than the current one.

The heterogeneous networks here refer to those using medium types specified by 3GPP and 3GPP2 as well as both wired and wireless media defined by the IEEE 802 standards. In general, both the mobile node and the network points of attachment such as base stations (BSs) and access points (APs) can operate in multi-modal. The handoff process can be initiated by measurement results reported by the data link layers of the mobile node, and both the mobile node and the network make decisions about connectivity [4].

This framework provides an MIH reference model for each different link-layer technologies defined by IEEE 802 family and 3GPP as well as 3GPP2. Accordingly, a set of MIH functions (MIHFs) have been defined within the protocol stacks. They are used to collect link information and control link behavior during handoff process with the following media independent services:

---

[1]Term "handover" is adopted in this standard, and "handoff" is used instead in this book when possible to align with the early discussion.

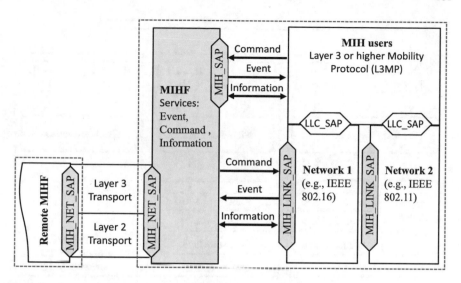

Fig. 7.10 General MIH function (MIHF) reference model and service access points (SAPs) [4]

- Event service to detect changes in link-layer properties and initiate appropriate events from both local and remote interfaces,
- Command service to provide a set of commands for the MIH users to control link properties for handoff and switch between links if necessary,
- Information service to provide the information on different networks and their services in order to make a more effective handoff decision across heterogeneous networks [4].

The interaction between MIHFs and other functional entities is carried out through service primitives, which are grouped in SAPs. As illustrated in Fig. 7.10, the MIH users can access these functions through a single media-independent MIH_SAP, while the MIHF obtains services from the lower layers via the media-specific MIH_LINK_SAP, which actually supports all communications between the MIHF and the lower layers of media specific protocol stacks. The MIH_NET_SAP is another media-dependent interface supporting the exchange of MIH information and messages with the remote MIHF. It uses the primitives specified by the MIH_LINK_SAP to support all transport services over the data link layer. The LLC_SAP is a conventional SAP for data transmission and reception between the data link layer and the layers above. A mobility management protocol stack, Layer 3 or Higher Mobility Protocol (L3MP), is logically identified within each network node [4]. Relationships between the MIHF and other functional components in the same network node are summarized in Fig. 7.11, including MIH signaling over the network, and local interactions between the MIHF and other layers in the same network, node and functional block.

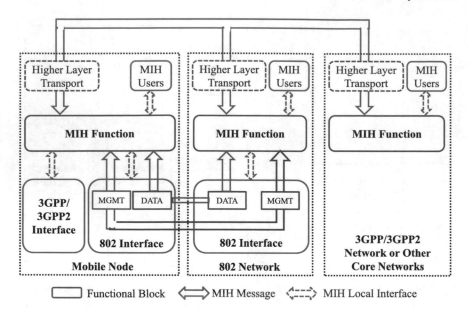

Fig. 7.11 Types of MIHF relationship [4]

7.3 Roaming

This issue is discussed for two types of networks: connection-oriented mobile cellular networks and connectionless IP networks.

7.3.1 Overview

As discussed earlier, routing is the key element for successful packet forwarding, and most routing protocols rely on certain identity information such as IP address and telephone number, which is often related to the geographical locations of the points that a node is attached to. Whenever a user moves out of the original service coverage into a new service area, this information should to be changed accordingly, and all potential source nodes that may transmit to it should be informed immediately so that the packets can be successfully forwarded to it. Apparently, this method is unscalable to a case with many users changing locations frequently. Furthermore, the network address may also be used as the identity of a node in communication, and changing such information will cause undesirable consequence.

Location management is proposed to ensure that, when a mobile user leaves its original location for a new place, the user can be still connected and use the subscribed services as usual. A general location management scheme operates as follows. Once a mobile node arrives at a new location, it automatically registers itself there to obtain a temporary identity information corresponding to this location, which is then transferred to the original location. The following methods can be used to maintain connectivity. (i) The sender learns the new identity information of the mobile node in the new location from the original location register, and sets a direct connection with the mobile node (e.g., GSM). (ii) The sender forwards packets according to the original identity, and the original location server relays data to the new location of the mobile user (e.g., mobile IP). Both GSM and mobile IP are discussed below.

7.3.2 Roaming in GSM

As illustrated in Fig. 7.8, the location management in GSM mainly consists of a home location register (HLR), a visitor location register (VLR) and an authentication center (AUC). The HLR is used to store the information of the registered users, such as the authentication information, and the VLR for the temporarily visiting users. The AUC is responsible to check the legitimacy of the mobile users. They operate as follows.

- Once a mobile node roams into a new location (called visitor), it is automatically registered at the VLR of the new location that it is visiting.
- The VLR of the visited location allocates a temporary identity to the visitor, which is transferred to the HLR of the visitor. The visitor can be reached during the dwell period in the new location through the temporary identity.
- When a user tries to call this visitor, the HLR of the visitor is signalled for connection setup since GSM is connection-oriented, so that a direct network connection between them can be set up.
- During the above process, the AUC needs to check the legitimacy of the visitor with possible certain measurements enforced for security if necessary.

7.3.3 Mobile IP

With connectionless IP, a potential data sender cannot have chance to obtain the new temporary identity of a mobile node roaming to a foreign network before data transmission. In this case, the packets need to be relayed through the home network to the foreign network that the mobile node is actually visiting. Such routing is called triangle routing, which is required for every packet forwarding with mobile IPv4, and can be avoided for sequentially packet forwarding with mobile IPv6. The following sections first discuss the major problems caused by mobility to IP, and then describe a solution adopted by mobile IPv4 and IPv6.

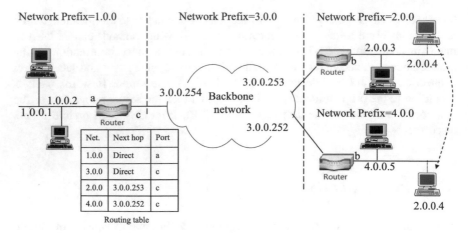

Fig. 7.12 Roaming effect on user connectivity

7.3.3.1 Problems for Roaming

With IPv4, a public IP address uniquely identifies an attach point in the Internet. Thus, a node assigned with an IP address must be located on the network indicated by this address so that it can receive the packet destined to itself; otherwise, the connectivity will loss. Due to the connectionless nature of IP as indicated in Fig. 1.21, no network connection is needed for any packet transmission. As illustrated in Fig. 7.12, when a user assigned with an address of 2.0.0.4 roams from the original network 2.0.0 to another network 4.0.0, if its potential senders cannot be informed immediately of the new network address, all packets destined to it will be still sent to the original network.

The IP address has a hierarchical structure corresponding to networks deployed in different geographical locations. Usually a maximal matching algorithm is adopted by routers for routing table lookup in determining to which network to forward packets. As illustrated in Fig. 7.12, in the routing table of router A, the network prefix in the column "Net" rather than the entire IP address is used as the index for table lookup. Matching operation starts from the left to the right of a given IP address and the network prefix, and the maximally matched one is selected. The next hop for the packet to be forwarded to is indicated by the corresponding outport. If the next hop is the current network, the corresponding part in the table is marked by "Direct"; otherwise, this part is marked by the network address of the next hop (e.g., 3.0.0.253). As illustrated in Fig. 7.12, if a mobile node assigned with an address of 2.0.0.4 and originally attached to the network 2.0.0 roams to the network 4.0.0, all the packets destined to 2.0.0.4 will be forwarded to the network 2.0.0 rather than 4.0.0 that it actually is visiting.

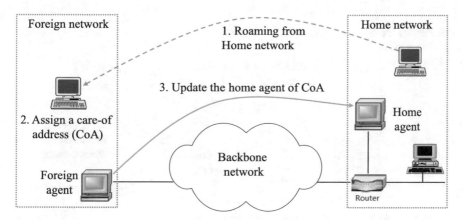

Fig. 7.13 Assignment of care-of-address (CoA) in mobile IPv4

7.3.3.2   Mobile IPv4 (MIPv4)

Networks are classified into home networks and foreign networks for roaming support. The home network is where a mobile node is registered, and has the same network prefix as its original IP address. The foreign network is any other network that the mobile node may visit. As illustrated in Fig. 7.13, there are following agents used to support roaming:

- Home agent stores the authentication information and maintains the updated location information of each mobile node registered in the network associated with this agent.
- Foreign agent assigns a temporal IP address, called care-of-address (CoA), to each visiting mobile node, and informs its home agent of the assigned CoA. It also secures the delivery of all packets relayed by the home network to the visiting mobile node.

Since IP is connectionless, a node can send any packet whenever it is available without signaling to the receiver. Thus, a sender cannot use the CoA to forward packets to a mobile node currently visiting a foreign network directly. Instead, its original IP address is always used to forward all packets destined to this node, resulting in a packet forwarding trace as illustrated in Fig. 7.14, which is explained below.

- Each packet from Node A to Node B is forwarded to B's home network according to its original IP address no matter where it is located.
- Upon receiving a packet destined to Node B, tunneling (See Sect. 7.3.3.3) is used to forward the packet to the foreign network that Node B is currently visiting following its CoA stored in the home agent.
- Upon receiving a tunneled packet, the foreign agent de-tunnels this packet first to obtain the original packet sent by Node A, and then delivers it to Node B according to the IP address carried by the original packet.

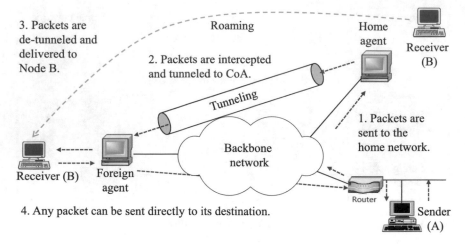

Fig. 7.14 Packet forwarding with triangle routing and tunnelling

- For Node B currently located in a foreign network to send packets to other nodes, the normal IP routing is used to deliver the packets to their destinations directly. If the source address is used for authentication or other security control (e.g., some routers may drop any packet whose source address does not match the prefix of the network that the packet comes from), tunneling has to be used in packet forwarding.

7.3.3.3 Tunneling

By tunneling, an IP packet is encapsulated into the payload of another IP packet without any changes to the encapsulated IP packet. It has been used widely in IP network to enable some new capabilities, such as running IPv6 over IPv4, IP multicast, Virtual Private Networks (VPNs) and mobility support.

The CoA of a roaming node is used to forward a packet from its home network to the foreign network that the node is currently visiting. Its original IP address will be used by the foreign network to secure the packet delivery to the visiting node, which is identified by its original IP address. Therefore, both the original address and the CoA have to be carried by the packet. However, an IP packet only has one field to carry the destination address, it is impossible for a packet to carry the two addresses. With tunneling, another IP packet with the CoA as the destination address is used to carry the original IP packet relayed by the home network to the foreign network indicated by the CoA.

7.3.3.4 Mobile IPv6 (MIPv6)

MIPv4 is an add-on feature built on the top of IPv4, while Mobile IPv6 (MIPv6) [12] is designed as a part of the IPv6 protocol stack to automatically support mobility as summarized below.

- The foreign agent used in MIPv4 is no longer required in MIPv6 because most related functions have been built in MIPv6. The major functions of the MIPv4 agent include provisioning of an CoA to a mobile node and de-tunneling the IP packets relayed by the home agent. With MIPv6, it is the mobile node that generates the CoA as described below.
- IPv6 uses a neighbor discovery protocol (NDP) to enable a mobile node to automatically detect the new network that it is going to move in, and obtain the network prefix. Then the node can combine this network prefix with its MAC address or IPv4 address to form an MIPv6 CoA, which is then transferred to its home agent by the mobile node.
- With MIPv6, no tunneling is required due to that a MIPv6 CoA also specifies the mobile node itself. Thus, the home agent can use this CoA to send IP packets directly to the mobile node.
- The triangle routing can also be avoided in MIPv6 through route optimization, with which, a mobile node can inform its correspondents of its CoA through binding update (BU). However, BU is vulnerable to attackers, who can also use BU to make spoofed binding. Thus, IP Security (IPSec) is embedded into IPv6.
- The IPv6 packet header can carry the home address of a mobile node beside its CoA so that its home address can be placed back to the source address field of the IP packet when necessary. This feature is useful because some applications use the source address as the user identity, whereas it is impossible in MIPv4 without tunneling.

7.4 Grade of Service (GoS)

The major performance indicators for handoff schemes include new call blocking probability (CBP), handoff call dropping probability (CDP) and call completion probability (CCP). CBP indicates the probability at which a new call is dropped, CDP is the probability for a handoff call to be dropped, while CCP is the probability for an admitted call to be completed without being either blocked or forcedly terminated [13]. A comprehensive performance indicator that reflects the interests of both users and operators, grade of service (GoS), is defined below:

$$GoS = CBP + \alpha \times CDP, \tag{7.1}$$

where α is a weight and often set to 10, which means that a dropped handoff call costs 10 times more than a blocked new call [14]. The lower the GoS, the better the performance. A consensus is that the user feels more boring with a handoff dropping than a new call blocking. Therefore, handoff calls are necessarily prioritized over new calls in network resource allocation. In the following sections, we will discuss some basic mechanisms for handoff support with simple performance analysis.

7.4.1 Approaches for Efficient Handoff

To achieve certain GoS, two levels of mechanisms are usually adopted: admission and handoff. The major admission level scheme is call admission control (CAC), which controls the maximum number of new calls to be admitted in order to maintain *CDP* below certain level. It can be stationary or dynamic. A stationary CAC, e.g., the Guard Channel (GC) scheme and its variants [14–17], determines a static admission condition according to statistical traffic characteristics. Its major advantage is simplicity for implementation without using realtime computing. However, it cannot adapt to instantly changing traffic load and may waste network resource. A dynamic CAC, e.g., Shadow Cluster (SC) [18] and Distributed CAC (DCAC) [19], can overcome this weakness but at the cost of more realtime computing and possible message exchange. These CAC schemes only consider inter-cell handoff and ignore intra-cell handoff in computing new call admission threshold, while the CAC scheme discussed in [20] takes both types of handoff into consideration.

A handoff level scheme aims to maximize successful handoff during handoff execution by leveraging channel allocation. A simple scheme is to queue handoff calls (QHC) when there is no free channel available to support the handoff. This scheme is suitable for handoff of non-delay sensitive applications such as data. Actually, GC is also a handoff level scheme because only handoff calls can use the reserved guard channels during handoff execution. Channel carrying (CC) [21, 22] is another handoff level scheme, allowing a handoff call to carry its currently occupied channel to the target cell that the call will hand over to for continuous use.

The performance of both call-level and handoff-level schemes depends on channel allocation schemes, which can be classified into Fixed Channel Allocation (FCA) and Dynamic Channel Allocation (DCA). With FCA, the set of channels allocated to each cell is fixed, whereas that with DCA can be adjusted dynamically. The above-mentioned GC and DCAC schemes are based on FCA, while the CAC scheme proposed in [20] is based on DCA. The CC and QHC schemes can be used with both FCA and DCA, and more research results can be found in the literature. The following sections will briefly introduce some of these schemes along with analysis on GC, DCAC and QHC.

7.4.1.1   Dynamic Channel Allocation (DCA)

DCA becomes more attractive than FCA to support ever-increasing demands for wireless channels. In general, DCA can provide more efficient solutions than FCA for TDMA-based cellular networks (e.g., UWC-136 based 3G networks [23]) due to its better adaptability to unevenly-distributed and changing traffic loads [24]. Since DCA allows any channel to be used by any cell subject to co-channel interference constraint, it is easier to implement a CC scheme with DCA than with FCA because a CC scheme with FCA has to either borrow a channel or extend channel reuse distance as discussed later.

DCA can be realized in either a centralized or distributed mode. With a centralized mode, all channels are grouped into a pool, which is managed by a central controller. For each call request, the base station will submit a request to the controller for channel allocation. In a distributed mode, a channel can be used in any cell provided that the signal interference constraint is satisfied [25]. A distributed DCA generates less signaling traffic especially with timid DCA, i.e., a randomly selected channel is assigned by a cell if it is not used by any of its interfering cells [26]. Its performance can be comparable with centrally administered FCA. An aggressive DCA, which suggests a mobile node to use a channel even if there is interference present, can reduce substantially CBP.

There are a lot of proposals for DCA implementation in the literature such as [24, 27]. One approach relies on communication and coordination between cells for channel assignment. This approach can provide optimal channel assignment to reduce or even eliminate co-channel interference. However, communications between cells generate extra traffic, and coordinations cause additional latency in handoff execution. A self-adaptive approach can overcome these weaknesses by deciding channel assignment mainly based on the local information but with possible increased co-channel interference. Due to its simplicity, this approach invoked lots of research. Among them, a signal strength measurement-based distributed DCA, e.g., the least interference scheme [28], is more attractive in terms of channel utilization because it can maximize the gain of channel spatial reuse [29].

7.4.1.2 Channel Carrying (CC)

CC allows a mobile node to continue the use of its currently occupied channel when it hands over to a new cell in order to improve handoff success probability. This approach is discussed first for a highway system in [30], where FCA jointly using channel borrowing is adopted [31]. That is, if no free channel is available for either a new call or a handoff call in a cell, this cell tries to borrow one from its neighbors. Particularly for a handoff call, if no channel can be borrowed, the cell lets the handoff call carry its currently used channel if the co-channel interference constraint is satisfied. A similar approach is discussed for the same environment in [32, 33], using DCA along with explicit information on the velocity of mobile nodes such that the same channel can be assigned to multiple mobile nodes moving in the same direction and separated by a certain distance, subject to the co-channel interference constraint. In this case, a mobile node may carry an assigned channel across different cells until the end of the call. CC has also been exploited differently in [21, 22] with FCA by simply extending the original channel reuse distance by one cell so that a channel can be carried once for handoff.

CC can be further classified into CC first (CCF) and CC last (CCL) according to carrying sequence. With CCF, a handoff call first tries to carry its currently used channel whenever it hands over to another cell. If this attempt fails, it then tries to occupy a free channel. With CCL, a handoff call tries to occupy a free channel first when it hands over. If no free channel is available, then it tries to carry its currently used channel. For both, a successful CC has to satisfy the co-channel interference constraint. Although the only difference between CCF and CCL is the carrying sequence, they may perform differently as discussed in [34], which conducts a performance study of different CC strategies in a DCA system. Some results are summarized below.

CCF may cause less channel switching operations than CCL, while CCL can have optimal channel assignment than CCF in terms of co-channel interference because CCL provides opportunities for cells to select optimal channels. Generally, both CC schemes outperform non-channel-carrying (NCC) in terms of handoff prioritization. They also perform differently against traffic loads, network topologies and handoff frequencies. (i) Both CCF and CCL can avoid the following undesirable situation: CDP > CBP, which appears in a DCA system without prioritizing handoff calls over new calls in channel assignment; (ii) CCL outperforms CCF in terms of handoff prioritization efficiency, and the gain given by CCL over NCC is remarkable and increases with traffic loads; (iii) DCA with CCL is comparable and even better than FCA jointly using GC [15] in terms of handoff prioritization.

7.4.1.3 Intra-cell Handoff with DCA

As mentioned earlier, there are two types of handoff in mobile cellular networks: intra-cell handoff and inter-cell handoff. The former refers to a process for a mobile node to switch its channel but it is still served by the same base station after channel switching, while the latter also changes its service base station [27].

There are some CAC schemes proposed for DCA such as [35–37] and an extension of the shadow cluster scheme originally proposed for FCA in [18]. They are based on the coordinated DCA, by which a channel is allocated dynamically to a node if the co-channel interference to other channels can be limited to certain level. Actually, they are similar to FCA in the sense that few intra-cell handoff calls are caused by new channel assignments and unnecessarily taken into account by CAC.

However, with measurement DCA, channel assignment is mainly determined according to the local information without considering the impact on existing calls present in other cells [38]. A channel may be assigned if its signal strength or the signal-to-interference ratio (SIR) satisfies certain threshold (e.g., 18 dB for SIR) so that a new call assignment may cause intra-cell handoff. As discussed in [20], as traffic load increases, intra-cell handoff becomes the dominant handoff. This is because failed intra-cell handoff eventually reduces the number of inter-cell handoff, which becomes less than that of intra-cell handoff. In this case, it is necessary for

Fig. 7.15 A queuing system for new calls and handoff calls without priority

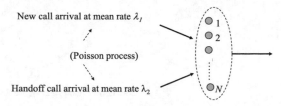

New call arrival at mean rate λ_1

(Poisson process)

Handoff call arrival at mean rate λ_2

N channels, whose service time each follows exponential distribution with mean μ^{-1}

CAC to consider both inter-cell handoff and intra-cell handoff to guarantee GoS, and such a CAC is investigated in [20].

7.4.2 Handoff Without Priority

The following analysis is based on results for queueing system $M/M/m$. The detailed discussion can be found in [39], and some results used here are listed in Appendix A.

7.4.2.1 Notation

The following notation defined in Fig. 7.15 and assumptions are adopted in the discussion.

- μ^{-1}: average channel holding time for both a new call and a handoff call, following exponential distributions.
- ψ^{-1}: average call lifetime also following exponential distribution.
- λ_1: mean new call arrival rate with Poisson arrival process.
- λ_2: mean handoff call arrival rate with Poisson arrival process.
- N: maximal number of voice channels, called system capacity.
- B: buffer size for queuing handoff calls in number of calls.

Let $\lambda \triangleq \lambda_1 + \lambda_2$. Since handoff calls are generated from admitted new calls, the following relationship between λ_1 and λ_2 is derived in [40]:

$$\lambda_2 = \frac{\mu(1 - CBP)\lambda_1}{\psi + \mu CDP}. \tag{7.2}$$

7.4.2.2 Performance Analysis

We first analyse *CBP* and *CDP* for a case without priority schemes used for handoff calls as illustrated in Fig. 7.15. The state transition diagram is illustrated in Fig. 7.16, where,

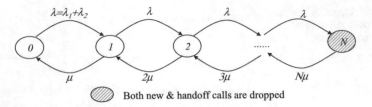

Both new & handoff calls are dropped

Fig. 7.16 State-transition without priority for handoff calls

$$\lambda(k) = \begin{cases} \lambda, \; 0 \le k \le N - 1, \\ 0, \; \text{otherwise}; \end{cases}$$

and

$$\mu(k) = \begin{cases} k\mu, \; 0 \le k \le N, \\ 0, \quad \text{otherwise}. \end{cases}$$

From (A.10), we can have the formula for the probability at which there are k calls in the system (P_k) as follows:

$$P_k = P_0 \prod_{i=0}^{k-1} \frac{\lambda}{(i+1)\mu} = P_0 (\frac{\lambda}{\mu})^k \frac{1}{k!} \tag{7.3}$$

for $k \le N$, and $P_k = 0$ otherwise. From (A.11), we have

$$P_0 = \left[\sum_{k=0}^{N} \left(\frac{\lambda}{\mu} \right)^k \frac{1}{k!} \right]^{-1}. \tag{7.4}$$

Both a new call and a handoff call will be dropped when all N channels have been occupied, i.e., $CBP = CDP = P_N$.

7.4.3 Queuing Handoff

As illustrated in Fig. 7.17, an arriving new call is dropped if all N channels are occupied, while an arriving handoff call is dropped only if both N channels and the buffer are fully occupied. That is, $CBP = \sum_{k=N}^{N+B} P_k$ and $CDP = P_{N+B}$.

The state transition diagram for this case is illustrated in Fig. 7.18.

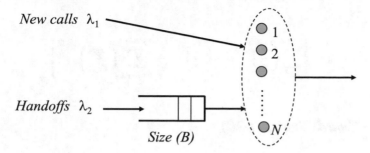

Fig. 7.17 Queuing handoff calls

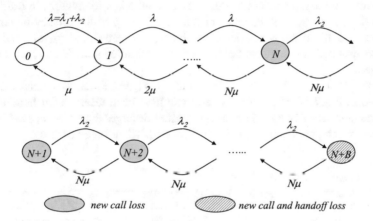

Fig. 7.18 State-transition for queuing handoff calls

We have

$$\lambda(k) = \begin{cases} \lambda, & 0 \le k \le N-1, \\ \lambda_2, & N \le k \le B+N-1, \\ 0, & \text{otherwise}, \end{cases}$$

and

$$\mu(k) = \begin{cases} k\mu, & 0 \le k \le N, \\ N\mu, & N+1 \le k \le B+N, \\ 0, & \text{otherwise}. \end{cases}$$

Then from (A.10), we have

$$P_k = \begin{cases} P_0(\frac{\lambda}{\mu})^k \frac{1}{k!}, & 0 \le k \le N, \\ P_0(\frac{\lambda}{\mu})^N \frac{1}{N!}(\frac{\lambda_2}{N\mu})^{k-N}, & N \le k \le B+N, \\ 0, & \text{otherwise}, \end{cases}$$

and

$$P_0 = \left\{ \sum_{k=0}^{N} \left(\frac{\lambda}{\mu} \right)^k \frac{1}{k!} + \left(\frac{\lambda}{\mu} \right)^N \frac{1}{N!} \sum_{k=N+1}^{N+B} \left(\frac{\lambda_2}{N\mu} \right)^{k-N} \right\}^{-1}. \qquad (7.5)$$

7.4.4 Guard Channel (GC)

GC is proposed for FCA-based cellular networks in [15]. With FCA, a fixed number of channels are assigned to one cell, and does not adapt to changes in traffic load. As illustrated in Fig. 7.19, its basic idea is to reserve g of the total N channels for handoff calls. If the number of occupied channels exceeds $(N - g)$, all arriving new calls are dropped, and only handoff calls can be admitted until N channels are fully occupied.

The original analysis for GC can be found in [15], with the state-transition diagram depicted in Fig. 7.20. Since g of N channels have been reserved for handoff calls, once the number of occupied channels is equal or larger than $N - g$, only handoff calls can be admitted, and new calls are dropped. Therefore, we have

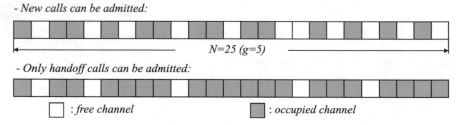

Fig. 7.19 Diagram of the guard channel scheme

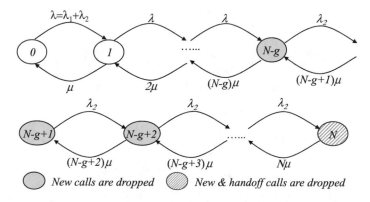

Fig. 7.20 State-transition with guard channels for handoff calls

$$\lambda(k) = \begin{cases} \lambda, & 0 \le k \le N - g - 1, \\ \lambda_2, & N - g \le k \le N - 1, \\ 0, & \text{otherwise}, \end{cases}$$

and

$$\mu(k) = \begin{cases} k\mu, & 0 \le k \le N, \\ 0, & \text{otherwise}. \end{cases}$$

Then, $CBP = \sum_{k=N-g}^{N} P_k$ and $CDP = P_N$, where

$$P_k = P_0 \left(\frac{\lambda}{\mu}\right)^{N-g} \left(\frac{\lambda_2}{\mu}\right)^{k-N+g} \frac{1}{k!} \tag{7.6}$$

with

$$P_0 = \left[\sum_{k=0}^{N-g} \left(\frac{\lambda}{\mu}\right)^{k} \frac{1}{k!} + \left(\frac{\lambda}{\mu}\right)^{N-g} \sum_{k=N-g+1}^{N} \left(\frac{\lambda_2}{\mu}\right)^{k-N+g} \frac{1}{k!} \right]^{-1} . \tag{7.7}$$

7.4.5 Distributed Call Admission Control (DCAC)

DCAC [19] is originally proposed for an one-dimensional (1D) cellular array as depicted in Fig. 7.21. Its basic idea is to allow a cell to take into account the number of active calls present in the cell itself and in its neighbors to determine a new call admission threshold for a time period (T) in order to guarantee CDP. T is assumed to be short enough so that the probability that a call hands over more than once during T is negligible, which is called T-assumption henceforth. DCAC is extended to two-dimensional (2D) cellular networks in [41].

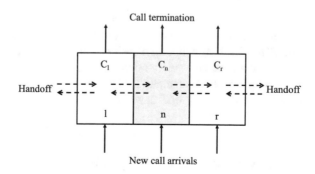

Fig. 7.21 Hypothetical topology of one-dimensional cellular networks

7.4.5.1 One-Dimensional Cellular Array

As illustrated in Fig. 7.21, the 1D cell array consists of cells C_l, C_n and C_r from the left to the right. With DCAC, the following two conditions have to be satisfied to carry out new call admissions in C_n during T:

1. A new call admission in C_n cannot affect *CDP* guarantee for the handoff calls coming from C_r and C_l to C_n.
2. A new call admission in C_n cannot affect *CDP* guarantee for the handoff calls from C_n to C_r and C_l because newly admitted calls in C_n may also hand over.

Let l, n and r denote the numbers of calls currently present in cells C_l, C_n and C_r at time t_0, respectively. Let p_m and p_s denote the probabilities that a call hands over to other cells and remains in the same cell during T, respectively. To guarantee *CDP*, the maximum number of calls to be present in C_n at $t_0 + T$ subjective to *condition 1* (n_1) is calculated below.

Following binomial distribution

$$B(x, X, p) = \binom{X}{x} p^x (1 - p)^{X-x}, \tag{7.8}$$

with mean and variance of Xp and $Xp(1 - p)$, respectively, $B(i, l + r, \frac{p_m}{2})$ and $B(j, n_1, p_s)$ calculate the probabilities that i calls out of $r + l$ calls hand over to C_n, and that j calls out of n_1 calls remain in C_n, respectively. The distribution for the number of calls (k) present in C_n at $t_0 + T$, $P_{t_0+T}(k)$, is calculated by the convolutional sum of the two binomial distributions for $i + j = k$. Thus, we have

$$CDP = \sum_{k>N} P_{t_0+T}(k). \tag{7.9}$$

7.4.5.2 Two-Dimensional Cellular Array

The DCAC is extended for a 2D cellular array as depicted in Fig. 7.22 [41]. The extended DCAC eliminates the original assumption of the Poisson CAC process. Following the T-assumption mentioned earlier, only the handoff calls generated by the existing calls and newly admitted calls in a cell are necessarily considered in admission threshold calculation. The sum of these two types of calls to be present in cell$_i$ between t_0 and $t_0 + T$ is bounded by

$$\mathcal{M}_i = max(h_i, \mathcal{N}_i). \tag{7.10}$$

Furthermore, unlike the 1D scheme [19], handoff calls generated by newly admitted calls in each cell are also counted in threshold calculation. Therefore, $\mathcal{M}_i$ instead of h_i and λT is used as the upper bound of the number of calls that affect *CDP*

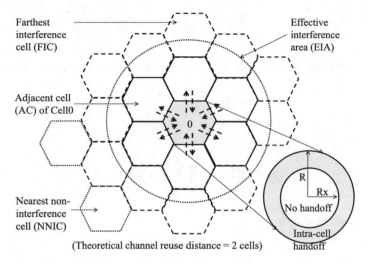

Fig. 7.22 2D cellular network: Adjacent Cell (AC), Farthest Interference Cell (FIC), Nearest Non
Interference Cell (NNIC) and Effective Interfering Area (EIA) [41]

guarantee to avoid the original assumption of Poisson CAC process. The following
additional parameters are used here.

- Φ_i: set of the neighbors of cell$_i$, and here $\Phi_0 = \{1, 2, 3, 4, 5, 6\}$.
- $p_{m,i}$: probability for a call to hand over to cell$_i$ within T.
- p_s: probability for a call to remain in the same cell within T.
- N_i: system capacity of cell$_i$.
- h_i: number of active calls present in cell$_i$ at time t_0.
- n_{ij}: preliminary new call admission threshold imposed by cell$_i$ to cell$_j$.
- $\mathcal{N}_i$: final call admission threshold of cell$_i$ for the period between t_0 and $t_0 + T$.

Consider that C_0 is calculating its call admission threshold for a period from t_0 to
$t_0 + T$. Following *Condition 1*, a preliminary admission threshold n_{00} is calculated
below. With binomial distribution $B(x, X, p)$, the probability that x calls out of n_{00}
calls remain in C_0 is

$$B_{s,0}(x) \triangleq B(x, n_{00}, p_{s,0}), \qquad (7.11)$$

and the probability that y calls out of $\sum_{\forall i \in \Phi_0} \mathcal{M}_i$ calls hand over to C_0 is given by

$$B_{m,0}(y) \triangleq B\left(y, \sum_{\forall i \in \Phi_0} \mathcal{M}_i, p_{m,0}\right). \qquad (7.12)$$

Then, the distribution for the number of calls to be present in C_0 at $t_0 + T$, $P_{t_0+T,0}(k)$,
is the convolutional sum of $B_{s,0}(x)$ and $B_{m,0}(y)$ with $x + y = k$. To bound *CDP*
below *CDP$_t$*, n_{00} must satisfy

$$\sum_{k>N_0} P_{t_0+T,0}(k) \leq CDP_t. \tag{7.13}$$

Similarly, according to *Condition 2*, n_{j0} ($j \in \Phi_0$) can also be calculated below with

$$B_{s,j}(x) \triangleq B(x, \mathcal{M}_j, p_s) \tag{7.14}$$

and

$$B_{m,j}(y) \triangleq B\left(y, \sum_{\forall i \in \Phi_j - \{0\}} \mathcal{M}_i + n_{j0}, p_{m,j}\right). \tag{7.15}$$

Then, n_{j0} must satisfy

$$\sum_{k>N_j} P_{t_0+T,j}(k) \leq CDP_t. \tag{7.16}$$

The same approximation method as used by the 1D DCAC [19] is adopted here. The convolutional sum of binomial distributions is approximated by Gaussian distribution $G(m, \sigma)$, where m and σ^2 respectively denote the sums of the mean and variance of all binomial distributions. Then, $\sum_{k>N_i} P_{t_0+T,i}(k)$ ($i = 0$ or j) is approximated by the integral over the tail of the Gaussian distribution expressed in terms of error function $Q(.)$, i.e.,

$$CDP_t \simeq Q\left(\frac{N_i - m}{\sigma}\right), \tag{7.17}$$

from which, n_{i0} can be derived by letting $Q(a) = CDP_t$. Obviously,

$$m = n_{00}p_{s,0} + \sum_{\forall i \in \Phi_0} \mathcal{M}_i p_{m,0},$$

$$\sigma^2 = n_{00}p_s(1 - p_s) + \sum_{\forall i \in \Phi_0} \mathcal{M}_i p_{m,0}(1 - p_{m,0})$$

for n_{00}. Then,

$$m = \mathcal{M}_j p_s + \left(\sum_{\forall i \in \Phi_j - \{0\}} \mathcal{M}_i + n_{j0}\right) p_{m,j},$$

$$\sigma^2 = \mathcal{M}_j p_s(1 - p_s) + \left(\sum_{\forall i \in \Phi_j - \{0\}} \mathcal{M}_i + n_{j0}\right) p_{m,j}(1 - p_{m,j})$$

for n_{j0}.

The remaining of the threshold calculation is conducted in the same way as used by the 1D DCAC [19], and we have

$$n_{00} = \frac{1}{2p_s}[a^2\mathcal{A}_0 + \mathcal{B}_0 - a\sqrt{a^2\mathcal{A}_0^2 + 4\mathcal{A}_0\mathcal{B}_0 + 4\mathcal{C}_0}]. \tag{7.18}$$

$$n_{j0} = \frac{1}{2p_{m,j}}[a^2\mathcal{A}_{j0} + 2\mathcal{B}_{j0} - a\sqrt{a^2\mathcal{A}_{j0}^2 + 4\mathcal{A}_{j0}\mathcal{B}_{j0} + 4\mathcal{C}_{j0}}]. \tag{7.19}$$

where, $\mathcal{A}_0 = 1 - p_s$, $\mathcal{A}_{j0} = 1 - p_{m,j}$, and

$$\mathcal{B}_0 = N_0 - \sum_{\forall i \in \Phi_0} \mathcal{M}_i p_{m,0},$$

$$\mathcal{C}_0 = \sum_{\forall i \in \Phi_0} \mathcal{M}_i p_{m,0}(1 - p_{m,0}),$$

$$\mathcal{B}_{j0} = N_j - \mathcal{M}_j p_s - \sum_{\forall i \in \Phi_j - \{0\}} \mathcal{M}_i p_{m,j},$$

$$\mathcal{C}_{j0} = \mathcal{M}_j p_s(1 - p_s) + \sum_{\forall i \in \Phi_j - \{0\}} \mathcal{M}_i p_{m,j}(1 - p_{m,j}).$$

Then the final call admission threshold is given by

$$\mathcal{N}_0 = \lfloor min\{n_{i0}, \forall i \in \Phi_0 + \{0\}\} \rfloor. \tag{7.20}$$

7.5 Summary

This chapter describes primary issues for mobility support in terms of roaming and handoff in mobile networks, focusing on mobile cellular networks and mobile IP networks. Actually, they share some similarities in the manner of supporting mobility although different terminologies are adopted. For example, a temporary network address or identity is assigned to each visitor by the foreign network that the visitor is visiting. They are different in terms of data forwarding process from a source node to the roaming node due to their connectionless or connection-oriented properties. However, mobility raises some challenging issues to network security [42] and vertical handoff in heterogenous mobile networks [43, 44], especially in large scale mobile networks consisting of satellites and various types of aeronautic vehicles as well as vessels, such as marine Internet [45].

References

1. Chai, R., Zhou, W.-G., Chen, Q.-B., Tang, L.: A survey on vertical handoff decision for hetero-geneous wireless networks. In: Prof. International Youth Conference on Informati Computing & Telecommunication (YC-ICT), pp. 279–282. Beijing, China (2009)
2. Edward, E.P., Sumathy, Dr. V.: A survey of seamless vertical handoff schemes for Wi-Fi/WiMAX heterogeneous networks. In: International Conference on Signal Processing & Communication. Bangalore, India (2010)
3. Ahmed, A., Boulahia, L.M., Gaïti, D.: Enabling vertical handover decisions in heterogeneous wireless networks: a state-of-the-art and a classification. IEEE Commun. Surv. Tutor. **16**(2), 776–811 (2014)
4. IEEE Std 802.21, IEEE Standard for Local and metropolitan area networks - Part 21: Media Independent Handover Services (2009)
5. Oliva, A.D.L., Banchs, A., Soto, I., Melia, T., Vidal, A.: An overview of IEEE 802.21: mediaindependent handover serviceS. IEEE Wirel. Commun. Mag. **15**(4), 96–103 (2008). Aug
6. Rappaport, T.S.: Wireless Communications. Principle & Practice. Prentice-Hall, New Jersey, USA (1996)
7. Lee, W.C.Y.: Mobile Cellular Telecommunications - Analog and Digital Systems, 2nd edn. McGraw-Hill Inc, New York City (1995)
8. Garg, V.K.: Wireless and Personal Communications Systems. Prentice-Hall, Upper Saddle River (2000). ISBN 0-13-234626-5
9. Raghavendra, R., Belding, E.M., Papagiannaki, K., Almeroth, K.C.: Understanding handoffs in large IEEE 802.11 wireless networks. In: Proceedings of ACM SIGCOMM Conference on Internet Measurement (IMC), pp. 333–338. San Diego, USA (2007)
10. Pack, S., Choi, J., Kwon, T., Choi, Y.: Fast handoff support in IEEE 802.11 wireless networks. IEEE Commun. Surv. Tutor. **9**(1), 2–12 (2007)
11. Agiwal, M., Roy, A., Saxena, N.: Next generation 5G wireless networks: a comprehensive survey. IEEE Commun. Surv. Tutor. **18**(3), 1617–1655 (2016)
12. Johnson, D., Perkins, C., Arkko, J.: Mobility support in IPv6, IETF RFC 3775 (2004)
13. Lin, Y.B.: Queueing priority channel assignment strategies for PCS hand-off and initial access. IEEE Trans. Veh. Tech. **43**(3), 704–712 (1994). Aug
14. Hong, D., Rappaport, S.S.: Traffic model and performance analysis for cellular mobile radio telephone systems with prioritized and non-prioritized handoff procedures. IEEE Trans. Veh. Tech. **35**(3), 77–92 (1986). Aug
15. Posner, E.C., Guerin, R.: Traffic policies in cellular radio that minimize blocking of handoff calls. In: Proceedings of International Traffic Congress (ITC) 11. Kyoto, Japan (1985)
16. Guerin, R.: Queueing-blocking system with two arrival streams and guard channels. IEEE Trans. Commun. **36**(2), 153–163 (1988). Feb
17. Ramjee, R., Nagarajan, R., Towsley, D.: On optimal call admission control in cellular networks. In: Proceedings of IEEE INFOCOM, pp. 43–50. San Francisco CA, USA (1996)
18. Levine, D.A., Akyildiz, I.F., Naghshineh, M.: A resource estimation and call admission algo-rithm for wireless multimedia networks using the shadow cluster concept. ACM/IEEE Trans. Netw. **5**(1), 1–12 (1997). Feb
19. Nagshineh, M., Schwartz, M.: Distributed call admission control in mobile/wireless networks. IEEE J. Sel. Areas Commun. **14**(4), 711–716 (1996). May
20. Jiang, S.M., Ling, X.H.: A CAC considering both intra-cell and inter-cell handoff for measurement-based DCA. IEEE Trans. Veh. Tech. **56**(2) (2007)
21. Li, J., Shroff, N.B., Chong, E.K.P.: Channel carrying: a novel handoff scheme for mobile cellular networks. In: Proceedings of IEEE INFOCOM, vol. 2, pp. 908–915. Kobe, Japan (1997)
22. Li, J., Shroff, N.B., Chong, E.K.P.: Channel carrying: a novel handoff scheme for mobile cellular networks. ACM/IEEE Trans. Netw. **7**(1), 38–50 (1999). Feb
23. TIA/EIA Standard, TDMA third generation wireless: introduction to channels. TIA/EIA-136-100-B (2000)

24. Katzela, I., Naghshineth, M.: Channel assignment schemes for cellular telecommunication systems: a comprehensive survey. IEEE Pers. Commun. Mag. 10–30 (1996)
25. Zhang, X., Zhuang, W.H.: A channel sharing scheme for cellular mobile communications. IEEE Pers. Commun. Mag. **9**, 149–163 (1999)
26. Cimini, L.J., Foschini, G.J., Chih-Lin, I., Miljanic, Z.: Call blocking performance of distributed algorithms for dynamic channel allocation in microcells. IEEE Trans. Commun. **42**(8), 2600–2607 (1994)
27. Li, V.O.K., Qiu, X.M.: Personal communication systems (PCS). IEEE Proc. **83**(9), 1210–1243 (1995). Sep
28. Chuang, J.C.-I.: Autonomous adaptive frequency assignment for TDMA portable radio systems. IEEE Trans. Veh. Tech. **40**(3), 627–635 (1991). Aug
29. Cheng, M., Chuang, J.C.-I.: Performance evaluation of distributed measurement-based dynamic channel assignment in local wireless communications. IEEE J. Sel. Areas Commun. **14**(4), 698–710 (1996). May
30. Kuek, S.S., Wong, W.C.: Ordered dynamic channel assignment scheme with reassignment in highway microcells. IEEE Trans. Veh. Tech. **41**(3), 271–277 (1992). Aug
31. Kuek, S.S., Wong, W.C.: Approximate analysis of a dynamic-channel assignment scheme with handoffs. IEE Proc. Commun. **141**(2), 89–92 (1994). Apr
32. Okada, K., Kubota, F.: A proposal of a dynamic channel assignment strategy with information of moving direction in micro cellular systems. IEICE Trans. Fundam. **E75-A**(12), 1667–1673 (1992)
33. Okada, K., Park, D.K., Yoshimoto, S.: A dynamic channel assignment strategy using information on speed and moving direction for micro cellular systems. IEICE Trans. Fundam. **E79-B**(3), 279–288 (1996)
34. Jiang, S.M., Ling, X.H., Chua, K.C.: Performance of channel carrying in DCA cellular networks. Wirel. Pers. Commun. **25**(3), 241–262 (2003). Jun
35. Wong, Y.M., Misic, J., Chanson, S.T.: Call admission control in DCA wireless network. In: Proceedings of IEEE Symposium Personal, Indoor & Mobile Radio Communication (PIMRC), pp. 665–671. Boston, USA (1998)
36. Re, E.D., Fantacci, R., Giambene, G.: Performance evaluation of different resource management strategies in mobile cellular networks. J. Telecom. Syst. **12**(4), 315–340 (1999). Dec
37. Tian, X.S., Ji, C.Y.: Bounding the performance of dynamic channel allocation with QoS provisioning for distributed admission control in wireless networks. In: Proceedings of IEEE INFOCOM, pp. 1356–1363. New York City, NY, USA (1999)
38. Egner, W.A., Prabhu, V.K.: Enhanced dynamic radio resource allocation performance using a gradient descent algorithm. In: Proceedings of IEEE Symposium Personal, Indoor & Mobile Radio Communication (PIMRC), pp. 1448–1452. Boston, USA (1998)
39. Kleinrock, L.: Queueing Systems, volume I: Theory. Wiley, New York (1975)
40. Lin, Y.B., Noerpel, A., Harasty, D.: The sub-rating channel assignment strategy for PCS handoffs. IEEE Trans. Veh. Tech. **45**(1), 122–130 (1996). Feb
41. Jiang, S.M., Li, B., Luo, X.Y., Tsang, D.H.K.: A modified distributed call admission control scheme and its performance. ACM Wirel. Netw. (WINET) **7**(2), 127–138 (2001)
42. Akyildiz, I.F., Jiang, X., Mohanty, S.: A survey of mobility management in next-generation all-IP-based wireless systems. IEEE Wirel. Commun. Mag. **11**(4), 16–28 (2004). Aug
43. Shenoy, N., Mishra, S.: Vertical handoff and mobility management for seamless integration of heterogeneous wireless access technologies. In: Hossain, E. (ed.) Heterogeneous Wireless Access Networks: Architectures and Protocols. Springer (2009)
44. Yan, X.H., Şekercioğlu, Y.A., Narayananb, S.: A survey of vertical handover decision algorithms in fourth generation heterogeneous wireless networks. Comput. Net. **54**(11), 1848–1863 (2010). Aug
45. Jiang, S.M.: On marine internet and its potential applications for underwater internetworking (extended abstract). In: Proceedings of ACM International Conference on Underwater Networks & Systems (WUWNet), pp. 57–58. Kaohsiung, Taiwan (2013)

Chapter 8
Network Security in RWNs

Abstract There are many security threats and attacks to the Internet (e.g., Fig. 1.29), which have caused immense economical cost and social damage. Therefore, network security becomes one of the upmost important issues, and is a key element of many critical applications. Due to the exposure and broadcast nature of wireless media, wireless networks are especially vulnerable to security attacks. This weakness is the major hurdle of further applying wireless networks into some important sectors. This chapter discusses fundamental network security technologies for wireless networks (The major references for this part include (Hunt, Total focus conferences, Singapore, 2003, [1], Boncella, Comput Secur (Elsevier) 9:269–282, 2002, [2], Hollingshead, 802.11 wireless security vs. basic network security principles 2004, [3])).

8.1 Overview

This section discusses briefly typical security attacks, basic security functions and the major security issues in wireless networks.

8.1.1 Typical Targets Under Attack

Security attacks to the Internet can be classified according to the attack targets as follows: (i) information transmitted through networks, (ii) systems attached to networks (e.g., servers), and (iii) network system itself (e.g., paralyzing network operation). This book will mainly focus on the first type of attack because insecure information transmission systems provide opportunities of other security attacks.

8.1.1.1 Information Transmission

Attacks on information transmitted through a network try to monitor, sniff or even alter the transmitted information, typically including

© Springer Nature Singapore Pte Ltd. 2018
S. Jiang, *Wireless Networking Principles: From Terrestrial to Underwater Acoustic*,
https://doi.org/10.1007/978-981-10-7775-3_8

- Interception: attackers try to capture the information transmitted on the network through eavesdropping or sniffing.
- Man-in-the-middle (or replay attack): independent connections between victims and an attacker are set up, and the attacker relays messages between the victims such that they believe talking to each other directly [4].
- Traffic analysis: attackers try to identify communication patterns even if they cannot read individual messages.
- Spoofing: attackers send messages with forged source addresses as identities.
- Data altering: the data transmitted over the network is modified or inserted with new data by attackers.

8.1.1.2   Systems Connected to Networks

Attacking systems connected to the network (e.g. servers) tries to achieve the following objectives:

- Denial of service (DoS): a large number of messages or a huge number of service requests are flooded to a victim system (e.g., a server) in a short time to make it out of service.
- Brute force attack: attackers try to figure out the passwords or other credential information to have illegal access to victim's system connected to a network through unlimitedly repeating trials.
- Session hijacking: attackers exploit valid communication sessions to obtain unauthorized access to the information or services in a computer system, such as TCP-hijacking [4].
- Configuration altering: the configuration of a system is modified so that attackers can operate it at their will for further attack. For example, with distributed DoS (DDoS), an attacker modify the configurations of many computers distributed world wide, which can be used to launch DoS attacks to one system simultaneously.

8.1.1.3   Network Operations

Attackers try to disable the normal operations of a network system, through

- Exhausting network resources: with which, attackers greedily consume network resources including bandwidth and buffer space, such as jamming wireless media.
- Changing network configuration: with which, attackers try to modify the configuration of networking units such as routing tables in routers.

8.1.2 Basic Network Security

Basically, network security should achieve the following objectives: (i) maintaining the normal network service and operation, (ii) assuring the privacy and integrity of the data transmitted over the network, (iii) guaranteeing the data delivery to the real receivers, and (iv) protecting the systems attached to the network from attacks as well as (v) identifying the attack sources. To this end, the following basic security functions should be implemented [1]:

- Authentication: ascertaining the user legitimacy of using network resource.
- Confidentiality: protecting the privacy of the data transmitted over the network.
- Integrity: identifying whether the data has been altered or not in the course of transmission.
- No-repudiation: preventing a sender from denying what has been sent by itself.

8.1.3 Wireless Network Security

The fundamental for network security is cryptography, and cryptographic algorithms usually are computationally complex and consume much energy. A ciphertext is usually larger than its original text, and takes more bandwidth in transmission

Many security schemes have been proposed for wired networks, but they cannot be applied to wireless networks directly due to critical differences between them as listed in Table 8.1. As illustrated in Fig. 8.1, wireless media are exposed to attackers, which make them more vulnerable to security threats than wired networks. These media are much more unreliable with less bandwidth than wired networks. Terminal mobility in wireless networks complicate the situation because it is more difficult to maintain mobile user's credential information for authentication. Furthermore, mobile terminals usually have lower computation capability and limited power supply than those in wired networks.

Table 8.1 Wireless networks versus wired networks for network security

Compared items	Wired networks	Wireless networks
Security complexity	Less complex	More complex
Terminal mobility	Not supported	Supported
Medium security	Secure	Insecure
Network bandwidth	Abundant	Scarce
Computation capability	Powerful	Weak
Power supply	Unlimited	Limited

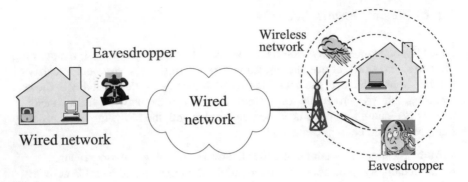

Fig. 8.1 Wireless and wired environments for network security

8.2   Security Primitives

The fundamental for network security is cryptography, based on which, several security primitives are constructed to secure communication and networking, typically including encryption and decryption, key management, digital signature and digital certificate as well as hash function.

8.2.1   Cryptography

Cryptography is a science for encrypting and decrypting of information, and is the fundamental of information security. As illustrated in Fig. 8.2, encryption is a process to scramble a plaintext (p) into a ciphertext (c) that cannot be easily understood. Decryption is a reverse process to restore a ciphertext to the original plaint text. For

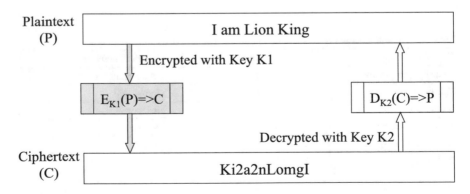

Fig. 8.2 Example on encryption and decryption

both encryption and decryption, cryptographic algorithms are used with some input variables usually called secret keys or key simply.

Let $E(p, K_e)$ denote an encryption function for an input plaintext p with encryption key K_e, while $D(c, K_d)$ as a decryption function for a ciphertext c with decryption key K_d. Then encryption process is mathematically expressed as $c = E(p, K_e)$, while decryption process as $p = D(c, K_d) = D(E(p, K_e), K_d)$.

Cryptography can be symmetric or asymmetric according to the relationship between the keys for encryption and decryption. With symmetric key cryptography, the decryption key (K_d) is the same as the encryption key (K_e), i.e., $K_e = K_d$. With asymmetric key cryptography, $K_e \neq K_d$, which means that two different keys are used for encryption and decryption, respectively. It is also called public key cryptography with the following important property. Given a public key pair (K_1, K_2), if K_1 is used for encryption, only K_2 can be used for decryption, and vice versa. Usually given a public key pair, one key is kept secretly, called private key, while the other is open to public, called public key. However, with an equivalent cryptographic strength, a symmetric key is much shorter than an asymmetric key. For example, a 40-56-bit symmetric key is equivalent to a 512-bit asymmetric key [1]. An example of public key cryptography is given in Appendix B.

8.2.2 Key Management

Key management is responsible for key generation, key distribution and key storage as well as key revocation. A set of algorithms are defined to generate keys. Key distribution consists of processes and protocols used to deliver keys to the corresponding users. The typical key distribution structure is Public Key Infrastructure (PKI), which is also widely used for digital signature. It consists of following components: (i) a security policy defining the direction on information security, the processes and principles for the use of cryptography, (ii) a certificate authority (CA) responsible for issuing and revoking certificates, (iii) a registration authority (RA), an interface between the user and the CA for authenticating the identity of the users and submitting the certificates to the CA, and (iv) a certificate repository and distribution system used to store and distribute certificates [1].

8.2.3 Hash Functions

A hash function, $H(x)$, can be thought of as a black box that accepts a digital object (x) and outputs an identifying number, $h = H(x)$, with the following properties [5, 6]:

- The same input will yield the same hash value, i.e., $H(x) = H(x')$ if $x = x'$; while different inputs should generate different hash values, i.e., $H(x) \neq H(x')$ if $x \neq x'$.

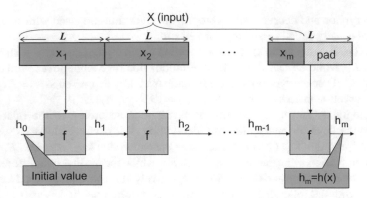

Fig. 8.3 Construction of a hash function [5]

- The input of a hash function $H(x)$ can be of any length, and it should be easy to compute $H(x)$ for any x, while the output has a fixed length for easy implementation.
- $H(x)$ should be one-way function, which means that it is computationally infeasible to find an input that can yield a hash value equal to an existing one.
- $H(x)$ should be collision-free, which means that it is computationally infeasible to find two different inputs x and y such that $H(x) = H(y)$.

Actually, the basis of a hash function is a compression function (f) that works on fixed size input blocks. As illustrated in Fig. 8.3, an input x of an arbitrary length is broken up into smaller blocks $x_1, \ldots, x_m$, and the last block has to be padded when necessary. The hash value of x is computed by repeatedly applying f as follows: given an initial value h_0, computing $h_i = f(x_i \oplus h_{i-1})$ for $i = 1, \ldots, m$, and taking h_m as the hash value of x, where $\oplus$ denotes concatenation operation.

Hash functions have many useful applications such as storing passwords and integrity protection since any modifications on the original data cause a change in the hash value. Computing the hash value requires no secret information so that anybody can create a valid hash for a given input. Particularly for long messages, it takes time and computation resource to make a direct comparison between them and digitally sign them, and hash functions can be used to generate short digests for comparison and digital signature.

8.2.4 Digital Signature

As illustrated in Fig. 8.4, a digital signature is created by encrypting a message with the sender's private key, and the recipient checks the signature through decrypting the message by using the sender's public key. Due to the characteristics of public key pair mentioned above, only a sender's public key can decrypt the ciphertext

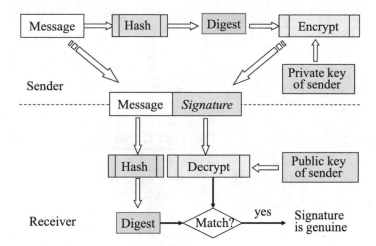

Fig. 8.4 Principle of digital signature

encrypted with its private key, which ascertains the uniqueness of a digital signature. However, the public key is inefficient to sign large messages so that a hash function is often used to generate a short digest. Then the digital signature is carried out to the digest rather than the original message as illustrated in Fig. 8.4 [6].

8.2.5 Digital Certificate

A digital certificate establishes a credential tie between a name or an identity and the information declared by the certificate such as public keys. As illustrated in Fig. 8.5, a certificate is issued by a certificate authority (CA) using the CA's private key to guarantee the authenticity of an issued certificate, while this CA's public key is published, and will be used by the user to verify the certificate available to the public. In other words, a CA is signed with the digital signature of the issuer [6].

A digital certificate system is required to issue and manage CAs, and usually consists of the following parts: (i) a certificate repository system that stores keys and certificates, (ii) a certificate revocation system to revoke some issued digital certificates due to mistakes in certificate vetting and key management, and (iii) a certificate distribution system used to distribute certificates [1].

8.2.6 Virtual Private Network (VPN)

A physical private network can be built up by using either private network infrastructure such as LANs or leased lines of a public network to link hosts located in

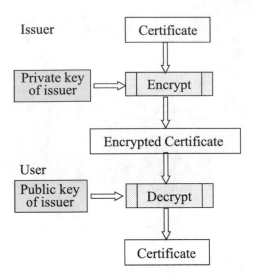

Fig. 8.5 Principle of digital certificate

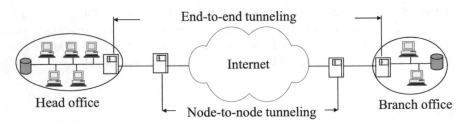

Fig. 8.6 Tunneling for virtual private network (VPN)

different geographical sites. The former is limited by small coverage, while the latter is expensive. Virtual private networks (VPNs) can provide private connection over a public network through tunneling. It can be used to (i) interconnect dispersed nodes located in different geographical locations, (ii) provide privacy and integrity for data transmitted across the public network, (iii) authenticate users before allowing communication with the VPN and (iv) provide the facilities of QoS and multicasting [1].

As illustrated in Fig. 8.6, there are two types of tunneling: end-to-end and node-to-node. With the former, VPN devices at each end of a VPN connection are responsible for tunnel creation, the encryption and decryption of data transferred between two sites. With the latter, the tunnel creation and termination occur at the gateway devices located at the network edges.

8.3 Standards for Network Security

As illustrated in Fig. 8.7, many network security standards have been developed to enforce network security for wired networks, and are the fundamental for wireless network security. The typical ones are introduced below, including authentication and security systems for the transport layer, network layer and data link layer.

8.3.1 Authentication Systems

A general authentication system is independent of low-layer networking technologies, typically including Remote Access Dial-in-User Service (RADIUS) and Diameter, which actually consist of Authentication, Authorization and Accounting (AAA) protocols.

8.3.1.1 Remote Access Dial-in-User Service (RADIUS)

RADIUS [7, 8] defines a protocol to perform authentication, authorization, and configuration information between a network access server (NAS) and a shared authentication server. It provides security, authorization and accounting services to dial-up links connecting the outside world through managing a single database of users, with the following properties [7]:

- A NAS operates as a client of RADIUS, and passes user information to the designated RADIUS servers, and acts on the returned response.
- The RADIUS server receives user connection requests, authenticates the user, and returns all configuration information to the client for service delivery. It can also act as a proxy client to other RADIUS servers or other types of authentication servers.

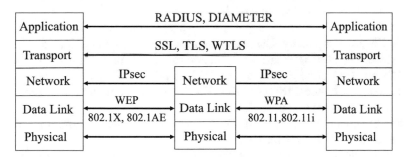

Fig. 8.7 Standards for network security [6]

- Transactions between the client and RADIUS server are authenticated with a shared secret that is never sent through the network. Any user passwords sent between the client and RADIUS server are encrypted.
- The RADIUS server can support several methods to authenticate a user, which typically include the Point-to-Point Protocol (PPP) [9], password pair (PAP) or challenge plus response pair (CHAP).
- All transactions are comprised of variable length attributes, and new attribute values can be added without disturbing existing implementations.

A user may initiate a PPP authentication request to the NAS, which either prompts the user for username and password with the PAP or challenge with the CHAP, and the user should reply correspondingly. A RADIUS client, i.e., the NAS, sends the username and encrypted password to the RADIUS server, which responds with a decision as follows: accept, reject or challenge. It has been shown that RADIUS cannot work well in large scale systems, and the Diameter protocol has been designed accordingly.

8.3.1.2 Diameter

It aims to provide an AAA framework for applications (e.g., network access or IP mobility) in both local and roaming situations through enhancing TACACS and RADIUS initially designed for dial-up PPP and terminal server access [10].

TACACS [11] is an access control protocol, allowing a TIP (a routing node that could accept dial up lines) to accept a username and password and send a query to a TACACS authentication server, which is a program running on a host. The host would determine whether to accept or deny the request with a response [11].

Diameter versus RADIUS

Diameter aims to fix the following issues that have not been addressed adequately by RADIUS [10]:

- Diameter supports application-layer acknowledgements and defines failover algorithms that are not defined in RADIUS.
- RADIUS does not provide support for per-packet confidentiality, and support for IPSec is not compulsory. Diameter provides universal support for transmission-level security with the Transport Layer Security (TLS) protocol over TCP and the Datagram Transport Layer Security (DTLS) protocol [12] over Stream Control Transmission Protocol (SCTP) [13]. DTLS provides communication privacy for datagram protocols based on TLS.
- Diameter runs over TCP and SCTP to provide reliable end-to-end transmission, whereas RADIUS runs over UDP.
- Diameter defines explicit agent behavior, which is not explicitly defined in RADIUS.
- The support of server-initiated messages in RADIUS is optional, while it is mandatory in Diameter. This allows some features to be implemented on demand

across a heterogeneous deployment, such as unsolicited disconnection or re-authentication/re-authorization.

Diameter also provides the following services that RADIUS does not: capability negotiation and peer discovery as well as configuration through DNS. It is backward compatible with RADIUS.

Protocol Overview

Diameter is a peer-to-peer protocol, with which, any node can initiate a request, and the communicating party may accept or reject it. Communication between peers begins with one peer sending a message to another, with the following units defined for protocol operation [10].

- A Diameter client is located at the edge of the network to perform access control, e.g., a NAS or a foreign agent. It generates Diameter messages to request AAA services for the user.
- A Diameter agent provides relay, proxy, redirect or translation services, but not local user authentication or authorization services.
- A Diameter server conducts authentication/authorization of the user.

The protocol message consists of a header followed by one or more Attribute-Value-Pairs (AVPs) to provide the following services [10]:

- Transporting user authentication information to enable the server to authenticate the user.
- Transporting service-specific authorization information to allow the peers to decide whether to grant a users access request or not.
- Exchanging resource usage information for accounting or capacity planning.
- Routing, relaying, proxying and redirecting of Diameter messages.

8.3.2 Transport Layer Security

Typical transport layer security systems include Secure Socket Layer (SSL) and its successor, TLS.

8.3.2.1 Secure Socket Layer (SSL)

SSL [14] was originally developed on the top of the transport layer (e.g., TCP) by Netscape to provides communication privacy over the Internet. It aims to secure end-to-end communications between browsers and servers by preventing eavesdropping, tampering, or message forgery with the following security services: (i) ensuring confidentiality and integrity of the transmitted data, and (ii) authenticating a server to assure that it is real as claimed. SSL uses both public keys and private keys for encryption and decryption. Now it has been replaced by TLS.

8.3.2.2 Transport Layer Security (TLS)

TLS [15], the successor of SSL, provides privacy and data integrity between two communicating applications over the Internet, with following properties: (i) an open standard-based solution with more no-proprietary ciphers, (ii) better error reporting capability, and (iii) using the Keyed-Hashing for Message Authentication (HMAC) protocol [16].

TLS is composed of record and handshake protocols. The record protocol layered on top of reliable transport protocols such as TCP provides connection security with the following basic properties [15]:

- The connection is private, using symmetric keys for data encryption. The keys are generated uniquely for each connection. The protocol can also be used without encryption.
- The connection is reliable due to using a message integrity check based on HMAC computed by secure hash functions.

This protocol also allows the server-client to authenticate mutually and negotiate encryption algorithms and keys before data transmission or reception [15].

The handshake protocol provides connection security with the following properties [15]:

- The peers identity can be authenticated using asymmetric/public keys.
- The negotiation is secure since the negotiated secret is not available to eavesdroppers.
- The negotiation is reliable since any modification to the negotiation communication will be detected.

8.3.2.3 Wireless Transport Layer Security (WTLS)

WTLS is an optional layer of the Wireless Access Protocol (WAP).[1] It provides security mechanisms based on public keys, similar to TLS in protecting integrity and confidentiality of information, authentication and protection against DoS attacks. It provides a secure session between a WAP device and a WAP gateway, and security facilities for encryption, strong authentication, integrity and key management. It is compliant with regulations on cryptographic algorithms plus key lengths in different countries [1]. However, the application of WAP is becoming less popular.

8.3.3 Network Layer Security

The typical network layer security system is IP Security (IPSec) for IP networks.

---

[1]http://www.wapforum.org.

8.3.3.1 Security Problems of IP

The IP was originally designed for use in well-established organizations such as governmental agencies, research institutions and universities, and network security had not been taken into account at the beginning. Its major security problems are listed below.

- No authentication is performed for either the sender or receiver of IP packets. The sender and receiver are identified simply by the IP address transmitted in plaintext, and can be easily spoofed and modified by attackers.
- The payload of an IP packet is also transmitted in plain text, and can be easily intercepted and understood by attackers without protection on information confidentiality or information integrity.

8.3.3.2 IP Security (IPSec)

IPSec [17, 18] is designed to provide cryptographically-based security for IPv4 and IPv6 at the IP layer, with the following security services: (i) access control to the IP layer, (ii) connectionless integrity (i.e., detecting modification of an IP datagram), (iii) data origin authentication, (iv) replay detection and replay rejection, and (v) limited traffic flow confidentiality. IPSec also specifies a minimal firewall functionality essential for the access control. Most of the security services are provided through the following two traffic security protocols: authentication header (AH) [19] and encapsulating security payload (ESP) [20], along with key management protocols [17].

The AH protocol protects data integrity and provides origin authentication, with optional anti-replay features. The ESP protocol additionally protects confidentiality to enable concealing packet length and facilitating generation and discarding of dummy packets. Both AH and ESP offer access control through key distribution and traffic flow management. The IPSec implementations must support ESP and may support AH because it has shown that ESP can almost provide all the requisite security services [17].

8.3.4 Link Layer Security

Typical link layer security systems include Extensible Authentication Protocol (EAP), IEEE 802.1AE and IEEE 802.1X.

8.3.4.1 Extensible Authentication Protocol (EAP)

Since PPP can only provide limited authentication methods, EAP is proposed to provide an authentication framework to support multiple authentication methods, and runs directly over the data link layer without using IP [21]. It improves previous authentication protocols such as PAP and CHAP through providing more sophisticated authentication methods such as token cards, Kerberos, PKI and symmetric key schemes. The authentication exchange is performed as follows [21]:

- The authenticator sends a request to authenticate the peer, typically an initial identity request.
- The peer sends a response in reply to a valid request.
- The authenticator sends an additional request, and the peer replies with a response.
- This conversation continues until either the peer cannot be authenticated after a certain number of retrials, or a successful authentication has been performed.

The EAP-TLS protocol provides a mutual authentication as well as support for fragmentation and reassembly unavailable in EAP, taking advantage of the cipher suite negotiation, mutual authentication and key management capabilities of the TLS protocol [22].

8.3.4.2   IEEE 802.1X

It defines a generic framework for authentication and authorization for IEEE LAN security. It authenticates a node before giving access to a LAN through port control, which allows a network administrator to restrict the use of IEEE 802 LAN service access points (SAPs), i.e., ports here, to secure communication between authenticated and authorized devices [23]. EAP is adopted to support authentication using a centrally administered authentication server, and EAP encapsulation Over LANs (EAPOL) is defined for necessary exchanges between peer ports. It is expected that no other protocols are needed to protect link communication with such physically secured communication [24].

A common architecture, functional elements and protocols are specified to support mutual authentication and secure communication. It requires PKI certificates on each client through a central RADIUS server that runs EAP, and enables a carrier for a secure delivery of session keys. The authenticator authenticates a client such as a network interface card (NIC) through authenticating it to a RADIUS server or Kerberos server [23]. Here, an authenticator is an entity facilitating the authentication of entities attached to the same LAN [24].

Figure 8.8 illustrates the components of controlled-port network access systems and the corresponding behaviors when an AAA server is used to centralize administration of authentication and authorization [24], and is explained below:

(a) Port, which is an entity through which communication is controlled and secured.
(b) Attached LAN, which provides MAC service to the ports client and its peers.
(c) Mechanisms, which define a secure connectivity association between the ports. This association is a security relationship comprising a fully connected subset of the SAPs in the associated stations attached to a single LAN.

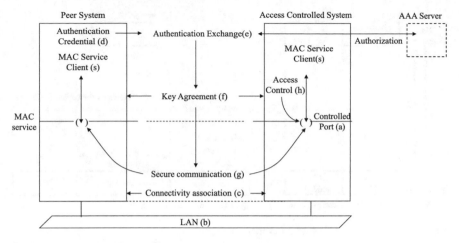

Fig. 8.8 Controlled-port network access components and processes [23]

The behaviors of a port's protocol entities engaged in access controlled communication include [24]:

(d) Possessing an authentication credential.
(e) Mutually authenticating peer ports.
(f) Communicating authorization data to the ports clients.
(g) Using the authentication results (i.e., success or failure) to agree on keys used to protect communication, and creating a secure connectivity association between peer ports.
(h) Protecting the data transferred within the secure connectivity association.
(i) Enforcing access control based on the authentication results.

8.3.4.3   IEEE 802.1AE

IEEE 802.1AE MAC Security Protocol (MACsec) [25] specifies cryptographic support of a controlled port for IEEE 802 medium access methods as illustrated in Fig. 8.9. It maintains the confidentiality and integrity of the transmitted data, and provides secure MAC services on a frame-by-frame basis according to security relationships maintained by the MACsec key agreement, which specifies the generation of the secure association key (SAK) used by MACsec [25]. However, MACsec protects communication only between trusted components, and cannot protect against inside attacks.

Basically, a secure communication over an insecure physical link with MACsec is provided with the following measures:

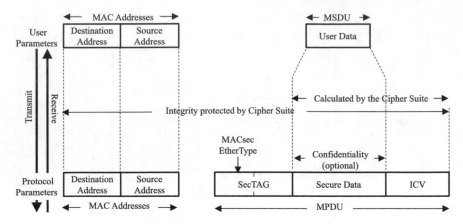

Fig. 8.9 Overview of MACsec architecture [25]

- A CA is created for connectivity between the stations attached to a single LAN, and is supported by unidirectional secure channels, which secure communication using symmetric key cryptography.
- Each secure channel is supported by an overlapped sequence of security associations, each of which uses a fresh SAK to guarantee the MACsec services and security for a sequence of transmitted frames.

8.4   Standards for Securing Wireless Links

Wireless local area networks (WLANs) have been widely used in practice today, such as the IEEE 802 WLAN family, Bluetooth, ZigBee and WiFi. Particularly, WiFi WLANs based on IEEE 802.11 can be found almost anywhere to provide pervasive access to the Internet, Therefore, we mainly focus on security issues of this network in the following discussion.

A WiFi WLAN usually consists of an access point (AP) and wireless stations, which collectively comprise a basic service set (BSS), and is also called infrastructure BBS. Differently, in an independent BBS, which is also called wireless ad hoc networks, no AP is used to coordinate communications between wireless stations. Multiple infrastructure BBSs can be linked through wired networks to form an extended service set (ESS), which can allow mobile stations to have transition mobility. A user is able to roam from one BBS to another without suffering from connectivity interruption.

This section focuses on security issues of WLANs that are based on the infrastructure BBS. Basically, a service set identifier (SSID) and an MAC address filtering scheme are jointly used to secure the access to a WLAN AP. They can only control which nodes can use a wireless medium, but does not protect confidentiality to the frames transmitted over air interface. Therefore, the Wired Equivalent Privacy (WEP) protocol is proposed.

The fundamental for wireless security is securing wireless links, and the following sections discuss security methods defined for wireless link layer security, which mainly include IEEE 802.11 (including WEP), WiFi Protected Access (WPA) and IEEE 802.11i. An end-to-end connection consisting of wireless links can be secured by WTLS discussed earlier.

8.4.1 IEEE 802.11

It specifies cryptographic methods to protect data transmitted over WLANs [26]. It also specifies key agreement and key distribution. A typical link layer security control provided by this standard comprises the SSID and MAC address filtering.

8.4.1.1 Service Set Identifier (SSID)

A large wireless network can be portioned into multiple sub-networks, each of which is covered and controlled by an AP and configured with an SSID. To access a sub-network, a wireless station should have a correct SSID that matches that of the AP. A wireless station can also be configured with multiple SSIDs so that it can access to different APs. This security measurement provides a minimum network security because it is easily compromised. An AP's SSID is shared by all wireless stations covered by this AP so that attackers can sniff the SSID from the traffic of an authorized user. A sub-network may also broadcast its SSID to indicate its existence, which is actually very popular practice in reality.

8.4.1.2 MAC Address Filtering

This scheme is also proposed to enhance network security jointly with the SSID matching. A wireless station can be identified by the unique IEEE MAC address of its network interface card (NIC). An AP can has a list of the MAC addresses of the authorized wireless stations. For a wireless station, only if its MAC address is included in this list and its SSID matches the AP's one, it is allowed to access this AP.

This scheme can be easily compromised by using a spoofed MAC address with a MAC address configurable NIC. In addition, the MAC address list must be kept updated especially for mobile stations. This raises a new challenge to network operation and causes the scalability problem. It is difficult for an AP to automatically update the MAC addresses of mobile stations because this operation requires secure information exchange between the AP and mobile stations. However, there is not yet a secure wireless channel at this stage. A manual intervention is required to update the list, which is not scalable for a large number of mobile stations.

8.4.1.3   Wired Equivalent Privacy (WEP)

The above SSID matching and MAC address filtering cannot secure the information transmitted in the air. A WEP protocol is defined by IEEE 802.11b to provide the following enhancements: (i) preventing eavesdropping through encrypting the payload of the transmitted frames, (ii) preventing unauthorized access to APs, and (iii) detecting modification to the received data.

Confidentiality

Rivest Cipher 4 (RC4) is adopted for encryption and decryption using a 40-bit (or 104-bit) shared symmetric key. As illustrated in Fig. 8.10, a key stream is generated by using one key plus a 24-bit initialization vector (IV), which however is sent in plaintext. The key stream tries to avoid two plaintexts from being encrypted with the same shared key. The payload of a frame along with its CRC value is encrypted with the key stream. The ciphertext along with the IV is transmitted to the receiver, which then decrypts the ciphertext to the plaintext with the same key stream generated in the same way as done by the sender.

Authentication

A shared symmetric key is also used for client authentication through the following process between an AP and a client. (i) The client first submits an authentication request to the AP. (ii) The AP returns a challenge phrase to the client. (iii) The client encrypts the challenge phrase using a shared symmetric key and transmits it to the AP. (iv) The AP compares the client's response with its stored phrase. If they match each other, the client is authorized and rejected otherwise.

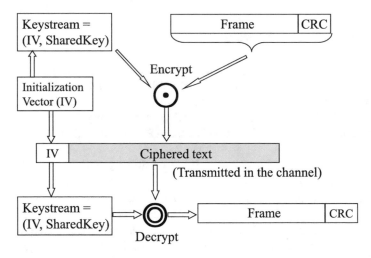

Fig. 8.10 Encryption and decryption of the WEP frame

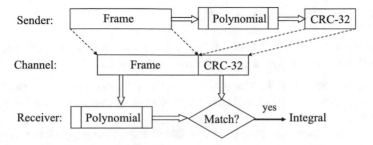

Fig. 8.11 Integrity protection with WEP

Integrity

As illustrated in Fig. 8.11, data integrity is protected by comparing the following two CRCs. One is received along with the data frame, and the other is generated for the received data frame using the same polynomial algorithm. If these two CRCs match each other, the data integrity is maintained.

Problems of WEP

Since an AP and all its covered stations share a symmetric key for encryption and decryption, an attacker can launch a dictionary-building attack to statistically analyze data traffic to realize a possible realtime automated decryption of all traffic. Although a 24-bit IV is adopted to prevent different stations from using the same key stream, the key length is computationally short so that the same key may be reused within a short time period.

The IV transmitted in plaintext facilitates figuring out the WEP key. An attacker can also crack down the authentication system without knowing the WEP key through the replay attack. That is, an attacker intercepts the encrypted credential information and then just reuses it. WEP simply uses CRC-32 for integrity protection, which is insecure because the CRC checksum itself can also be modified for a modified frame using the same algorithm.

The above security mechanisms can only provide loose network security, which can be easily compromised by using some free tools available on the Internet. They provide fast and effective statistical attacks through collecting as many frames as possible from a WEP-protected wireless network. Thus, an interim solution namely WiFi Protected Access (WPA) is defined by the WiFi alliance, and a long-term solution IEEE 802.11i is defined by IEEE.

8.4.2 WiFi Protected Access (WPA)

There are two versions of WiFi Protected Access (WPA) protocols: the first generation WPA (WPA) and the second generation WPA (WPA2) [27].

8.4.2.1   WPA (1st Generation WPA)

WPA aims to improve IEEE 802.11 security through software upgrading of the existing hardware mainly with the following enhancements.

- It utilizes a so-called Temporal Key Integrity Protocol (TKIP) to enable more data protection through changing encryption keys for every frame, which is called per-packet key (PPK).
- It provides a message integrity check through Michael algorithm, and uses an extended initialization vector (IV) to provide a stronger integrity protection than the CRC used in WEP.
- It implements an IEEE 802.1X server and EAP to further strengthen its authentication capability.

WPA provides two application modes: personal mode and enterprise mode. The personal mode is designed for use in home and offices, where it is not cost-effective to implement and maintain an authentication server such as RADIUS. The enterprise mode is designed to provides industrial strength security for business sectors through using the IEEE 802.1X and RADIUS servers [27].

Personal Mode

TKIP is used to replace the WEP encryption to provide more secure encryption. To maintain compatibility with WEP, TKIP is designed to use the same hardware-based calculation mechanisms as used by WEP. This mode has the following properties [27]:

- Using pre-shared keys (PSKs) for authentication: Each user must enter a shared static key or passphrase to access the network. This information is usually stored on the user device and needs to be entered only once. A passphrase must be at least 8 characters long, and a 20-character long one is recommended, containing numbers and special characters.
- TKIP is used to create dynamic keys for encryption and mutual authentication. A PSK automatically changes the keys at a preset time interval, making it much difficult for an attacker to figure out them. It also sets a unique starting key for each authenticated client using PSK. Refer to Sect. 8.4.3.1 for more detail.
- Message integrity: It is used to provide a better data integrity protection than the CRC used by WEP. Refer to Sect. 8.4.3.1 for more detail.

Note that the passphrase should be changed whenever an individual user able to access the network is no longer authorized to use it, or when a device configured to use it is lost or compromised.

Rekeying refers to a process that changes the encryption key of an ongoing communication to control the amount of data encrypted with the same key [4]. However, rekeying encryption keys used by WEP can be only carried out manually, which is a tedious process and impractical for large organizations. If an AP's key is rekeyed in this way, no clients can access this AP until the encryption keys of these clients are also rekeyed. With WPA, the rekeying of encryption keys is required for every frame. TKIP allows the key change between the AP and the corresponding wireless clients to be automatically synchronized on a frame basis [27].

Enterprise Mode

The Extensible Authentication Protocol (EAP) is used for message exchange during the authentication process, and a RADIUS server is used to provide industrial strength security. There are tools available on the Internet that can modify a WEP-encrypted frame payload and the corresponding integrity check value (ICV). Thus, Michael algorithm described in Sect. 8.4.3.1 is used to replace the CRC-based ICV for a stronger integrity protection [27].

8.4.2.2 WPA2 (2nd Generation)

WPA2 has not been designed to overcome the weaknesses of WPA but to support the new mandatory security features standardized by the IEEE 802.11i. It maintains backward compatibility with WPA. The major difference between WPA and WPA2 is that the Advanced Encryption Standard (AES) [28] is used as the cipher in WPA2. WPA still uses RC4 adopted by WEP but is enhanced by the PPK scheme supported by TKIP. AES can provide enough security to meet the high level standards of many federal government agencies [1].

Like WPA, WPA2 also supports both enterprise applications and home applica tions with the following additional features. The enterprise mode supports authentication in the following two phases: an open system authentication, with which only the MAC address of the NIC is used as the identification, and a joint use of the authentication methods adopted by IEEE 802.1X and EAP. The personal mode supports the use of PSK.

8.4.3 IEEE 802.11i

It is an amendment to WPA2, aiming to provide a long-term solution to secure IEEE 802.11 WLANs with the following objectives: (i) providing commercial grade security, (ii) addressing all known issues of WEP and (iii) maintaining backward compatibility with WEP and forward compatibility with all existing and planned amendments [24]. To fully support IEEE 802.11i, most existing IEEE 802.11 WLAN devices need to be replaced in order to process AES cipher. AES needs more powerful processing capability and processing time with more energy consumption than the cipher used in WEP.

A Robust Security Network Association (RSNA) is introduced to realize the above objectives. A security association is a set of policy(ies) and key(s) used to protect information, and an RSNA is an association used by a pair of stations if they use the 4-way handshake defined by the standard to establish authentication or association between them [24]. In an RSNA, IEEE 802.11 provides functions to protect data frames, and IEEE 802.1X provides authentication along with an authentication server and a controlled port, while they collaborate to provide key management [24]. RSNA security comprises the following components:

- Temporal Key Integrity Protocol (TKIP), a cipher suite enhancing the WEP protocol on pre-RSNA hardware. A cipher suite includes a set of algorithms designed for data confidentiality, data authenticity, data integrity and replay protection.
- An AES based counter mode with CBC-MAC protocol (CCMP) used between stations and APs.
- Key management providing key hierarchy including pairwise key hierarchy and group key hierarchy.
- Security capability discovery and security negotiation.

8.4.3.1 Temporal Key Integrity Protocol (TKIP)

The following modifications to WEP are conducted by TKIP for more robust data protection [24]:

- To protect against forgery attacks, the transmitter calculates a keyed cryptographic message integrity code (MIC) over the MAC source and destination addresses, the priority, and the plaintext data of the MAC service data unit (MSDU) (i.e., the payload of the MAC frame). This MIC is appended to the MSDU data, as illustrated in Fig. 8.12.
- A countermeasure is used to overcome the weakness of the TKIP MIC to defeat forgery message integrity attacks.
- A per-MPDU TKIP sequence counter (TSC) is used to sequence MAC protocol data units (MPDUs) so that the out-of-order MPDUs will be dropped by the receiver.
- A cryptographic mixing function is used to combine a temporal key, the Transmit Address (TA) and the TSC into the WEP seed (i.e., per-MPDU key) by concatenating an encryption key to an IV. The key mixing function is used to compute per-MPDU key in order to defeat weak-key attacks against the WEP key.

More discussions on key mixing and MIC are given below.

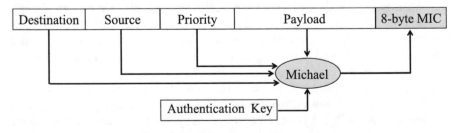

Fig. 8.12 Parameters for message integrity code (MIC) [29]

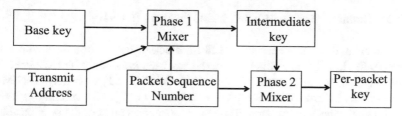

Fig. 8.13   Key mixing [29]

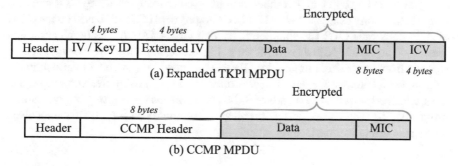

(a) Expanded TKPI MPDU

(b) CCMP MPDU

Fig. 8.14   MPDU formats for TKPI [29]

Key Mixing

The encryption key is 128-bit long, and the data integrity key is 64-bit long. The encryption key length is mainly constrained by some WEP off-load hardware, while the length must be long enough to avoid key reuse. An AP and a wireless station use different keys for transmission [30]. As illustrated in Fig. 8.13, a PPK is mixed with the base key, the transmit MAC address and the packet sequence number. Such mixing aims to build a better per-packet encryption key, prevent weak-key attacks, de-correlate the WEP IV from each frame, and defend against address hijacking as well key reuse [30].

Message Integrity Code (MIC)

The MIC is similar to the ICV used by WEP, with an additional MIC for the MPDU. The MIC is computed according to the destination address, the source address, the payload of the data frame and an authentication key as illustrated in Fig. 8.12. As illustrated in Fig. 8.14a, an 8-byte MIC is inserted between the payload and a 4-byte ICV. The MIC contains a frame counter to detect forgery attempts caused by wireless replay attacks. The wireless device computes an MIC using the same mechanism as used to compute the ICV. The MIC is encrypted along with the payload and the ICV to make it much more difficult to tamper with the data [24].

8.4.3.2   Counter Mode with CBC-MAC Protocol (CCMP)

The counter mode (CTR) encrypts a 128-bit block of data at one time with a 128-bit encryption key that is different from the key used previously. The Cipher Block Chaining (CBC) Message Authentication Code $(\mathbb{MAC})^2$ [31], or simply CBC-MAC, constructs a $\mathbb{MAC}$ from a block cipher (also refer to [32]), and a chain of blocks is created such that each block depends on the proper encryption of the previous block.

The CTR with CBC-MAC (CCM) [33] combines the CTR and CBC-MAC. CCM is a symmetric key block cipher mode providing confidentiality using CTR and data origin authenticity using CBC-MAC. The CCM protocol (CCMP) based on the CCM of AES combines CTR for confidentiality and CBC-MAC for authentication and integrity. The CCMP MPDU is depicted in Fig. 8.14b, and the CCMP header is constructed from the packet number, key ID and extended IV [24]. The confidentiality, data authentication, and replay protection are enhanced by using fresh cryptographic keys, which are provided by means of the 4-way handshake and group key handshake protocols [24]. The replay protection allows a station to detect whether a received data frame is an unauthorized retransmission.

8.4.3.3   Key Management

A 4-way handshake is used to manage pairwise keys to confirm mutual possession of a Pairwise Master Key (PMK) by two parties and distribute a group temporal key (GTK) [24, 30].

4-Way Handshake

It is based on IEEE 802.1X EAPOL-Key frames, and completes the IEEE 802.1X authentication process to achieve the following objectives [24, 30]:

- Confirming that a live peer holds the current PMK and the cipher suite selection.
- Deriving a fresh pairwise transient key (PTK) from the PMK.
- Installing the pairwise encryption and integrity keys into IEEE 802.11.
- Transferring the GTK and its sequence number from the authenticator to the supplicant (i.e., peer).
- Installing the GTK and its sequence number in the station and the AP if not already installed.

After a successful handshake, the authenticator and the supplicant have authenticated each other. Then, the IEEE 802.1X controlled ports are unblocked to permit general data traffic [24, 30].

Group Key Handshake (GKH)

The GKH is used to issue a new GTK to peers who have already established security associations with the local station through the 4-way handshake protocol. A PTK

$^2\mathbb{MAC}$ is artificially used here to distinguish it from the abbreviation for Medium Access Control.

Fig. 8.15 Pairwise key architecture [29]

is used to protect a GTK during its transferring to the receiving station. The GKH process can be triggered by the supplicant through sending an EAPOL-key frame [24].

Pairwise Key

As illustrated in Fig. 8.15, the pairwise key hierarchy takes a pairwise master key (PMK) to generate a PTK, which is partitioned into a key confirmation key (KCK), a key encryption key (KEK) and temporal keys (TKs) used by the MAC to protect unicast communication between the receive stations of the authenticator and supplicant. PTKs are used between a single supplicant and a single authenticator. Here a master key is a secret key used to derive other cryptographic keys for protecting data transfer [24].

The above construction of a PTK guarantees fresh session keys because with a random 256-bit string, there is a statistically insignificant chance that a PTK will ever repeat. The PTK is bound to the station and the AP [30].

8.4.3.4   Discovery and Negotiation

The discovery process and negotiation process is used to enable an IEEE 802.11i-based device to be inter-operable with already-deployed and non-802.11i device, and create a mechanism to extend the IEEE 802.11i framework to permit authenticated key management (AKM) protocols and cipher suites that are not defined by IEEE 802.11i. As illustrated in Fig. 8.16, during the discovery process, an AP tries to find the security policy adopted by available WLANs, such as the available AKM protocol unicast and multicast cipher suites. During the negotiation process, both the AP and the wireless station try to enable each other to agree on the security policy to be used with an association [30].

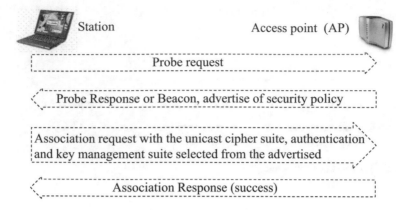

Fig. 8.16 Discovery and negotiation diagram [29]

8.5 Summary

This chapter briefly discusses the major security issues for wireless networks, and introduces the principles of the major standards available so far. More reading for wireless security can be found in surveys such as [34]. Many solutions are proposed for infrastructured wireless networks mainly used as access networks, and may not perform well in other types of wireless networks such as mobile ad hoc networks (MANETs) and wireless sensor networks (WSNs) as well as underwater acoustic networks (UWANs) because they are different in several aspects that influence the operations of these solutions. Surveys for MANETs and WSNs can be found in [35, 36], respectively, and more discussion on security for UWANs will be provided in Chap. 13.

References

1. Hunt, R.: Network security – systems and architecture. In: Total Focus Conferences, Singapore. http://www.cosc.canterbury.ac.nz (2003). Accessed 26–27 March 2003
2. Boncella, R.J.: Wireless security: an overview. Comput. Secur. (Elsevier) **9**, 269–282 (2002)
3. Hollingshead, T.: 802.11 wireless security vs. basic network security principles. SANS Institute 2004, available on WWW (2004)
4. The free encyclopedia: Wikipedia. http://en.wikipedia.org/wiki/
5. Menezes, A.J., van Oorschot, P.C., Vanstone, S.A.: Handbook of Applied Cryptography. CRC Press, Boca Raton (1996)
6. Jiang, S.M.: Securing underwater acoustic networks: a survey. IEEE Commun. Surveys Tutor, Submitted (2017)
7. Rigney, C., Willens, S., Rubens, A., Simpson, W.: Remote authentication dial in user service (RADIUS). IETF RFC 2865 (2000)
8. DeKok, A., Lior, A.: Remote authentication dial-in user service (RADIUS) protocol extensions. IETF RFC 6929 (2013)

9. Simpson, W.: The point-to-point protocol (PPP). IETF RFC 1661 (1994)
10. Fajardo, V., Arkko, J., Loughney, J., Zorn, G.: Diameter base protocol. IETF RFC 6733 (2012)
11. Finseth, C.: An access control protocol, sometimes called TACACS. IETF RFC 1492 (1993)
12. Rescorla, E., Modadugu, N.: Datagram transport layer security version 1.2. IETF RFC 6347 (2012)
13. Stewart, R.: Stream control transmission protocol. IETF RFC 4960 (2007)
14. Freier, A., Karlton, P., Kocher, P.: The secure sockets layer (SSL) protocol version 3.0. IETF RFC 6101 (2011)
15. Dierks, T., Allen, C.: The transport layer security (TLS) protocol, version 1.2. IETF RFC 5246 (2008)
16. Krawczyk, H., Bellare, M., Canetti, R.: HMAC: keyed-hashing for message authentication. IETF RFC 2104 (1997)
17. Kent, S., Seo, K.: Security architecture for the internet protocol. IETF RFC 4301 (2005)
18. Bellovin, S.: Guidelines for specifying the use of IPsec version 2. IETF RFC 5406 (2009)
19. Kent, S.: IP authentication header. IETF RFC 4302 (2005)
20. Kent, S.: IP encapsulating security payload (ESP). IETF RFC 4303 (2005)
21. Aboba, B., Blunk, L., Vollbrecht, J., Carlson, J., Levkowetz, H.: Extensible authentication protocol (EAP). IETF RFC 3748 (2004)
22. Aboba, B., Simon, D.: PPP EAP TLS authentication protocol. IETF RFC 2716 (1999)
23. IEEE Std 802.1X, Port based network access control (2004)
24. IEEE Std 802.11i: Amendment 6: medium access control (MAC) security enhancements (2004)
25. IEEE Standard 802.1AE: IEEE standard for local and metropolitan area networks: media access control (MAC) Security (2006)
26. IEEE Std 802.11: Medium access control (MAC) sub layer and 3 physical layer specifications (1997)
27. Wi-Fi Alliance: Deploying wi-fi protected access (WPA^{TM}) and WPA2^{TM} in the enterprise (2005)
28. Heron, S.: Advanced Encryption Standard (AES). Netw. Secur. **2009**, 8–12 (2009)
29. Mangir, T.: IEEE 802.11i (PPT). http://www.doc88.com/p-9905244343279.html (2007)
30. Chaplin, C., Qi, E., Ptasinski, H., Walker, J., Li, S.: 802.11i overview,. IEEE 802.11-04/0123r1 (2005)
31. Bellare, M., Kiliany, J., Rogaway, P.: The security of the cipher block chaining message authentication code. J. Comput. Syst. Sci. **61**(3), 362–399 (2000)
32. International Standard Organization (ISO): Information technology C security techniques C message authentication codes (MACs) C part 1: mechanisms using a block cipher, 1999, ISO/IEC 9797-1 (1999)
33. Whiting, D., Housley, R, Ferguson, N.: Counter with CBC-MAC (CCM). IETF RFC 3610 (2003)
34. Shin, M.H., Ma, J., Mishra, A., Arbaugh, W.A.: Wireless network security and interworking. Proc. IEEE **94**(2), 455–466 (2006)
35. Djenouri, D, Khelladi, L., Badache, A.N.: A survey of security issues in mobile ad hoc and sensor networks. IEEE Commun. Surv. Tutor. **7**(4), 2–28 (2005) (Fourth Quarter)
36. Butun, I., Morgera, S.D., Sankar, R.: A survey of intrusion detection systems in wireless sensor networks. IEEE Commun. Surv. Tutor. **16**(1), 266–282 (2014) (First Quarter)

Part II
Underwater Wireless Acoustic Networks (UWANs)

Chapter 9
Overview of Underwater Acoustic Communication

Abstract Since radio signals are strongly attenuated in saltwater and often scattered by suspended particles (similar for optical signals), acoustic wave becomes major communication medium used in underwater environments (Preisig, Proceedings of the ACM international WS, underwater networks (WUWNet), Los Angeles, USA 2006, [1]). However, some characteristics of underwater acoustic channels raise new challenges to underwater network protocol design, which are summarized below.

9.1 Underwater Acoustic Environments

Underwater acoustic communications are mainly affected by the following aspects as illustrated in Fig. 9.1: (i) physical properties of underwater acoustic wave, (ii) propagation environments and (iii) characteristics of the propagation medium, i.e., seawater discussed here.

A typical underwater propagation environment consists of the following components: seawater propagation medium, various interfaces and objects moving over and under water, such as vessels and whales. The typical interfaces include one between water and air, and the other between water and seafloor. These interfaces may reflect acoustic signals, especially in shallow water areas. A dynamic water-air interface consisting of moving wave driven by wind may create a scattering surface of acoustic signal. Moreover, moving objects may not only become reverberation surfaces, but also generate noise as discussed later.

The primary factors that influence acoustic wave propagation in seawater medium include temperature gradients and salinity of the medium. Changes in temperature and salinity from the surface to the bottom make the medium density to decrease with depth, resulting in density stratification. When a wave propagates through an uneven density medium, not only its propagation speed varies with the density, but also it will be refracted, resulting in ray bending. Changes in temperature will further contribute to change in propagation speed.

The major physical properties of underwater acoustic wave include its low propagation speed and signal-to-heat conversion, which increases with signal frequency and propagation distances. Thus, only low frequency acoustic signal can propagate over long distances in underwater environments.

© Springer Nature Singapore Pte Ltd. 2018

233

S. Jiang, *Wireless Networking Principles: From Terrestrial to Underwater Acoustic*,
https://doi.org/10.1007/978-981-10-7775-3_9

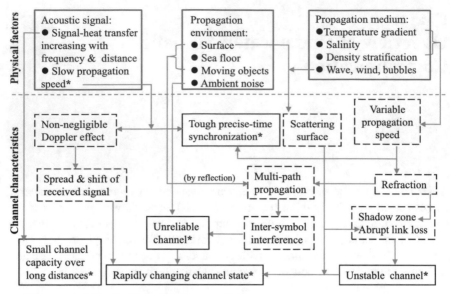

Fig. 9.1 Underwater propagation environments for underwater acoustic channels

9.2   Peculiar Features of Underwater Acoustic Channels

As discussed below, due to the above properties, the underwater acoustic channel is unreliable with small capacity and long latency, which along with technological and operation limitations raise new challenges for the network protocol design in UWANs.

9.2.1   Long and Changing Propagation Delays

The propagation speed (V in m/s) of acoustic waves in seawater environments is approximately 1500 m/s (i.e., 0.67 s/km), which is five order of magnitude slower than that of light. It is further affected by the temperature (T in Celsius), the salinity (S in unit *ppt*) and the depth (D in meter), as shown by the following empirical formula [2]:

$$V = 1449.2 + 4.6T - 0.055T^2 + 0.00029T^3$$
$$+ (1.34 - 0.01T)(S - 35) + 0.016D. \tag{9.1}$$

Long and changing propagation delays cause some problems that degrade the performance and efficiency of many network schemes commonly used in RF wireless networks. The typical effects are summarized below.

9.2.1.1 Non-negligible Protocol Overhead

The ratio of the time used to transmit a packet (t) to the time used to access the channel (τ, i.e., the time interval between the packet's arrival time and the start time of being transmitted), $\frac{\tau}{t}$, can be used to evaluate MAC protocol efficiency as follows:

$$\frac{\tau}{t} \approx \frac{x \times \frac{\iota}{r} + y \times \frac{d}{V}}{\frac{l}{r}} = x \times \frac{\iota}{l} + y \times \frac{r}{l} \times \frac{d}{V}, \tag{9.2}$$

where, r is transmission rate in bits/s, ι is protocol overhead in bits, l is packet length in bits, and d is the sender-receiver distance in meter. Both x and y are a positive coefficient, and their settings depend on protocol design.

If a MAC protocol originally designed for an RF wireless network (i.e., $V_r = 300,000$ km/s for radio wave) is directly applied to an UWAN (i.e., $V_a = 1.5$ km/s for acoustic wave), we only need to look at the second item in (9.2), i.e., $y \times \frac{r}{l} \times \frac{d}{V}$ (denoted by Δ henceforth), which is determined by the propagation speed in this case. Then the ratio of Δ_a for acoustic wave to Δ_r for radio wave, $\frac{\Delta_a}{\Delta_r} = \frac{V_r}{V_a} = 2 \times 10^5$. This means that Δ may be negligible in an RF wireless network but may not in an UWAN, particularly using handshake protocols.

9.2.1.2 Non-negligible Doppler Distortion

Due to the slow propagation speed of acoustic wave, Doppler effect becomes more severe in UWANs because its magnitude is proportional to the ratio $\alpha \triangleq \frac{v}{V_x}$, where v is the relative speed between transmitter-receiver pair and V_x is propagation speed. This effect causes motion-induced distortion by spreading the bandwidth (B) of the received signal to $(1 + \alpha)B$ (called Doppler spreading) and shifting the reception frequency (f) by an offset of αf (called frequency shifting) [3]. In an RF wireless network, $\alpha = 1.48 \times 10^{-7}$ for $v = 160$ km/h, the above distortion can be neglected by the receiver. However in UWANs, $\alpha = 2 \times 10^{-3}$ for an underwater vehicle moving at $v = 3$ m/s (which is typically used in the literature for protocol evaluation). Such a motion-induced Doppler distortion contributes to dynamics of channel status as discussed in Sect. 9.2.2.3, and requires frequency synchronization to capture the right frequency and bandwidth for efficient reception.

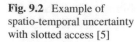

Fig. 9.2 Example of spatio-temporal uncertainty with slotted access [5]

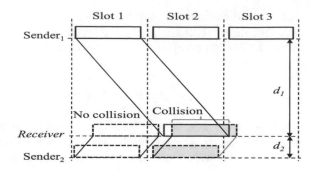

9.2.1.3 Spatio-Temporal Uncertainty

As discussed in [4], long propagation delays may cause spatio-temporal uncertainty, which causes design principle of MAC protocols different from those with negligible propagation delay. In this case, signals transmitted at different times may arrive at the same receiver simultaneously, while signals sent at the same time may arrive at the same receiver at different times, due to differences in signal propagation delay between different sender-receiver pairs. Figure 9.2 shows an example: Sender$_1$ transmits in slot 1. If Sender$_2$ also transmits in Slot 1, two transmissions will arrive at the receiver without collision. However, if Sender$_2$ transmits in Slot 2, collision will happen at the receiver.

The major objective of MAC protocols is to avoid collision at the receiver. If multiple simultaneous transmissions are not allowed at the same frequency, a typical method to avoid collision is to prevent the receiver from receiving multiple signals simultaneously at the same frequency. When the signal propagation delay is negligible, this can be achieved by allowing different source nodes to transmit at different times so that their signals can arrive at the receiver at different times too. However, when propagation delay is not negligible, such treatment may lead to the spatio-temporal uncertainty as mentioned above.

9.2.1.4 Difficulties for Precise Time Synchronization

Time synchronization is usually used to enable a group of nodes to have the same timing reference in order to design more efficient communication and network protocols, such as those used in CDMA and TDMA systems. Since GPS cannot work well in underwater environments to provide globally unique clock, time synchronization in UWANs can be realized through inter-node message exchange. Due to variable underwater acoustic propagation speed and long propagation delay, it is very difficult to implement a precise time synchronization [6, 7], and only relatively precise clock synchronization can be realized when the propagation delay is predictable and static for short periods of time [8]. Furthermore, channel status may change in very short time scale [9] (Please also refer to Sect. 9.2.2.3 for more information), and the probability of such change increases with the propagation delay.

9.2.2 Primary Characteristics

Besides slow and variable propagation speed of acoustic signals, underwater acoustic channels are also characterised by low capacity and poor reliability as well as wideband as discussed below [5, 10].

9.2.2.1 Small and Busy Channel

As discussed in [11], underwater acoustic channels are both distance and frequency selective fading. That is, path loss depends not only on the distance between the transmitter and receiver, but also on the signal frequency, which determines the absorption loss caused by the transfer of acoustic energy into heat. This loss increases with frequency and distance.

The underwater acoustic channel is busy because not only a limited amount of bandwidth is available for communication but also the channel is used by navigation. The acoustic bandwidth available for underwater communication depends on transmission distances and transmission frequencies. This is due to both spreading loss and signal-heat transfer increase with transmission distances, while signal-heat transfer also increases with transmission frequencies.

Let A_0 denote the unit-normalizing constant, $a(f)$ the absorption coefficient, and k the spreading factor that describes the geometry of propagation. Attenuation occurring in an underwater acoustic channel over a distance l for a signal of frequency f is given by [11]

$$A(l, f) = A_0 l^k a(f)^l, \qquad (9.3)$$

which can be expressed in dB as follows:

$$10log\frac{A(l, f)}{A_0} = k \times 10logl + l \times 10loga(f), \qquad (9.4)$$

where the first term is the spreading loss, and the second term is the absorption loss.

A popular empirical formula used to estimate absorption coefficient, $a(f)$ in dB/km (f in kHz [12]), for frequencies from 100 Hz to 1 MHz is Thorps formula as follows [11, 13]:

$$10loga(f) = 0.11\frac{f^2}{1 + f^2} + 44\frac{f^2}{4100 + f^2} + 2.75 \times 10^{-4}f^2 + 0.003 \qquad (9.5)$$

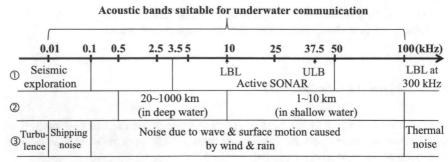

Fig. 9.3   Underwater channel characteristics versus frequency

for the scenario with seawater's salinity of 35% *ppt*, $pH = 8$, temperature of 4 °C and depth equal to 0 m. From this formula and we can have $a(10) = 1.19$ dB/km, $a(100) = 34.07$ dB/km, which means that with $f < 10$ kHz, long range communication is feasible [11, 13]. For lower frequencies, the following formula may be used [11, 13]:

$$10loga(f) = 0.002 + 0.11\frac{f^2}{1 + f^2} + 0.011f^2. \tag{9.6}$$

Furthermore, the spreading loss due to the expansion of transmitted energy over a large surface also increases with transmission distances as the signal propagates as shown by the first term in (9.4), which furthers limit the underwater acoustic channel capacity over long distances [1, 3]. Factor k in this term is commonly set to 2 for spherical spreading, 1 for cylindrical spreading, and 1.5 for practical spreading.

As summarized in Fig. 9.3, for a long range communication (i.e., 20 ∼ 2000 km) in deep water, the upper frequency limit ranges from 500 Hz to 10 kHz. For a medium-range communication (i.e., 1 ∼ 10 km) in shallow water, the upper-frequency limit is 10 ∼ 100 kHz [13, 14]. For short range communication less than 1 km, available bandwidth becomes much wider, especially at ranges less than 100 m. Note that underwater laser-optical communication may also be feasible for much higher data rates at ranges less than 100 m [14].

Theoretically, for data rates of underwater acoustic channels, a short-range communication can have more than 100 kbits/s. For a medium-range communication, the maximum achievable data rate is 50 kbits/s. For example, communication of voice and still images of 10 kbits/s from a submarine to an 4.6-km apart Unmanned Underwater Vehicle (UUV) at frequency 20 ∼ 30 kHz is reported in [15]. For a long range communication, the maximum achievable data rate is 10 kbits/s for 20 km. At 2 ∼ 4 kHz, it was possible to establish a communication channel of 2.4 kbits/s between an underwater vehicle and a surface ship 65 km away [15].

As illustrated in Fig. 9.3, the above-mentioned small bandwidth of underwater acoustic channels is also used by other applications such as active SONAR using frequency ranging from $3.5 \sim 50$ kHz [16]. For underwater positioning systems, underwater locator beacon (ULB) operates at 37.5 kHz, and long baseline (LBL) uses 10 kHz. For scientific research, frequencies less than 100 Hz are used for seismic exploration.

9.2.2.2 Unreliable and Unstable Channel

The unreliability of underwater acoustic channel is mainly reflected by high BERs and link failure. High BER is mainly caused by multipath propagation and ambient noises in underwater environments. The major courses leading to link failure include the refraction and surface scattering of acoustic signal.

As illustrated in Fig. 9.1, multipath propagation is caused by acoustic signal reflected from surfaces, seabeds and other objects. In this case, a signal from a transmitter may arrive at the same receiver in different paths probably with phase shift [13]. These out-of-phase simultaneously arriving signals may cause severe Inter-Symbol Interference (ISI), with which a signal for one symbol may interfere with those for subsequent symbols. Different from an RF receiver, in which the ISI may involve only a few symbols, the ISI in a single-carrier UWAN may span tens or even hundreds of symbol intervals due to long propagation delays [3], which makes it much more difficult to resolve ISI in demodulation phase [15].

The number of multipaths depends on characteristics of environments in which the signal propagates. In shallow water or inshore area, both the surface and seabed can reflect signal, while in deep water or offshore area, the number of reflectors becomes much less. Moreover, surface scattering contributes dynamics in multipath propagation effect, causing signal arrival not only to fluctuate in time but also to spread at the receiver [1]. The time spread increases with range, frequency as well as wind speed. Also spatial coherence is mainly determined by the characteristic of surface scattering, which makes such effect with a horizontal structure to be more severe than with a vertical one [17].

As illustrated in Fig. 9.1, another factor that influences the quality of underwater acoustic channels is plentiful noises present in underwater environments, which typically include ambient noise of the ocean and self-noise of vessels [8, 17–19], as well as ocean animals such as whales and even shrimps [20]. As illustrated in Fig. 9.3, there are four types of noise sources that affect acoustic communication at different frequency bands [19]. These noises can be described by Gaussian statistics and a continuous power spectral density estimated with the following empirical formulaes in dB as a function of frequency in kHz [11, 13]:

$$10logN_t(f) = 17 - 30logf$$
$$10logN_s(f) = 40 + 20(s - 0.5) + 26logf - 60log(f + 0.03)$$
$$10logN_w(f) = 50 + 7.5w^{\frac{1}{2}} + 20logf - 40log(f + 0.4)$$
$$10logN_{th}(f) = -15 + 20logf,$$

where, w is the wind speed in m/s, and

- $N_t(f)$: Turbulence noise that influences only the very low frequency region of less than 10 Hz
- $N_s(f)$: Shipping noise is dominant in the frequency region between 10 and 100 Hz.
- $N_w(f)$: Wave and other surface motions caused by wind and rain contribute to the noise in the region of 100 Hz ~ 100 kHz. However, this region is used by the majority of acoustic systems.
- $N_{th}(f)$: Thermal noise is dominant for frequency over 100 kHz.

On the other hand, an acoustic link may suffer failure in a shadow zone, where almost no acoustic signal presents as explained below. An acoustic signal is refracted during its propagation in uneven density seawater, which may cause the signal path to be blended in a way that unusually high transmission loss occurs in certain areas. In shallow water, the location of a shadow zone varies due to the vertical movement of water [13, 21]. Similarly with surface scattering, very high intensity and rapidly fluctuating arrivals in channel impulse response may happen, leading to abrupt link loss [1].

9.2.2.3 Rapidly Changing Channel States

In [22], a measurement result over a 3000-m link in a 100-m deep water column shows the following oscillations of the average SNR of the output of a channel equalizer: about 9 ~ 5.7 dB within about 0.5 min, and about 9 ~ 3.5 dB within less than 1.5 min. As discussed in [1], channel coherence time can be in order of 40 ms, while Doppler distortion may further reduce such time. The experimental results reported in this paper also show large oscillations of BER in short time intervals. For example, BER drops from 0.11 (with one equalizer) and 0.06 (with four equalizers) to 8.1×10^{-4} within about 2 s, and both show periodic increases.

From networking point of view, one-hop link latency, which is the sum of medium access delay (ξ), transmission time and propagation delay along the link that an incoming packet will be transmitted over, affects network protocol performance. ξ largely depends on the adopted MAC protocol as well as traffic load and traffic pattern, and is quite dynamic by nature. The transmission time and propagation delay can be expressed respectively by $\frac{l}{r}$ and $\frac{d}{v}$ in (9.2). Due to low transmission rates and slow propagation speeds in UWANs, it is easy for the latency to reach seconds. For example, to transmit a 256-byte data packet at 10 kbits/s over a 4.6-km link as reported in [15] (also mentioned in Sect. 9.2.2.1), the sum of $\frac{l}{r}$ and $\frac{d}{v}$ is already 3.27

seconds, and medium access delay may make the latency much larger. Within such a large link latency, the quality of an one-hop link may change significantly.

9.2.2.4 Wideband Channel

As discussed in [23], systems whose bandwidth is smaller than 1% of the center frequency of the signals are called narrowband, and those between 1 and 20% are called wideband, while the others are called ultra-wideband (UWB) [10]. Popular frequency bands used by acoustic communication vary with communication ranges. For example, a popular frequency band is about $8 \sim 14$ kHz for ranges up to a few kilometers, while the upper-frequency limit is $10 \sim 100$ kHz [13, 14] as mentioned earlier. Relatively, UWA channels can be qualified as wideband at least [23], which makes the popular narrowband channel model inappropriate [24]. The shortcomings of narrowband channel models in this case are demonstrated by the measurements and analysis of acoustic propagation effects conducted in [23, 24].

As discussed in [24], wideband UWA channels are typically characterized by frequency-dependent fading statistics such as the mean reception power, time-varying delays and frequency-dependent fluctuation rates, which causes the framework of wide-sense stationary uncorrelated scattering developed for narrowband systems using correlated scattering to mean correlated taps not applicable in this case. Frequency-dependent fading statistics includes frequency-dependent path losses due to bottom loss and surface loss, absorption of seawater and scatter in the water column. Frequency-flat and frequency-selective fading may occur in multipath environments, where a channel may have paths with different Doppler shifts, resulting in frequency-dependent fluctuation rates [10, 24].

9.2.3 Technological and Operational Limitations

With the limitation of the available acoustic communication technologies, current underwater acoustic communication systems are also characterised by the following technological and operational limitations.

9.2.3.1 Difficult Full-Duplex Communication

A transducer conducting acoustic-electronic signal conversion is used for underwater acoustic communication. Similar to antenna design, the size of a transducer should also be proportional to wavelength. For a typical frequency suitable for underwater acoustic communication, say 10 kHz, the size of a transducer used in seawater should be at least 7.5 cm, i.e., a half of wavelength. Thus, it is difficult to spatially separate transducers using different frequencies far enough in a space-constrained platform. In this case, with duplex communication, a transmitted signal will saturate the receivers. Therefore, underwater acoustic communication is often in a half-duplex mode [25].

9.2.3.2   Asymmetrical Transmission/Receiption Rates

As discussed in [25], for a small self-propelled mobile platform such as AUVs, it is easy for them to transmit at high data rates but not for them to receive at high rates due to propulsion noise and difficulties in mounting receiver arrays therein. With currently available acoustic communication technologies, higher transmission rates can be achieved with a single transducer typically using phase-shift keying. However, for reception, to combat the multipath interference, it often requires an array of spatially separated traducers, which however are difficult as mentioned above.

9.2.3.3   Large Power Consumption

Power consumptions of both transmission and reception in underwater acoustic transceivers are much higher than those in RF transceivers. For example, transmission/reception power consumptions are about 50W/3W for an underwater acoustic transceiver, while 80mW/30mW for an RF transceiver [26]. The transmission power consumption dominates, and further depends on channel condition and transmission distances. As mentioned in [14], the transmission power is usually larger than 100 watts for long range communication.

In a self-propelled platform such as AUVs, the propulsion power dominates, and energy efficiency for communication and networking is not a primary concern. However, for battery-operated devices such as stationary underwater sensors, the energy efficiency is still a major concern, and similar for those powered by renewable energy (e.g., wave-gliders) because such kind of energy supply is not stable. Sufficient energy saving is important to maintain sustainable network operations in this case.

9.2.3.4   Costly Device and Deployment

Today, the cost of communication devices and sensors and network deployment as well as maintenance in underwater environments are much higher than those in terrestrial environments, especially AUVs equipped with GPS and satellite communication facilities to be used when it floats on the water surface. As reported in [25], an acoustic modem with a rugged pressure housing without underwater sensors currently costs roughly $3k, while both repairs and recovery are also expensive. In this case, dense development of UWANs is not an cost-effective option, which results in sparse UWANs.

Table 9.1 RWNs versus UWANs

Compared items	RFWNs	UWANs
Maximum channel capacity	Gbit/s	kbit/s ($<$ 10km)
Bit error rate (BER)	10^{-2}-10^{-6}	Higher
Propagation delay	Negligible*	Large and variable
Feasibility of full duplex	Larger	Smaller
Feasibility of time synchronization	Larger	Smaller
Frequency-selective loss	No	Yes
Communication power consumption	Smaller	Larger
Node distribution in networks	Dense	Sparse
Asymmetric link situations	Less	More
Occasional link failures	Less	More

*An exception is satellite networks especially GEO satellite networks discussed in Chap. 1

9.3 Summary

Table 9.1 compares wired networks, RF wireless networks (RWNs) and UWANs. The differences between these two types of networks listed in this table remind us that protocols developed for RWNs might not be applied in UWANs directly, and these peculiar features have to be taken into account in UWAN protocol designs.

References

1. Preisig, J.: Acoustic propagation considerations for underwater acoustic communications network development. In: Proceedings of the ACM International WS, Underwater Networks (WUWNet), Los Angeles, USA (2006)
2. Etter, P.C.: Underwater Acoustic Modeling, Principles, Techniques and Applications, 2nd edn. E & FN Spon, (1996)
3. Stojanovic, M.: Underwater acoustic communications: design considerations on the physical layer. In Proceedings of the Annual Conferences Wireless on Demand Network Systems and Services (WONS), Garmisch-Partenkirchen (2008)
4. Syed, A.A., Ye, W., Krishnamachari, B., Heidemann, J.: Understanding spatiotemporal uncertainty in medium access with ALOHA protocols. In: Proceedings of the ACM International WS, Underwater Networks (WUWNet), Montreal, Canada (2007)
5. Jiang, S.M.: State-of-the-art medium access control (MAC) protocols for underwater acoustic networks: a survey based on A MAC reference model. IEEE Commun. Surv. Tutor. **20**(1), 1 (2018) (Quarter 2018)
6. Pompili, D., Akyildiz, I.F.: Overview of networking protocols for underwater wireless communications. IEEE Commun. Mag. 97–102 (2009)
7. Melodia, T., Kulhandjian, H., Kuo, L.-C., Demirors, E.: Advances in underwater acoustic networking. In: Basagni, S., Conti, M., Giordano, S., Stojmenovic, I. (eds.), Mobile Ad Hoc Networking: the Cutting Edge Directions, pp. 804 – 852. Wiley-IEEE Press (2013)
8. Syed, A.A., Ye, W., Heidemann, J.: Time synchronization for high latency acoustic networks. In: Proceedings of the IEEE INFOCOM, pp. 1–12. Barcelona, Spain (2006)

9. Kredo, K. II, Djukic, P., Mohapatra, P.: STUMP: exploiting position diversity in the staggered TDMA underwater MAC protocol. In: Proceedings of the IEEE INFOCOM, Rio de Janeiro, Brasil (2009)

10. Jiang, S.M.: On reliable data transfer in underwater acoustic networks: a survey from networking perspective. IEEE Commun. Surv. Tutor. **PP**(99) (2018)

11. Stojanovic, M.: On the relationship between capacity and distance in an underwater acoustic communication channel. In: Proceedings of the ACM International WS, Underwater Networks (WUWNet), Los Angeles, USA (2006)

12. Sozer, E.M., Stojanovic, M., Proakis, J.G.: Underwater acoustic networks. IEEE J. Ocean. Eng. **25**(1), 72–83 (2000)

13. Catipovic, J.A.: Performance limitations in underwater acoustic telemetry. IEEE J. Ocean. Eng. **15**(3), 205–216 (1990)

14. Catipovic, J., Brady, D., Etchenmendy, S.: Development of underwater acoustic modems and networks. Oceanography **6**(3), 112–119 (1993)

15. Lanzagorta, M.: Underwater communications. MORGAN & CLAYPOOL (2012)

16. Dạfmico, A., Pittenger, R.: A brief history of active sonar. Aquat. Mamm. **35**(4), 426–434 (2009)

17. Stojanovic, M.: Underwater acoustic communication. Wiley Encyclopedia of Electrical and Electronics Engineering, vol. 9, pp. 98–101. Wiley, Mew York (1999)

18. Makhija, D., Kumaraswamy, P., Roy, R.: Challenges and design of MAC protocol for underwater acoustic sensor networks. In: Proceedings of the Modeling & Optimization in Mobile, Ad Hoc & Wireless Net (2006)

19. Burrowes, G.E., Khan, J.Y.: Investigation of a short-range underwater acoustic communication channel for MAC protocol design. In: Proceedings of the Signal Processing and Communication Systems, pp. 1–8 (2010)

20. Mahmood, A., Chitre, M.: Ambient noise in warm shallow waters: a communications perspective. IEEE Commun. Mag. **55**(6), 198–204 (2017)

21. Quazi, A.H., Konrad, W.L.: Acoustic underwater communications. IEEE Commun. Mag. 24–30 (1982)

22. Tomasi, B., Toso, G., Casari, P., Zorzi, M.: Impact of time-varying underwater acoustic channels on the performance of routing protocols. IEEE J. Ocean. Eng. **38**(4), 772–784 (2013)

23. van Walree, P.A., Otnes, R.: Ultrawideband underwater acoustic communication channels. IEEE J. Ocean. Eng. **38**(4), 678–688 (2013)

24. van Walree, P.A.: Propagation and scattering effects in underwater acoustic communication channels. IEEE J. Ocean. Eng. **38**(4), 614–631 (2013)

25. Partan, J., Kurose, J., Levine, B.N.: A survey of practical issues in underwater networks. In: Proceedings of the ACM International WS, Underwater Networks (WUWNet), Los Angeles, California, USA (2006)

26. Kredo, K.B., II, Mohapatra, P.: A hybrid medium access control protocol for underwater wireless networks. In: Proceedings of the ACM International WS, Underwater Networks (WUWNet), Montreal, Canada (2007)

Chapter 10
MAC for UWANs

Abstract The peculiar features of underwater communications cause MAC protocols proposed for RWNs unable to be used directly in UWANs, and many research
results on UWAN MAC protocols are reported in the literature. This chapter reviews
typical UWAN MAC protocols following the MAC reference model discussed in
Chap. 3. The main reference for this chapter is (Jiang, IEEE Commun Surv Tutor,
2018, [1]).

10.1 Overview

In RWNs where the signal propagates almost at light speed of 300,000 km/s, the
signal propagation delay between senders and receivers in small networks or differences in propagation delays between different nodes can be negligible. In this case,
how to provide collision-free reception at a receiver becomes an issue on how to
enable collision-free transmissions from different senders [1]. We call this kind of
MAC protocol as sender-centric MAC protocol henceforth. The major MAC design
strategies for such protocol is to arrange transmissions from different nodes to occur
at different times, different frequencies or different spreading codes. This strategy is
adopted by most terrestrial wireless networks, and can significantly simplify MAC
protocol design. However, this simplification may not make sense with long propagation delay in UWANs because a collision-free transmission does not always lead to
a collision-free reception, while a collided transmission may lead to a collision-free
reception, as illustrated in Fig. 9.2. This phenomena happens due to differences in
propagation delays between senders and receivers, and is also called spatio-temporal
uncertainty [2]. However, we can exploit the long propagation delay as an opportunity because it can provide larger space-time volume than smaller delay [3] to allow
more concurrent transmissions that can be successfully received.

On the other hand, the very small underwater acoustic channel capacity maximally
at kbps-level [4, 5] and highly dynamic underwater communication environments
[6, 7] require more bandwidth-efficient and adaptive UWAN MAC protocols. Furthermore, unreliable underwater acoustic channel makes it undesirable to frequently
exchange messages for MAC operation. Therefore, some physical-layer techniques

© Springer Nature Singapore Pte Ltd. 2018   245
S. Jiang, *Wireless Networking Principles: From Terrestrial to Underwater Acoustic*,
https://doi.org/10.1007/978-981-10-7775-3_10

that allow concurrent communications at the same frequency, such as CDMA and MIMO, are used in UWAN MAC protocol design. However, for CDMA in multi-user systems, precise time synchronization between senders and receivers [8] and power control to avoid the near-far effect are difficult in UWANs [9, 10]. For MIMO, how to obtain the realtime channel state information (CSI) and install antenna array in space-limited nodes such as autonomous underwater vehicles (AUVs) are two major challenging issues.

10.2   Challenges and Categories

The major challenging issues on UWAN MAC protocol design and a category of typical UWAN MAC protocols are summarized below.

10.2.1   Medium Utilization

It measures the efficiency of a medium used for user data transmission. The guard time and guard band used by multiplexing schemes are two major factors that degrade this efficiency. The protocol overhead used by a node to obtain medium access opportunity (e.g., control messages used by reservation and handshaking) also affects the efficiency. In very small capacity UWANs, maintaining high medium utilization is especially important but becomes more difficult with long propagation delays. This results because spatio-temporal uncertainty makes sender-centric MAC protocols unable to avoid collision at the receiver. On the other hand, receiver-centric MAC protocols often require cooperation between the senders and receiver, and long propagation delays affect the efficiency of such cooperation [1].

10.2.2   Energy Efficiency

Due to excessive medium attenuation, underwater acoustic communication usually requires much high transmission and reception powers, which are much larger than those of RF communication. For example, tens of Watts are typically required for transmission, depending on transmission distances, while tens of mWatts up to a few Watts for reception, depending on the type of processing [11]. In battery-powered UWANs, it is very difficult and costly to recharge and re-deploy underwater nodes, and minimizing transmission and reception activities is necessary. On the other hand, transmission power has to be large enough to achieve acceptable data rates in noisy underwater acoustic channels over long transmission distances [1].

10.2.3 Quality of Service (QoS)

The MAC protocol is a key factor in QoS provisioning to satisfy application requirements in terms of medium access delay and effective throughput. The former is the
time that data has to wait before being transmitted, and the latter is the amount of
data successfully received per time unit. However, long propagation delays cannot
provide a large space for medium access delay, while the very small channel capacity
does not allow large protocol overhead used by MAC to provide QoS [1].

10.2.4 Mobility

In mobile UWANs, Doppler effect affects communication quality as mentioned in
Sect. 9.2.1.2. MAC protocol design has to take into account the effect of terminal
mobility on accuracy of time synchronization and localization. Particularly for an
AUV, the impact of its own noises on communication quality is an additional issue
necessarily to be addressed, and its narrow space also limits the application of some
technique for better performance such as acoustic arrays [1].

10.2.5 Fairness

When propagation delays or differences in propagation delays between different
nodes are negligible, location-dependant unfairness can be neglected so that a simple
fairness police such as First-in-First-out (FIFO) can work well. In the case of long
propagation delays, an earlier departure may arrive at a node later than its followers,
resulting in earlier sent packets may not be served earlier accordingly due to their
different journeys. Particularly, MAC protocols based on carrier sensing or frame
sensing are favorable to nodes near the signal source so that they can have more
access opportunities [1, 12].

10.2.6 Protocol Validation

Besides mathematical analysis for simple scenarios, computer simulation has been
extensively used to validate UWAN MAC protocols because sea-field trial is very
expensive for UWANs. How well a simulation tool can properly simulate the peculiar
features of underwater acoustic channels and high dynamics of UWANs affects the
credibility of simulation results. Prototyping with laboratory test can also be used
to verify a UWAN MAC protocol, and can check protocol complexity, but the test
environment is still different from the reality [1].

Fig. 10.1 A classification of
the reviewed UWAN MAC
protocols

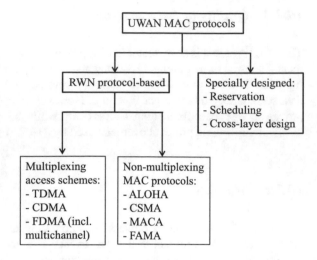

10.2.7 Classification of UWAN MAC Protocols

There are many UWAN MAC protocols reported in the literature as surveyed in
[1], which can be roughly categorized into two as illustrated in Fig. 10.1: UWAN
MAC protocols based on popular RF ones and deliberately designed UWAN MAC
protocols. It is relatively straightforward to modify popular RWN MAC protocols
by making them adaptive to peculiar features of UWANs. This kind of protocol
include those based on multiplexing-based access schemes, i.e., TDMA, CDMA and
FDMA as well as multi-channel schemes, and those modifying the popular RF MAC
protocols without multiplexing such as ALOHA, CSMA, MACA and FAMA. The
deliberately designed UWAN MAC protocols include those based on scheduling and
reservation schemes as well as cross-layer design. More details of these protocols
can be found in Sects. 10.3 and 10.4.

10.3 UWAN MAC Based on RWN Protocols

This chapter will first review some typical UWAN MAC protocols that are based on
multiplexing access schemes. As mentioned in Sect. 3.3.1, a MAC protocol based on
multiplexing schemes can simplify its design, and the major remaining MAC design
issues include transmission request submission and allocation result notice. Then,
some UWAN MAC protocols that modify well-established RWN MAC protocols
without using multiplexing schemes are introduced. These RWN MAC protocols
include ALOHA, S-ALOHA, CSMA, CSMA/CA, MACA and FAMA, which are
listed in Table 4.1. Table 10.1 summarizes these reviewed UWAN MAC proposals
following the MAC reference model described in Sect. 3.2.

Table 10.1 Modified popular RWN MAC protocols

Protocol name and reference	MAC reference model (Fig. 3.6)			Major features		
	Oper. cycle	Access unit	MAC mechanism	Topology	Synchronization	Validation method
Seaweb [13]		SB,TS	HS	CL	✓	T
UW-MAC [14]		CO	FA,ME	CL		S
[15]		CO	HS	TR		
[16]	SF	TS	PS	ST		A,S
[17]		SB,SL	SC,SA,ME,CS	CL	✓	S
ALOHA-CS [18]			CS,BO	MH		S
ALOHA-RB [19]			CS,BO	ST		S
CS-MAC [20]		SL	FA,SA,HS,CS	MH	✓	S
CSMA-ALOHA [21]			CS,BO	MH		S
DACAP [22]			FA,HS,BO	MH		S
PCAP [23]			FA,HS	OH		S
S-ALOHA-GB [2]		SL	SA	ST	✓	A,S
S-FAMA [24]		SL	HS	MH	✓	A,S
CUMAC [25]			HS,CS	MH		S

Refer to Sect. 3.2 for the descriptions of abbreviation

10.3.1 Multiplexing Access Protocols

As discussed earlier, the typical multiplexing access protocols include TDMA, CDMA and FDMA. Multichannel protocols are treated as a special case of FDMA.

10.3.1.1 TDMA-Based MAC Protocols

There are many TDMA-based MAC protocols proposed for UWANs. This kind of protocol usually jointly uses other MAC mechanisms such as reservation and scheduling for time slot allocation, e.g., ST-MAC [26], CT-MAC [27] and TFO-MAC [28], which will be discussed in Sect. 10.4. Here we introduce one scheme that does not use the above MAC mechanisms.

A TDMA-based MAC protocol using a superframe structure for energy efficiency was discussed in [16] for a star-topology UWAN with a sink. Instead of using the fixed-size time slot, a defer time which is defined as a time interval between when a node receives the last bit of the superframe and when it begins to send a frame, is used for channel allocation in order to make allocation adaptive to propagation delays. The defer time for each node is calculated by the sink, which arranges the transmission sequence of each transmitting node according to their distances to it in an ascending order so that the sink can receive data frames sent by each node back-to-back without gap to maximize bandwidth utilization. Particularly for a node with a transmission sequence i (called node i), it starts its transmission from the ith defer time $(T_d(i))$ calculated as follows:

$$T_d(i) = (i - 1)T - \frac{2(d_i - d_1)}{v}, \tag{10.1}$$

where T indicates the frame length in time unit, d_i is the distance between node i and the sink, and v is the signal propagation speed.

With such transmission sequence, the node closest to the sink first starts immediately data transmission, which is then followed by the second closest node and so on. To implement this protocol, the sink should have the information on the distance between itself and every node, and there are not two nodes that have the equal distance to the sink. The sink needs to broadcast the transmission schedules to all nodes, which must be synchronized with the same timing reference.

However, (10.1) cannot allow the sink to receive data frames sent by each node back-to-back without gap between them as expected by the proposed MAC protocol. We think that the correct one should be as follows:

$$T_d(i) = T_d(1) + (i - 1)T - \frac{d_i - d_1}{v} \tag{10.2}$$

for $i > 1$. Figure 10.2 illustrates the relationship between $T_d(i)$, d_i and T. We can have $T_d(2) = T_d(1) + (T - \frac{d_2 - d_1}{v})$ and $T_d(3) = T_d(2) + (T - \frac{d_3 - d_2}{v})$. After manipulation, $T_d(3) = T_d(1) + 2T - \frac{d_3 - d_1}{v}$. Similarly, we can get (10.2).

Fig. 10.2 Principle of a TDMA-based MAC protocol

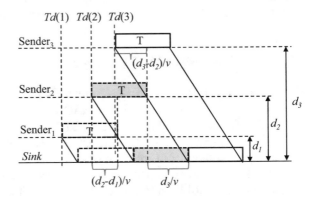

10.3.1.2 CDMA-Based MAC Protocols

CDMA-based MAC can reduce the medium access delay of protocols based on handshaking or negotiation in the case of long propagation delay in UWANs. Such kind of protocol can leverage CDMA's capability against channel fading in underwater environments as mentioned in Chap. 9. Here CDMA-based MAC protocols for UWANs are based on asynchronous CDMA, i.e., spreading codes are not transmitted in parallel with the data. The major issues to be addressed by CDMA-based MAC protocols include spreading code assignment, power control and energy efficiency, which are similar to those for terrestrial RWNs. Several such kinds of MAC protocols are described below according to the network topologies that the protocols are designed for [1].

Tree Topology

The CDMA-based MAC protocol for a tree-topology UWAN in [15] adopts a topology probe process for code and power assignment. This process is launched by the master node periodically through a dedicated channel that is configured with a specific code. The process broadcasts a topology discovery message carrying the following information: a set of codes selected randomly from the entire set of spreading codes, the identity of the sender and the transmission power level. Each node within the range of the master node's probe selects one code from the provided set, and sends a response to the sender with the selected code. The response message includes the signal strength of the received probe, and a set of codes randomly chosen from the entire collection. It is possible for more than one node receiving the probe to select identical codes. In this case, the probe source has to send an amended probe to notify the affected nodes or even assign a code directly to them.

Each node will respond the topology probes and the responses of other nodes that it receives. If no response is received within a given time period, the probe sender will resend the probe with an increased power level subject to the pre-determined level of the initiation probe. If a timeout for the expected response with the increased transmission power happens again, the node assumes that it is on the edge of the network and wraps up the topology discovery process by transmitting a completion message to the master node.

This protocol can allow new nodes to be added and dead nodes to be removed dynamically. The central node periodically updates spreading codes and power levels to control the transmission range of each node, and permit a reuse of spreading codes in different geographical areas of the network.

Distributed Topology

A transmitter CDMA-based MAC protocol with single carrier (UW-MAC) is investigated for a distributed topology UWAN in [14]. It aims to achieve the following objectives: high throughput efficiency, low medium access delay and low energy consumption. The throughput efficiency is defined as the ratio between the bit rate delivered to the sink (correct bits) and the bit rate offered to the network. UW-MAC adopts a closed-loop distributed algorithm to set transmission power and code length [10].

A node, say node i, randomly accesses the channel by broadcasting a short message called Extended Header (EH) with a common chaotic code that is known by all nodes. The EH contains the information on the final destination (i.e., a surface station), the chosen next hop and the parameters to be used by node i to generate the chaotic code to be used to transmit the data packet. After the EH transmission, node i immediately transmits the data packet with the optimal transmission power and code length set by a power-code self-assignment algorithm. If the EH is successfully decoded by the chosen next hop, it locally generates the spreading code for decoding the upcoming data packet. Once the receiver successfully decodes the received packet, it sends an ACK to node i to avoid end-to-end retransmissions. If node i does not receive the ACK after the timeout, it retransmits the packet subject to a predefined maximum retransmission times.

The power-code self-assignment algorithm and the corresponding optimization algorithm with a closed-loop solution are carried out in a distributed manner based on the channel models for both deep and shallow water environments. Accordingly, the MAC protocol periodically collects the channel state information (CSI) from the neighborhood and feeds the algorithm with the required information [14].

The simulation results show that UW-MAC can simultaneously achieve the above-mentioned three objectives in the cases of deep and shallow water because it can dynamically find the optimal trade-off among these objectives according to the application requirements [14]. However, to run the protocol by setting the transmission power and spreading factor, a node needs to periodically collect and leverage CSI from the neighborhood, such as the multiple access interference (MAI) and normalized receiving spread signal of neighboring nodes. Active nodes are required to broadcast this information periodically [10].

10.3.1.3   FDMA-Based MAC Protocols

According to Shannon theory, the effective data rate (R) depends on both bandwidth (B) and signal-to-noise ratio (SNR) as follows: $R = B \log_2(1 + \text{SNR})$. Since the acoustic frequency band suitable for long distance underwater transmission is very small, the bandwidth per sub-band is further shrunk, which makes it difficult to combat frequency-selective fading and multipath propagation in UWANs mentioned earlier [9, 10, 29]. Moreover, increasing transmission power for high data rates is undesirable in many situations especially for battery-powered nodes [30]. Therefore, there are few MAC protocols simply using FDMA.

Instead, there are several MAC protocols based on Orthogonal FDMA (OFDMA), with which, a shared medium is divided into multiple orthogonal sub-channels, similar to OFDM, to enable different nodes for simultaneous access to the medium. The major issues to be addressed include subcarrier allocation, protocol adaptability and energy efficiency as discussed below. Here, the multi-channel approach is treated as a special case of FDMA.

An Implementation

An FDMA-based MAC protocol is implemented in the Seaweb'98 system [13]. It uses three interleaved FDMA sets, each of which has 40 Multi-Frequency-Shift Keying (MFSK) tonals and two codewords for a cluster structure. All nodes in the same cluster are assigned the same FDMA carrier set for reception. However, only 20 MFSK tonals and one codeword are associated with each cluster because half the bandwidth is used as guard bands. It can permit simultaneous transmissions in all clusters without MAI. Within a cluster, TDMA is used for the intra-cluster bandwidth sharing. The filed testing was based on a 300-bps modulation to yield a net FDMA bit-rate of just 50 bps, resulting in low bandwidth efficiency. Thus, FDMA was abandoned in the subsequent Seaweb systems [31].

A MACA-like RTS/CTS handshake protocol is also adopted in the Seaweb system to exchange information about addressing, ranging, channel state estimation, adaptive modulation and power control, rather than for collision avoidance as used by many MAC protocols. Collision avoidance is achieved by using FHSS or DSSS with pre-assigned hopping sequence or spreading codes, working as follows. An initiating node transmits an RTS waveform with an FHSS series or DSSS pseudo-random codes to the uniquely intended receiver. Upon detecting the request, the addressed node awakens to demodulate the signal in order to capture an estimate of the channel scattering function and signal excess, and sends a CTS also using FHSS or DSSS to acknowledge the receipt and specify appropriate modulation parameters. Finally, data packets are transmitted by the initiating node after the RTS/CTS handshake [13].

Orthogonal FDMA (OFDMA)

An OFDMA-based MAC protocol for a stationary UWAN is investigated in [17, 32]. Underwater nodes are grouped into clusters, and in each cluster, the nodes transmit data to the cluster head. A multi-hop path relaying is required for nodes far away from the cluster head. Each cluster head is connected to a surface station. The channel is divided into a number of orthogonal sub-channels to enable simultaneous communication for a pair of neighboring nodes. An allocated sub-channel can be reserved by a granted node pair until they relinquish it. The time is further slotted such that data transmission is allowed only at the beginning of each time slot.

The basic protocol operation is based on pilot messages broadcast during the initiation process, which has to be conducted by each node to its neighbours via all sub-channels. A node receiving a pilot message from a sub-channel measures the SNR, which is then informed to the sender. After receiving the SNR information from all its neighbors, a node selects the best sub-channel as a control sub-channel between itself and all its neighbors. At the end, each node can have a SNR list for all sub-channels, which is used for proper sub-channel selection to minimize transmission power for the required transmission rate.

To avoid collision for pilot message broadcast, a CSMA-style protocol is further used. That is, each node has to back off a random time period before sending its pilot signal. At the end of this period, the node senses the carrier. If all sub-channels are free, it sends its pilot message; otherwise it backs off another random time period. When more than one node selects the same sub-channel for data transmission, scheduling is further used to differentiate their transmission sequences to avoid collision.

Multichannel Protocols

There are several multichannel MAC protocols proposed for UWANs, and here we introduce one proposal aiming to solve the triple hidden terminal problem in a multichannel UWAN with single-transceiver. This problem also appears in terrestrial RF ad hoc networks [33].

A control channel is used to coordinate data channel allocation. Since a node can work either on the control channel or a data channel, but not on both simultaneously, a node communicating in a data channel cannot learn about the channel assignment undergoing in the control channel. Once this node switches to communicate with other nodes, it may select a data channel that has just been allocated to other nodes, resulting in collision.

Particularly with long propagation delays in UWANs, even if all relevant nodes are listening to the control channel at the same time, a CTS frame sent by the RTS receiver that contains the information on a data channel allocation may arrive later at another node, e.g., the CTS frame sent by node B to node A arrives later at node D in Fig. 10.3. If this node also conducts the same data channel allocation and has just sent out a CTS frame to another RTS sender for this allocation, e.g., node D also allocated to node C the same data channel as allocated by node B to node A. In this case, the same data channel will be used by two nodes to transmit data simultaneously, probably leading to data collision at the receivers. In Fig. 10.3, both nodes A and C use the same data channel for transmission, resulting in collision in node B.

To handle the triple hidden terminal problem in multi-hop long-delay UWANs, the Cooperative Underwater Multichannel MAC (CUMAC) protocol [25] adds a beacon in the original RTS/CTS handshake process. The beacon broadcast by the intended receiver of an RTS frame through the control channel aims to seek cooperation

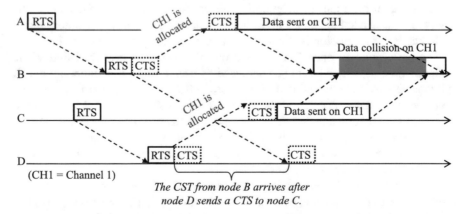

The CST from node B arrives after node D sends a CTS to node C.

Fig. 10.3 Tipple hidden terminal problem due to large propagation delay [1, 33]

from its neighbors for collision detection. The whole RTS/Beacon/CTS negotiation process over the control channel works as follows:

- A node wishing to transmit data first sends an RTS frame, which carries some useful information such as a set of data channels available at the RTS sender.
- Upon receiving the RTS frame, the intended receiver selects a data channel probably to be used for the data transmission requested by this RTS according to the information carried therein and its own perceived channel conditions. Then the receiver broadcasts a beacon message to seek cooperation from its neighbors for collision detection. This message carries the information on the data channel selected for the RTS request.
- The RTS receiver starts a timer and waits for the responses from its neighbors. If no collisions are reported by its neighbors before the timeout, it replies the RTS sender a CTS frame to inform it of the selected data channel; otherwise, the receiver selects another data channel and broadcasts a new beacon message again, and so on.
- It is possible for multiple neighbors to send responses simultaneously to congest the control channel. Thus, instead of sending a message when a neighbor detects a collision, it sends a tone pulse [34] sequence at a specific time, which is a predefined bit sequence consisting of multiple periodic tone pulses. This sequence is calculated based on the location information such that the sequence can arrive at the intended receiver on the expected detection time point. With such a small duration and randomly chosen periodicity of tone pulse sequences, a very low probability for one sequence destined for one node to overlap with that for another node is expected.

10.3.1.4 Protocol Evaluation

Some multiplexing-based MAC protocols ignore potential difficulties in providing basic operation conditions to run the MAC proposals [1]. For example, several code assignment schemes for CDMA do not discuss how to assure practically adequate orthogonality of the theoretically orthogonal codes assigned for nodes with different large time offsets. Actually, signals may propagate alone different paths, some of which cause long and variable delays [35]. For TDMA and slotted access MAC protocols, a primary problem is how to tradeoff well between minimizing collision and maximizing bandwidth utilization in setting guard time. Those proposals relying on precise time synchronization and message exchanging (e.g., handshaking) will suffer the same problems as faced by other types of protocols to be discussed later.

Multichannel-based MAC protocols can prevent transmission collision between data and control messages, actually they suffer the same problem as FDMA-based protocols in terms of low bandwidth utilization due to guard bands inserted between sub-channels. However, this issue has not been addressed by this kind of protocol in performance evaluation. Furthermore, the use of a dedicated control channel (DCC) leads more waste of the very small acoustic channel capacity. It can simplify synchronizing a sender-receiver pair to the same data channel, but may lead to the triple hidden terminal problem. Therefore, research on non-DCC based protocols such as those for RWNs [36] is necessary for UWANs.

10.3.2 MAC Protocols Without Multiplexing

The major modifications to ALOHA, CSMA, MACA and FAMA in order to make them suitable for UWANs are summarized below:

- Changing protocol configurations to make a protocol aware of propagation delay and node locations. These major configurations include contention window (CW), guard time for slotted access and inter-frame space (IFS).
- Modifying protocol operations to support concurrent handshaking and transmissions, centralize access control and allow nodes to transmit or handshake during the waiting period set by the original protocol.
- Using slotted access to replace the random access adopted by the original protocol.

10.3.2.1   ALOHA

A modified ALOHA protocol proposed in [19] requires a node to randomly back off before transmission once a packet arrives in order to reduce collisions, jointly using a congestion window (CW) and a timer. When a packet arrives at a node for transmission, it randomly chooses a starting value between 0 and $CW - 1$, and then sets a timer accordingly to count down. The timer will pause once the channel is found busy, and the node starts its transmission when the timer reaches 0.

To reduce collisions caused by transmissions from senders with different distances to the same receiver, a modified S-ALOHA protocol proposed in [2] inserts a guard time to each time slot to avoid collision at the receiver as illustrated in Fig. 10.4. Thus, the slot duration is increased to $T + \beta \times P_{max}$, where, T is the frame transmission time, β is the fraction of the maximum propagation delay (P_{max}) that nodes have to wait from the end of the last transmission. If $\beta = 1$, it means no overlap between transmissions in different slots. As illustrated in Fig. 10.4, senders 1 and 2 transmit at slots 1 and 2, respectively. To avoid collision at the receiver, the guard time should be $(d_1 - d_2)/v$. The worst case is that d_1 is equal to the maximum distance, while sender 2 is so close to the receiver that $d_2 \approx 0$.

Since the setting of β affects bandwidth utilization and the collision probability at the receiver, an optimal setting of β is also studied. As shown by a mathematical analysis in [37], which renames this protocol as Propagation Delay Tolerant ALOHA (PDT-ALOHA), the throughput can be $17 \sim 100\%$ higher than the original S-ALOHA with a proper parameter setting. However, the simulation study in [2] also shows that without time synchronization, this modified S-ALOHA cannot get any benefit.

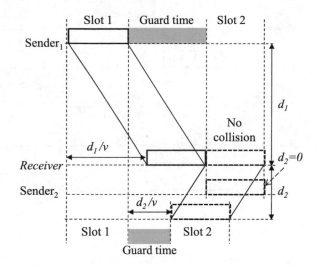

Fig. 10.4 Guard time setting for S-ALOHA in UWANs [1]

10.3.2.2 CSMA

The ALOHA with carrier sensing (ALOHA-CS) [18] is actually a variant of CSMA, which uses a maximum backoff window ranging between twice and five times the maximum propagation delay (P_{max}). A node keeps sensing the channel and stays silent before data transmission until the channel becomes idle. Once the channel is sensed idle, the node starts its transmission. If a transmission error occurs, the sender backs off a random time for a new attempt, similar to CSMA with collision detection (CSMA/CD). If consecutive transmission occurs, the maximum backoff time is set up to $5P_{max}$. A similar version is called CSMA-ALOHA discussed in [21], where the sensing duration is randomly set shorter than the time required so that the signal can propagate over the sensing range.

The Propagation-delay-tolerant Collision Avoidance Protocol (PCAP) [23] conducts the following modification to MACA in order to reduce the inefficiency caused by a long waiting time for the CTS in UWANs. The transmission of the CTS frame is deferred such that it will reach the RTS sender after $2P_{max}$ from when the RTS frame was sent out. That is, once the RTS intended receiver receives the RTS frame, it has to defer its transmission of the CTS frame by $2(P_{max} - Psr)$, where P_{sr} is the sender-receiver propagation delay. The above modification tries to allow the RTS sender or its neighbours to take other actions during the waiting period, such as transmitting another data frame or handshaking for the next transmission plan, while the expected CTS frame can be received at the proper time. However, how to arrange these actions is not discussed with PCAP but with the channel stealing MAC protocol (CS-MAC) in [20].

With CS-MAC [20], a node follows the DCF of IEEE 802.11 (i.e., CSMA/CA) to compete for channel access. Once a node wins, it sends immediately an RTS frame. To handle the hidden terminal problem, upon receiving the RTS frame, the

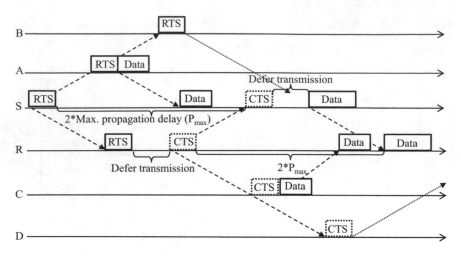

Fig. 10.5 Principle of CS-MAC and PCAP [20]

receiver has to postpone the transmission of the CTS frame for a duration equal to $2(P_{max} - Psr)$ as illustrated in Fig. 10.5, where the intended receiver R receives the RTS from the source S, and postpones returning the CTS frame to it. Similarly, the source node also defers the same amount of time for data transmission from when it receives the expected CTS frame. Other nodes that overhear the RTS or CTS frames are allowed to transmit data frames in order to minimize bandwidth waste due to the waiting period. For example, upon receiving the RTS, node A transmits a data frame immediately to R, which can be received successfully within this period. Similarly, node C can also send a data frame to node R after receiving the CTS. However, neither node C nor node D transmits to avoid collision. To reduce collision in this case, the time interval between when an RTS frame is sent and when the expected CTS frame is received (equal to $2P_{max}$) is slotted. Each slot lasts for a duration of a data frame, and a guard time is inserted between adjacent slots to prevent collisions from consecutively received data frames.

10.3.2.3   MACA and FAMA

Some modifications to enhance MACA try to leverage long propagation delays and improve efficiency and success of handshaking. Note that some proposals claim themselves as modifications of FAMA, actually they are variants of MACA because they do not sense carrier before sending RTS frames. As mentioned earlier, the major difference between MACA and FAMA is that FAMA uses carrier sensing while MACA does not before sending an RTS frame.

With the original FAMA protocol in the case of long propagation delays, the transmission of RTS/CTS frames may collide with an ongoing data transmission.

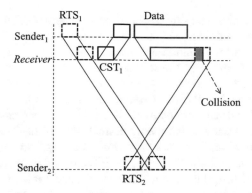

Fig. 10.6 Collision between RTS and data frames with the original FAMA [1, 24]

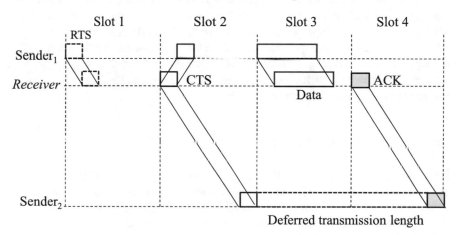

Fig. 10.7 Principle of S-FAMA [1, 24]

For example illustrated in Fig. 10.6, sender$_1$ first submits its RTS frame RTS$_1$, and the receiver returns CTS$_1$ accordingly. However, sender$_2$ also submits RTS$_2$ before the arrival of RTS$_1$/CTS$_1$, and RTS$_2$ collides with the data frame from sender$_1$ at the receiver. To overcome this problem, the slotted FAMA (S-FAMA) [24] suggests that a frame is allowed to be transmitted only at the beginning of one slot. The slot length is set to P_{max} plus the transmission time of a CTS or an RTS frame as illustrated in Fig. 10.7, where sender$_2$ defers its transmission according to the information carried by the CTS. An ARQ protocol is also used to send ACK or NACK frames to acknowledge the data receipt. To accommodate for clock drift in time synchronization, a guard time, which increases the slot length, should be inserted. However, the slot size proportional to the maximum propagation delay reduces bandwidth utilization in large UWANs.

The Distance Aware Collision Avoidance Protocol (DACAP) [22] adopts an MACA-like RTS/CTS handshake to maximize throughput with less energy consumption and collision avoidance using a spacial reuse scheme. For the same receiver, a frame coming from a far-away sender will not corrupt another frame from a much closer sender if they are transmitted at the same power level. Thus, handshaking aims to avoid collisions from nodes closer than a certain distance, and the length of the idle period is set according to the distance between the nodes, which is measured by the sender according to the RTS/CTS round-trip time (RTT).

The following handshake protocol is defined. Upon receiving an RTS frame, the receiver sends a CTS frame to the sender immediately. If the CTS sender overhears an ongoing transmission, it sends a very short warning frame to the RTS sender. Upon receiving the expected CTS frame, the RTS sender waits some time before its data transmission. During this period, if it overhears another transmission or receives a warning frame from the intended frame receiver, it defers its transmission.

The impact of packet sizes on the performance of CSMA and DACAP is analyzed through simulation study in [38], with the following observations. An improper setting can result in a high performance penalty. The performance on throughput, latency and energy consumption in multi-hop UWANs can be greatly improved by a proper parameter setting. The best packet size depends on data generation rate, transmission rate and BER. Such observation is also initially found in terms of throughput for a single-hop network in [39].

10.3.2.4   Performance Evaluation

It is shown analytically that ALOHA has better robustness than S-ALOHA, and may even outperforms it [40, 41]. The simulation study in [42] shows that the normalized throughput of ALOHA changes little for different depths; but in deep water scenario, this throughput of nodes located near the surface is about half that in deeper water. With a string topology [43], it is shown analytically that saturation may occur in less than five hops, and within three hops for an optimal load. Packets from upstream nodes have very small opportunity to reach the gateway due to collision. Similarly, for p-persistent ALOHA, without packet dropping, throughput increases with traffic load, but the latency also increases significantly if nodes along the path defer transmission to avoid collisions at their downstream nodes [44]. A similar observation is also reported for p-persistent S-ALOHA in [45].

For a large sparse UWAN, the simulation study in [46] shows that ALOHA and S-ALOHA perform better in the case of very low traffic load, while the handshaking protocol (HP) is better in a dense UWAN in the case of higher traffic load. However, the HP cannot work well with large transmission range, which however does not affect ALOHA and S-ALOHA. The HP can perform better for large packets with higher throughput for bursty traffic. A slight modified S-ALOHA also suggests using smaller slots to improve throughput [41]. The simulation study in [47] shows that ALOHA-CS [18] consumes most energy with less delay, while FAMA consumes less energy with largest delay per successful reception, and MACA is just in between them.

The analysis in [48] shows that S-FAMA can efficiently avoid hidden terminal problem (HTP) with considerably reduced throughput especially for large networks, and T-Lohi (to be discussed in Sect. 10.4) is better in resilience against congestion and energy-throughput tradeoff. In a deep water sparse network, the simulation study in [21] shows that CSMA-ALOHA outperforms all, and T-Lohi is better than DACAP for network throughput as explained below. Both DACAP and T-Lohi suffer large delay for handshaking and contender detection. For DACAP, an RTS transmission may also interfere with data transmission, while T-Lohi suffers from the HTP. The sea-field trial reported in [49] shows that CSMA yields the best throughput efficiency, but DACAP decreases it quickly. For packet latency, which is defined as the average time between data packet generation and data packet reception at the sink, T-Lohi is the best, while DACAP is worse than CSMA.

Reference [50] derives the closed-form expressions for mean service time and throughput against propagation delay, high detection and decoding errors for MACA. Analytical results match the simulation results using preliminary sea trial settings. Several MACA-like protocols are analyzed in [51], showing that an optimal protocol parameter setting can improve performance (e.g., throughput) significantly. A sea trial for a small UWAN shows that the performance at sea is slightly worse than the results predicted by the analysis and simulation.

10.4 Newly Designed UWAN MAC Protocols

This section discusses some typical MAC protocols that are specifically designed for UWANs by taking into account the peculiar features of UWANs in MAC protocol design. These protocols may also jointly use popular MAC protocols or other MAC mechanisms proposed for RF wireless networks discussed in Chaps. 3 and 4, e.g., signal tone used in the Busy Tone Multiple Access (BTMA) protocol. The discussion is organized according to the UWAN MAC protocol categories specified in Fig. 10.1, each of which is defined below [1].

- Reservation-based MAC protocols: A sender has to submit a request to reserve medium access opportunity for data transmission. This request has to be granted by other nodes such as the receiver.
- Scheduling-based MAC protocols: Different from the reservation-based protocol, a sender itself computes its transmission schedule to make transmission decisions by itself.
- Cross-layer designed MAC protocols: MAC protocols jointly use information and/or capabilities available at other layers, such as the physical layer, the network layer and the transport layer, in design to improve further the performance, bandwidth utilization and energy efficiency.

Table 10.2 summarizes these protocols following the MAC reference model discussed in Chap. 3.

Table 10.2 New UWAN MAC protocols

Protocol name and reference	MAC reference model (Fig. 3.6)			Major features		
	Oper. cycle	Access unit	MAC mechanism	Topology	Synchronization	Validation method
DC-MAC [52]			FA,RE	MH		S
COD-TS [53]	SF	MS,SL	HS,SC	CL	✓	A,S
ST-Lohi [34]	GF	SL	FA,SI,CS,BO,ME	MH	✓	S
DOTS [54]			HS	*MH*	✓	S,P
ST-MAC [26]		TS	ME	TR	✓	A,S
UAN-MAC [55]	LF		FA,PS	MH		S
CT-MAC [27]		TS	ME,PR,MIMO	ST	✓	A,S
FDA/AFDA [56]			FA,Coding	ST		A,S
TFO-MAC [28]	SF	TS	RE,OFDM	CL	✓	S

Refer to Sect. 3.2 for the descriptions of abbreviation

10.4.1 Reservation-Based MAC Protocols

With a reservation-based MAC protocol, a sender has to request transmission opportunity by submitting a short request usually through random access. The request needs to be granted by other nodes with possible resource allocation. With a receiver-centric MAC protocol, the request has to be granted by the receiver as a MAC decision. Most reservation protocols use messages for request submission and MAC operations, and they are called message-based reservation. There are also some protocols using signal for request submission, which are called signal-based reservation.

10.4.1.1 Message-Based Reservation

The complexity of reservation process depends on network topologies, and centralized ones can simplify the process because the central unit can coordinate the operations. Therefore, in the following, we review the protocols proposed for centralized and distributed topologies, respectively.

Centralized Topology

Two features were found in some commercial modem-based real systems [53, 57], i.e., low transmission rates and long preambles, which both severely degrade MAC protocol performance. Based on this finding and relevant analyses, a time sharing-based MAC protocol called Cluster-based On-Demand Time Sharing MAC (COD-TS) is proposed. COD-TS organizes nodes into clusters by using a distributed clustering algorithm proposed in [58]. A cluster head is responsible for arranging

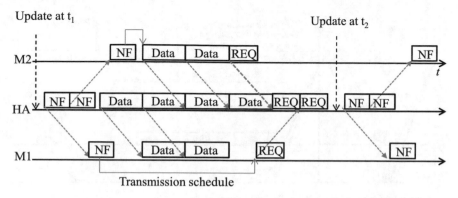

REQ = Request frame, Data = Data frame, NF = Notification frame

Fig. 10.8 A communication round for single hop in COD-TS [1, 53]

sending time of both data and requests for each cluster member, which actually is its one-hop neighbor. A member needs to submit a request first to the head before sending data through the time slot allocated to this node in each communication round (CR). A CR is defined as a sequence of data/control packet exchanges between two successive schedule updates of the cluster head. The head notifies the requesting nodes of the data sending time in the next CR. Thus, intra-cluster communication is fully coordinated by the cluster head with a collision-free schedule based on a superframe structure and mini-slots. The nodes are also synchronized in order to measure the propagation delay and implement scheduling [1].

An example is given in Fig. 10.8, where the time interval between t_1 and t_2 is a CR for the cluster controlled by the head HA. Each CR starts with a notification packet (NF) sent by the cluster head to its cluster members following the previously sent schedule update. Each NF informs the intended member of when to send request and data packets. The current schedule update arranges the slots for data and request transmissions to incur from t_1, e.g., data packets from M2 to HA and request packets from M1 to HA. The next schedule update with NFs in the next CR starts from t_2.

COD-TS is also extended for multi-hop networks through organizing the cluster heads to form an ad hoc network, in which, the heads communicate with each other directly. Therefore, the problems of RF wireless ad hoc networks also exist here, such as collision. To solve these problems, COD-TS requires a cluster head to know the up-to-date schedule of each cluster member's neighbors within two hops for generating schedule through exchanging the up-to-date schedule with its neighboring cluster heads (NCHs). To this end, a cluster head may increase its transmission power, and exchanged packets are sent with a CDMA-based scheme. To avoid schedule update conflicts, a cluster is not allowed to generate or notify its new schedule during the schedule update conflict intervals, which is the time interval between when a NCH generates new schedule and when the corresponding schedule exchange packet arrives at the corresponding node.

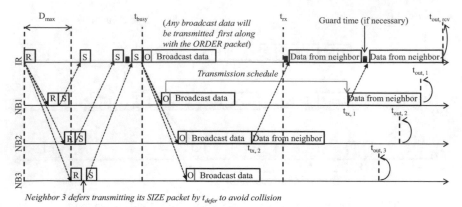

Fig. 10.9 The 4-way handshaking with multiple node polling in RIPT [59]

Distributed Topology

To copy with long propagation delay, a Receiver-Initiated Packet Train (RIPT) MAC protocol is proposed for a multi-hop UWANs in [59]. It jointly uses handshake-based reservation and packet train transmission schemes. A receiver initiates the reservation process and arranges packets from different neighbors to arrive in a packet train within each round of handshake, using a multiple-node polling scheme. Therefore, multiple nodes are allowed to transmit data frames to the same receiver per handshaking. To this end, every node has to know the propagation delays between itself and its neighbors so that their transmissions can be scheduled to avoid collision at the receiver. Since a receiver can efficiently reduce collision therein, RIPT can alleviate the hidden/exposed-terminal problems.

Figure 10.9 illustrates a 4-way (i.e., RTR/SIZE/ORDER/DATA) handshake protocol adopted by RIPT, which is explained as follows:

- An initiating receiver (IR) broadcasts an RTR (ready-to-receive) frame to inform through polling all its neighbors of that it is ready to receive for a certain duration in the upcoming time.
- Upon overhearing this RTR, any neighbors of the receiver (e.g., Neighbors 1 ~ 3) wishing to transmit to it during this handshaking period sends a SIZE frame to inform the IR about the number of frames to be sent. If a node receives the RTR within the pre-defined time interval, it replies immediately; otherwise, it needs to defer its transmission by t_{defer} to avoid collision.
- After receiving the requests from its neighbors, the receiver broadcasts an ORDER frame to inform its neighbors of their relative transmission orders and the number of frames allowed to transmit.
- Each granted neighbor transmits its data frames following the received order, e.g., Neighbor 2 sends first at time $t_{tx,2}$ and Neighbor 1 follows up at time $t_{tx,1}$ as depicted in the figure.

An intended receiver triggers the above handshaking process randomly and blindly, which may cause the following problems. More delay will be introduced if the receiver does not initiate the process immediately when a node has data to send. This situation will become worse if another node close to the receiver triggers the process earlier because such a process will block the receiver from doing the same thing. A similar impact will be also caused by RTR frame transmission, which is triggered randomly and may cause collision. Similarly, for SIZE frame transmission, one polling may trigger multiple nodes to send their SIZE frames simultaneously. Although the data frame collision can be avoided by utilizing the receiver-initiated reservations mentioned above, the control frame collision may become a performance bottleneck.

Thus, an enhancement with multichannel proposed in [52] divides the channel into two subchannels: a control channel (CCH) used for channel reservation, which is similar to that carried by RTR/SIZE/ORDER, and a data channel (DCH) used for data transmission, similar to DATA. It tries to eliminate the hidden terminal problem by grouping nodes into different collision domains with adequate space separation between them to avoid frame transmission interference. A receiver-initiated handshake with polling for precise time-space determination tries to provide multiple collision-free transmissions. However, it cannot eliminate handshaking failure of RIPT although the handshaking process is carried in the CCH, because both schemes adopt random access schemes for handshaking. Another problem caused by channel division is that the probability for the CCH becoming a performance bottleneck is much higher than that in RIPT due to the smaller CCH's capacity. Furthermore, a dedicated CCH cannot be used for data frame transmission, resulting in low channel utilization.

10.4.1.2 Signal-Based Reservation

A tone-based contention MAC protocol (T-Lohi) discussed in [34, 60] exploits proactive tone to reserve the channel, and uses carrier sensing to verify the reservation result. A node wishing to transmit a data frame first sends a short tone, and then listens to the channel for the duration of a contention round (CR). If the node dose not overhear any other tones at the end of the CR, which means a successful channel reservation, it can transmit a data frame; otherwise it backs off and will try again in a later CR. However, if a reservation tone is sent out without any control, it is likely that reservation tones may collide with ongoing transmissions. Therefore, both synchronized T-Lohi (ST-Lohi) and conservative unsynchronized T-Lohi (UT-Lohi) have been discussed.

ST-Lohi synchronizes each CR by distributing the reference time so that every node can know the boundaries of each frame and its reservation period. In this case, each node can decide when to send a reservation tone and when to send its data frames to avoid collision after it wins the contention. It synchronizes contention period and data transmission into slots. A CR lasts for a duration of $CR_{ST} = P_{max} + T_{tone}$, where P_{max} is the maximum one-way propagation delay, and T_{tone} is the tone

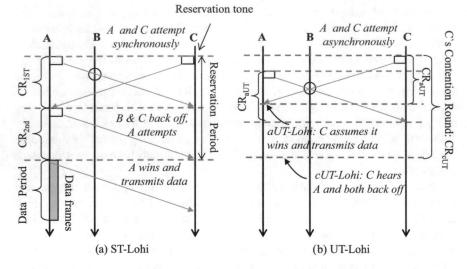

Fig. 10.10 Synchronous T-Lohi versus asynchronous T-Lohi [1, 34]

detection time. As illustrated in Fig. 10.10a, during the first CR, nodes A and C contend simultaneously by sending reservation tones but fail, and node B overhears A's tone first. During the second CR, only A sends a reservation tone while the other nodes back off. In this case, A wins the reservation and starts its transmission.

With UT-Lohi, a reservation tone can be sent out anytime, and carrier sensing is used to detect the channel. Only when the channel is sensed still idle at the end of a pre-defined period, the sender wins the reservation. As illustrated in Fig. 10.10b, nodes A and C both send out a reservation tone, and each senses the channel for a duration equal to $CR_{cUT} = 2(P_{max} + T_{tone})$ with conservative UT-Lohi (cUT-Lohi). At the end of this period, both nodes A and C receive the tones from each other, backs off certain time, and contend again later. With aggregative UT-Lohi (aUT-Lohi), the detection period is shorten to $CR_{aUT} = P_{max} + T_{tone}$. As depicted in the figure, no other tones are received by node C at the end of this period, which means that node C wins the reservation.

T-Lohi also uses a low-power wake-up tone receiver to reduce energy consumption, i.e., a node is activated only when a tone is detected by the receiver. A contender counting is also used to improve fairness and stability under heavy traffic load. A Markov analysis in [3] quantifies the bound of convergence time for MAC protocols using an exact contender counting, showing that such counting can make contention to converge quickly with an asymptotic limit of 3.6 contention rounds on average, which is independent of network density. It explains the load-stability of T-Lohi [1]. Some comparison with other protocols can also be found in Sect. 10.3.2.4.

10.4.1.3   Remarks

Reservation-based MAC protocols are more suitable for centralized topologies, in which a central unit (which is often the receiver too) arranges transmission with less information exchanged between nodes, and collision-free reception can be assured. Actually, this kind of MAC protocol is a sort of receiver-centric MAC protocol, and the reservation procedure provides cooperation between senders and receivers. Since no central unit is available in distributed topologies, the reservation operation becomes more complex [1].

A message-based reservation protocol is more informative than a signal-based ones, but the former is more vulnerable to low channel quality and request collision. A short signal-based reservation can avoid these problems but with only inexplicit reservation, e.g., no explicit information on how long and to whom the reservation is made for. Furthermore, the delay from request submission to decision arrival is a large protocol overhead for short message transmission for both types of reservation protocols, especially in large UWANs [1].

10.4.2   Scheduling-Based MAC Protocols

Different from reservation-based MAC protocols, with a scheduling-based MAC protocol, the sender computes its transmission schedule and makes a MAC decision for collision-free transmission by itself. To this end, some information is needed, such as propagation delay, traffic load and even the schedules of other nodes. The performance of such protocols depends on the accuracy of the used information and time synchronization used by many such protocols. Note that, scheduling for QoS is used to differentiate medium access sequences of different packets according to their QoS requirements [1].

10.4.2.1   Scheduling Constraint Formulation

A scheduler may use some scheduling constrains to make scheduling decisions, and some examples are introduced below.

Spatial-Temporal Conflict Graph (ST-CG)

To handle spatio-temporal uncertainty in a tree-topology UWAN for data gathering application, a TDMA-based MAC protocol discussed in [26, 61], called spatial-temporal MAC scheduling (ST-MAC), uses a Spatio-Temporal Conflict Graph (ST-CG) for schedule computation. In an ST-CG, a vertex indicates a transmission link, and an edge between two vertices represents the conflict relationship between the two transmission links with extra information on the edges to describe spatial uncertainty. An ST-CG is constructed according to topology, mutual interference

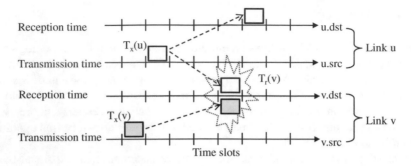

Fig. 10.11 Conflict relationship in ST-CG for ST-MAC [1, 26]

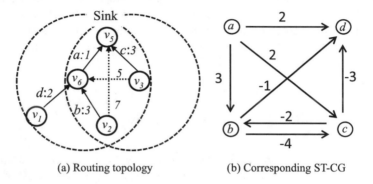

| (a) Routing topology | (b) Corresponding ST-CG |

Fig. 10.12 An example on a routing topology and the corresponding ST-CG for ST-MAC [26]

between any two transmissions and propagation delay, and used to compute schedules that allow each node to know when to sleep, transmit and receive.

For a unicast routing topology $G(V, E)$, V denotes a set of nodes and E is the set of transmission links. Let $PD(v_i, v_j)$ indicate the propagation delay between two nodes with the time slot used as the time unit, and '-1' indicates no interference between them. Let $G'(V', E')$ denote an ST-CG corresponding to a $G(V, E)$, where $V' = E$, and E' is the set of conflict relationships between any two transmissions. Then a conflict relationship, $Conflict(u, v)$ with $u, v \in V'$, exists if there is a transmission over link u affecting the reception over link v as depicted in Fig. 10.11, where $u.src$ and $u.dst$ indicate the sender and receiver of link u, $T_x(u)$ and $T_x(v)$ are transmission times over links u and v, and $T_r(v)$ for reception time over link v, respectively. Obviously, the propagation delay, which can be ignored in RWNs, has to be considered here because it causes spatio-temporal uncertainty.

Figure 10.12a depicts an example of $G(V, E)$, in which the letters denote transmission links, and the arrows indicate the propagation directions, i.e., the link set $E = \{a, b, c, d\}$, the node set $V = \{v_1, v_2, v_3, v_4, v_5\}$, and the number associated with each link is $PD(v_i, v_j)$. The dashed lines with arrows represent interference relationships. Figure 10.12b is the corresponding $G'(V', E')$, where the number associated with each edge with arrows pointing to the conflicted vertex v represents a conflict delay between links u and v, $c_{u,v}$ in time slots, which is calculated as follows (See Fig. 10.11):

$$c_{u,v} = PD(v.src, v.dst) - PD(u.src, v.dst). \qquad (10.3)$$

For the reception at the same destination $v.dst$, if link u transmits with $c_{u,v}$ later (for $c_{u,v} > 0$, see Fig. 10.11) or earlier (for $c_{u,v} < 0$) than link v, u will conflict with v. In this case, edge (u, v) is added to E'. Only when $c_{u,v} = 0$, the transmission will not cause conflict at $v.dst$.

We can calculate $c_{u,v}$ for each of the following conflicts: (i) two links are jointed with a common node, and (ii) two disjoint links interfere with each other, as discussed below.

The first category can be further divided into the following three cases for calculation:

- Two links have the same destination, i.e., $u.dst = v.dst$. Then, $c_{u,v} = PD(v.src, v.dst) - PD(u.src, u.dst)$. For example, link pairs (a, c) and (b, d) in Fig. 10.12a yield $c_{a,c} = 2$ and $c_{b,d} = -1$ in Fig. 10.12b.
- The destination of one link is the source of the other link, i.e., $u.src = v.dst$. Then, $c_{u,v} = PD(v.src, v.dst)$ since a node usually cannot transmit and receive simultaneously. For example, link pairs (a, b) and (a, d) in Fig. 10.12a yield $c_{a,b} = 3$ and $c_{a,d} = 2$ in Fig. 10.12b.
- Two links have the same sources, i.e., $u.src = v.src$. Then, $c_{u,v} = 0$ since a node cannot transmit simultaneously to more than one destination with unicast.

For the second category, we need to consider whether a links reception is interfered by another link's transmission, and there are also two cases as discussed below:

- Interfering in one direction, i.e., the transmission of link u affects the reception of link v, but not vice versa. For example in Fig. 10.12a, link c's transmission from node v_3 will affect link d's reception at node v_4, thus following (10.3), $c_{c,d} = PD(v_1, v_4) - PD(v_3, v_4) = -3$ as illustrated in Fig. 10.12b.
- Mutual interfering, i.e., the transmission of link u affects the reception of link v, and vice versa. In Fig. 10.12a, link c's transmission from node v_3 will affect link b's reception at node v_4, thus following (10.3), $c_{c,b} = PD(v_2, v_4) - PD(v_3, v_4) = -2$ as depicted in Fig. 10.12b. Similarly, link b's transmission from node v_2 will affect link c's reception at node v_5, resulting in $c_{b,c} = PD(v_3, v_5) - PD(v_2, v_5) = -4$.

A Solution for Vertex Coloring Problem

Based on the ST-CG discussed above, the time slot assignment becomes an issue on a vertex (i.e., a transmission link) coloring problem. However, the traditional heuristic solution for vertex-coloring does not take into account propagation delay, and cannot work well in UWANs with long propagation delay. As mentioned earlier, when propagation delay is negligible, a collision-free transmission means a collision-free reception. Thus, a color vertex corresponding to a time slot assignment indicates an available time slot for either transmission or reception. However, a large propagation delay causes spatio-temporal uncertainty so that transmission and reception should be scheduled separately. To this end, a color is used to represent the transmission time, and the corresponding reception time is just the sum of the transmission time and the propagation delay between the sender and receiver.

To solve the NP-complete vertex coloring problem in ST-CG with polynomial time, a centralized heuristic approach that takes both traffic load and routing information into account, called Traffic based One-Step Trial Approach (TOTA), is investigated. With TOTA, the sink first collects the necessary information, and calculates the transmission and reception schedule, which is then broadcast to all nodes. It uses an one-step-trial scheme to search proper time slots without real transmission. That is, a vertex may be colored during its available time slot in the ST-CG, and a colored vertex indicates an available collision-free time slot for transmission. When a vertex is colored at a time slot, adjacent vertices are interfered and are not allowed to transmit packets at the corresponding time slot due to conflict delays, which are called conflict states. These states are updated by performing the one-step trial without real transmission, and this information is used to determine the schedule assignment. The example illustrated in Fig. 10.13 demonstrates an example of TOTA operation [61], which is described below. All vertices attempt to color their first available time slots in the ST-CG simultaneously, which means that all links try to transmit at their first available time slots. Which link has the highest transmission priority depends on conflict states that are marked by the coloring operation as discussed below.

- Vertex a: it has 3 conflicting vertices, i.e., b, c and d as illustrated in Fig. 10.13a. If a transmits at its first slot, its transmission will collide with the transmission of b to start 3 time slots later (i.e., on the $4th$ time slot) and that of c to start 1 time slot later (i.e., reversing the direction toward a so that '-1' becomes '1'). Since

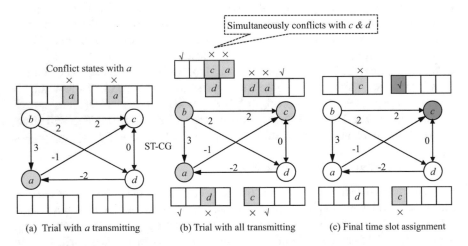

(a) Trial with a transmitting   (b) Trial with all transmitting   (c) Final time slot assignment

Fig. 10.13 Example on coloring operation with TOTA for possible transmissions on the first time slot with a trial carried out for each link (i.e., vertex) corresponding to an ST-CG [61]

no transmission can occur earlier than that on the first time slot, confliction with d will not happen in this round.

- Vertex b: It also has 3 conflicting vertices, i.e., a, c and d, its transmission on the first time slot will not lead to any collision with the transmissions from these links in this round since no transmission can occur earlier than that on the first time slot.
- Vertex c: Similar to the above, its transmission on the first slot will collide with that of b to start 2 time slots later and that of d starting on the same time slot, as illustrated Fig. 10.13b.
- Vertex d: Similarly, its transmission on the first slot will collide with those of both a and b started 2 time slots later and that of c started on the same time slot. Note that the Fig. 10.4 in [61] only shows the conflict relationship from d to c. In order to match the final trial result marked for c, there should be a conflict relationship from c to d too as depicted in Fig. 10.13b.

All possible conflicting states are marked by ⊠, and the remaining time slots marked by ☐ are available for assignment as depicted in Fig. 10.13b. Now each vertex identifies its earliest available time slot that is located at slot position l as follows: the 1st time slots for both a and b (i.e., $l = 1$), 3rd time slot for c and 2nd time slot for d, which are ticked (✓) in Fig. 10.13b. Reference [61] suggests that l is used to evaluate the condition of a conflict state because after a vertex attempts to allocate as compactly as possible the transmission time slots in its temporal conflict state, where l is the position of the last allocated transmission slot. Thus the vertex with the largest l has the highest priority to assign its first time slot for transmission since it has fewer marked time slots in its final conflict state. Accordingly, vertex c has the highest priority for time slot assignment as illustrated in Fig. 10.13c, and the conflict states corresponding to this assignment are used in the next iteration following the same procedure used above. Actually, if the first time slot is assigned to vertex b, it will not cause any conflict state as illustrated Fig. 10.13b.

To satisfy different traffic demands, the number of time slots assigned to a node for data transmission is set proportional to its data rate, which means that nodes with higher traffic load will be assigned more time slots. Furthermore, a distributed TOTA (DTOTA) is also studied, with which, a node calculates its time slot assignment locally without using global information.

10.4.2.2 Leveraging of Long Propagation Delay

Long propagation delay can be exploited to enable more concurrent transmissions particularly in the following scenarios: (i) multiple senders transmit to the same receiver simultaneously, and (ii) multiple concurrent transmissions are allowed between any nodes.

A propagation delay-aware opportunistic transmission scheduling (DOTS) based MAC protocol is discussed in [54, 62]. It adopts a MACA-like RTS/CTS protocol and a scheduling scheme to realize concurrent collision-free transmissions in the case of large propagation delay. The RTS/CTS handshake is carried out according to the following conditions [63]:

- The RTS wait time should be greater than the maximum propagation delay (P_{max}) to allow a transmitted frame to reach its maximum transmission range.
- The CTS wait time should be greater than the sum of the RTS transmission time and $2(P_{max} + T_{tr})$, where T_{tr} denotes the transmit-to-receive transition time of hardware.

Based on a network-level time synchronization, a node can locally calculate a distributed transmission and reception schedules to perform concurrent transmission. To this end, every node has to overhear neighboring transmissions to obtain relevant information in order to build a delay map database, which is updated for every overheard MAC frame header. It contains the following information: the source and destination addresses of each overheard MAC frame, the time stamp at which a MAC frame was sent out and the estimated propagation delay between the source and the destination of the MAC frame. The database is used by a node having a frame to send to estimate collisions at the intended receivers, and make intelligent scheduling decisions for transmission to increase collision-free concurrent transmissions.

As illustrated in Fig. 10.14, node X first transmits an RTS frame to node Y, and node U also transmits an RTS frame to node V, while X and U are exposed to each other, following the schedules. Node Y replies with a CTS frame after waiting for a time period equal to the sum of packet transmission time and P_{max}. During this waiting period, node U also receives the RTS frame sent by Node X. Suppose that node U also has data to send. Since collision only occurs at receivers, it can start its transmission without any harm to receptions at other nodes if its transmission will not collide with any ongoing and prospective receptions at the nodes located within its interference range. For example, although the data frame from node X will arrive

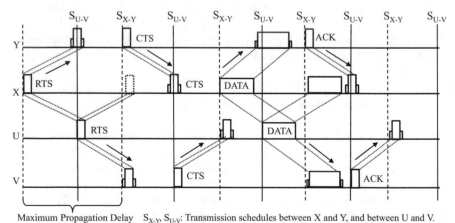

Fig. 10.14 Concurrent transmission with DOTS: multi-sender to multi-receiver [54]

at node U when it is transmitting a data frame to node V, such overlapping does not affect the corresponding receptions at both nodes Y and V. To this end, it is necessary for a node to compute precisely the transmission and reception schedules of each neighbor based on the delay database.

10.4.2.3 Energy Efficiency

To save energy, letting a node sleep during an idle period is a method extensively adopted, and scheduling can arrange the time for nodes to sleep and wake up. For example, in [55, 64], a distributed energy-efficient MAC protocol, called Underwater Wireless Acoustic Networks - Medium Access Control (UAN-MAC), is discussed for delay-tolerant application. With UAN-MAC, each node announces its scheduled transmission cycle before going to sleep so that its neighbors can know when to wake up to listen to its transmission. During the sleeping period, a node turns off its transceiver circuit to save energy.

Figure 10.15 explains how a node can achieve a locally synchronized schedule without knowing the propagation delay. Node A first broadcasts a SYNC packet at the beginning of its cycle period, which announces its transmission cycle period of T_A, then goes to sleep. When a node joins the network, say node B, it has to listen to the channel until it receives a SYNC packet for frame synchronization with the SYNC sender. After receiving the SYNC packet from node A, node B can learn the lengths of node A's sleeping state (i.e., T_A) and prospective transmission cycle. Then it also goes to sleep for the same duration T_A from when it decodes out the SYNC packet, and will wake up at the end of this period to listen to the channel for reception. Similarly, for the transmission in the wake-up state, the initial transmission

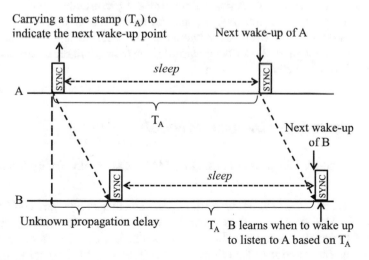

Fig. 10.15 Basic idea of the UAN-MAC protocol [1, 55]

time can be selected randomly within the first cycle. The same transmission time will be used locally in the consecutive cycles with an identical time offset between two transmissions in consecutive cycles [1].

This protocol does not rely on network-level time synchronization without using any information on the propagation delay, as long as the variance of propagation delay is negligible for one cycle to the next, and the clock drift is not significant per cycle. A separate simulation study conduced in [65] shows that since UAN-MAC strongly relies on the synchronization between the nodes' schedules, network performance drops fast as synchronization drift increases, and periodic exchange of synchronization messages can be a solution for this problem.

10.4.2.4   Remarks

In UWANs, scheduling is typically used to arrange collision-free receptions and sleeping/wake-up to improve bandwidth utilization and save energy. Its most abstractive feature is to leverage long prorogation delay to improve channel utilization by enabling more concurrent transmissions. Many scheduling-based MAC protocols usually use some information such as prorogation delay and even schedules of other nodes. How to get such valid information is a key issue, while long propagation delay may obsolete carried information, and transmission error may worsen such situation. Particularly, when time slots are used as MAC access units, they also need to handle similar problems faced by TDMA-based protocols mentioned earlier, and inaccurate time synchronization affects the efficiency of calculated scheduling decision [1].

The complexity of a scheduler depends on network topologies and the information to be used, and information gathering is a major overhead. However, many protocols do not state clearly what information is used and how to obtain it without discussing the impact of information gathering on the protocol performance. For example, to obtain more and accurate information, specific message exchange is necessary at the cost of more protocol overhead. Although a piggyback scheme can reduce such overhead, it cannot guarantee sufficient information to be available timely for a node to make a correct scheduling decision.

10.4.3   Cross-Layer Designed Protocols

Typical techniques jointly used to design UWAN MAC protocols are listed below [1].

- Coding: Usually collided frames are dropped immediately by many UWAN MAC protocols. However, these frames can be recovered by using coding technique subject to certain coding conditions. For example, a collided frame is usually retransmitted, which may lead to a second collision. These two collided frames can be used jointly to recover the collided frame with successive interference cancellation (SIC) [66].

- Multi-input multi-out (MIMO): With MIMO, antenna arrays are installed at both transmitter and receiver to allow multiple nodes to transmit concurrently at the same frequency without causing reception failure. MIMO can significantly improve channel capacity through exploiting propagation spatial diversity without using additional transmission power or bandwidth. In UWANs, it can further exploit rich scattering and multipath fading to provide higher spectral efficiency [67, 68].

- Orthogonal Frequency Division Multiplexing (OFDM): With OFDM, a channel is divided into multiple subchannels such that data symbols modulated over these subchannels can be transmitted in parallel. Adjacent subchannels overlap in such a way that their information-bearing waveforms are orthogonal without causing inter-channel interference (ICI).

10.4.3.1 Encoding and Decoding

It is shown that if collisions follow some patterns, a collided frame can be effectively decoded [69]. Thus, a Zigzag decoding scheme is investigated in [66] to combat the hidden terminal problem as described below. Usually, once a collision due to hidden terminals happens, the sender will trigger retransmissions with a random backoff, probably resulting in a second collision because the backoff delay is typically much shorter than the packet transmission time. Instead of dropping the collided frames, the Zigzag decoding tries to exploit pattern difference between two collisions to recover them through iterative decoding. That is, for a collided frame, its clear chunk (say C1) in the first collision is used to decode another chunk (say C2) in the second collision by subtracting C1 from C2, and so on, until the two collided frames are fully decoded [1].

As illustrated on the left side of the dashed line in Fig. 10.16, two frames, A and B, collide with each other. However, chunk C_{a1} of frame A and chunk C_{b5} of frame B are clear for decoding. Therefore, they can be used to decode some collided chunks in the retransmitted frames as shown on the right side of the dashed line. That is, C_{a1} is used to decode C_{b2}, C_{b5} to decode C_{a4}, and so on, until all collided chunks are decoded as illustrated in the figure.

To make Zigzag decoding successful, two already collided frames should collide each other again in the subsequent retransmissions, which however is difficult for implementation in reality. Thus, a distributed MAC protocol, Flipped Diversity ALOHA (FDA), is studied in [56] to handle this problem, which is renamed as Asynchronous FDA (AFDA) in [70]. This protocol tries to improve ALOHA's performance by using the Zigzag decoding scheme in UWANs with long and varying propagation delay, while source nodes have no reception capability.

As illustrated in Fig. 10.17a, a super frame consists of the original frame and its flipped replica, in which S and E denote the beginning and the end of the original frames, respectively. These two frames are transmitted back-to-back without any

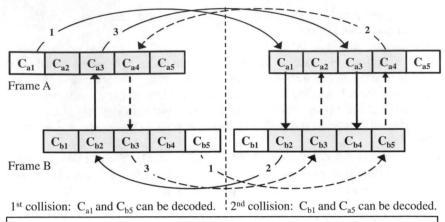

1st collision: C_{a1} and C_{b5} can be decoded. ¦ 2nd collision: C_{b1} and C_{a5} can be decoded.

> The remaining blocks are calculated following the lines below one-by-one: i) Solid lines from C_{a1}: C_{b2}, C_{a3} and C_{b4}, ii) Dashed lines from C_{b5}: C_{a4}, C_{b3} and C_{a2}.

Fig. 10.16 Principle of Zigzag decoding [1, 66]

Fig. 10.17 Basic idea of the FDA protocol [70]

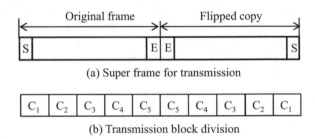

(a) Super frame for transmission

(b) Transmission block division

interval between them. Figure 10.17b shows the corresponding transmission block division for Zigzag decoding.

10.4.3.2 Multi-input Multi-out (MIMO)

Reference [27] investigates a distributed MAC protocol, called Coordinated Transmission MAC (CT-MAC), for uplink sharing in a MIMO-based UWAN, where signals from underwater nodes are transmitted to a surface buoy. It exploits MIMO to enable MAC-level simultaneous transmissions to combat long propagation delay, which is similar to the logical MIMO-based MAC for mobile cellular networks introduced in [71, 72]. That is, with such kind of MAC, m receive-antennas can allow m nodes to transmit simultaneously at the same collision domain, which can shorten medium access delay with a transmission rate equal to that of a Single-Input-Single-Output (SISO) channel.

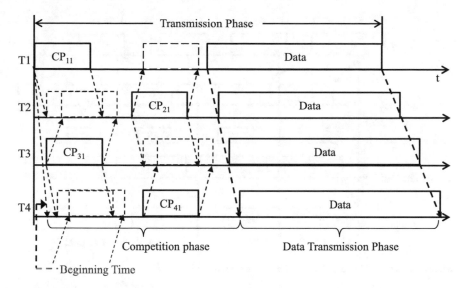

Fig. 10.18 The sequence diagram of 1D CT-MAC: solid lines refer to transmitted packets and dash lines represents received packets [27]

With CT-MAC, the base station tries to select active nodes to transmit simultaneously and ensure (i) their transmissions to arrive at the base station at the same time, and (ii) the number of the selected nodes approaching to $\frac{M_r}{M_t}$, where M_r is the number of hydrophones available at the base station and M_t the number of transducers equipped at each node [1]. The operation is divided into two stages: initialization and continuous transmission. In the first stage, nodes exchange control messages to get the global network information, and in the second stage data packet are transmitted without collisions. The transmission phase consists of competition and data transmission with equal lengths as illustrated in Fig. 10.18.

During the competition phase, every competing node first sends out a competition packet (CP), which contains the priority level (PL) information. To obtain the global competition information, every node forwards the new PL information received previously. This process continues until all nodes obtain such information. To reduce CP collision, nodes with even IDs send first and then starts receiving CPs from their neighbors, which is followed by nodes with odd identity (ID). The only m nodes with highest PLs win the competition, where m indicates the maximum number of parallel transmissions allowed by the MIMO uplink. As illustrated in Fig. 10.18, the beginning of the transmission cycle for each node is adjusted by an offset to compensate for the propagation delay to assure data packets from different nodes to arrive at the base station simultaneously. The time offset is estimated during the allocation process of node identity.

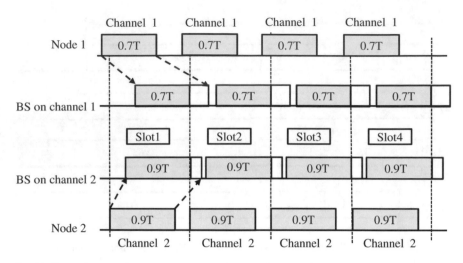

Fig. 10.19 TFO-MAC: fixed channel assignment with OFDM [28]

10.4.3.3   Orthogonal Frequency Division Multiplexing (OFDM)

Reference [28] investigates an MAC protocol for the OFDM based uplink communication in a cellular-like UWAN, called TDM with FDM over OFDM MAC (TFO-MAC). In this network, a set of nodes are covered by a base station, which synchronizes every node with it for transmission. The available bandwidth is divided into multiple subbands with FDM, and OFDM is used in every subchannel for data transmission. Each channel is further divided into equal-length time slots, which are organized into superframes. In this case, it is expected that a large available bandwidth especially for short range communication can be fully used to overcome the relatively small bandwidth provided by OFDM modems currently available for UWANs [1].

However, multipath delay spread in UWANs greatly affects the performance of OFDM modem because the guard time between OFDM blocks should be set larger than the propagation delay in order to eliminate inter-block interference (e.g., 25 ms as reported in [73]). In this case, after one block transmission, the sender has to wait for at least one guard time to transmit the next block. Since the delay depends on a node's location, nodes at different locations experience significantly different delay spread, which will make their throughput different as explained below.

Figure 10.19 depicts a system working on channels 1 and 2 each associated with different base stations each with a data rate of R in bps. Suppose nodes 1 and 2 requesting a service with a data rate 0.8R and the multipath delays for nodes 1 and 2 are $0.3T$ and $0.1T$, respectively, where T is the slot length. With a fixed channel assignment, the maximal data rate that node 1 can get is 0.7R since only $(T - 0.3T)$ can be used for data transmission, while node 2 can get a data rate of 0.9R (i.e., $T - 0.1T$), as shown in Fig. 10.19. In this case, node 1 cannot be satisfied, whereas

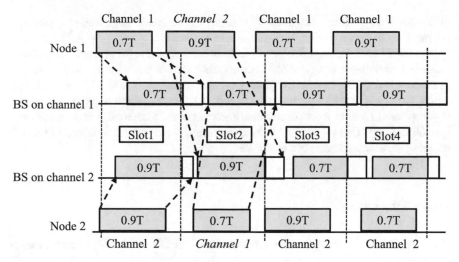

Fig. 10.20 TFO-MAC: dynamic channel assignment with OFDM [1, 28]

node 2 is over-provisioned. Such situation can be avoided with a dynamic channel assignment, which can allow nodes 1 and 2 to use channels 1 and 2 alternatively as shown in Fig. 10.20. Then, both nodes 1 and 2 can be satisfied in terms of data rates on average.

To achieve the above objective, TFO-MAC tries to adjust adaptively channel access for each node by jointly taking into account channel assignment, transmission mode and transmission power with an iterative algorithm [1]. This algorithm first initializes the transmission mode and power of every node on every channel, and then optimizes the channel assignment in every slot as well the transmission mode and power. At the initial stage, the base station first sends a channel measurement message sequentially on every available channel at the beginning of each superframe. Each node follows the sequence of this message to measure the quality of the downlink channel, and randomly selects a channel to send out a channel request to the base station if it has data to send. This request indicates the node's minimal data rate requirement and the channel measurement results. Then, the node listens to the well-known control channel to learn the channel assignment result. Once the node obtains a channel assigned by the base station, it will transmit data with the specified transmission mode at the specified transmission power in the assigned time slots.

10.4.3.4 Remarks

Cross-layer design can address several issues related to the peculiar features of UWANs as summarized below. Both MIMO and OFDM can improve channel reliability and utilization, and MIMO enables MAC layer simultaneous transmission. Recovering collided frames with coding technique can further improve bandwidth utilization and protocol efficiency [1]. Interference mode may improve protocol adaptability to dynamic environments, while routing information can help a node to make better MAC decision. However, technique limitations for practical implementations of cross-layer designed MAC protocols have not been addressed adequately. For example for MIMO, with a typical acoustic frequency around 10 kHz, the wavelength (λ) is 15 cm with a propagation speed of 1.5 km/s, which requires an antenna (hydrophone) sized at least several centimeters because antenna size depends on λ (pp168 in [74]). This requires a big antenna array, which makes it difficult to install it in a space-limited node such as AUVs.

10.5   Discussion

Here we discuss some major problems remaining for UWAN MAC protocol designs as summarized in [1].

- The widely used protocol validation method is computer simulation, whereas prototyping test in laboratory and sea-trial are seldom used. This may cause the investigated performance of some UWAN MAC protocols questionable in real environments. As reported in [49] for CSMA, T-Lohi and DACAP, there is a significant gap between sea trial and simulation results with an inadequate acoustic channel model or without considering the overheads and delays caused by the specific hardware. Only the sea trial for a small UWAN using MACA-like protocols shows that the performance at sea was slightly worse than that predicted by the analysis and simulation [51]. Special attention should be paid for those protocols using time synchronization, localization and information collection as discussed below.
- Time synchronization is used to enable nodes to have the same timing reference for communication and networking, which is widely used in CDMA-based, TDMA-based and slotted access protocols. Due to difficulties in realizing precise time synchronization underwater as discussed in Sect. 9.2.1.4, some synchronization-based UWAN MAC protocols such as STUMP and [75] have considered synchronization error in performance evaluation, but many protocols ignore such impact, and may be only theoretically sound but ineffective in reality. A similar problem also exists for difficult underwater localization [19] since localization error may also happen.
- Many MAC protocols need to collect information not available locally for MAC operations, using short messages like RTS/CTS. However, as mentioned in [27], a long preamble sequence (e.g., $0.49 \sim 1.5$ s, depending on particular modem) will be added for signal synchronization and channel estimation for actual transmission in a complex underwater acoustic channel, whereas a 100-bit packet transmitted at 1 kbps [24] takes about only 0.1 s. Thus, a short message may be actually transmitted in much longer time than expected. Furthermore, underwater acoustic channel quality may change in very short time scale [76], and more changes may happen over long transmission time. This factor will impact successful reception and fur-

ther affect MAC protocol efficiency. Furthermore, energy consumption for short message transmission should be also taken into account since transmission and reception consume lot of energy. However, these issues have not been considered adequately in performance evaluation.

• Some MAC protocols adopt several MAC mechanisms to handle the problems in UWANs for better performance, which however makes the protocols more complex and costly for implementation. The simpler a MAC protocol, the more robust for its operation. Actually, there were also a large number of MAC protocols proposed for terrestrial RWNs, and many of them have been modified for UWANs. What is widely used in practice today is CSMA/CA-based IEEE 802.11, while RTS/CTS is only an optional part sitting on top of it. Particularly for scheduling-based MAC protocols, realtime scheduling decision requires sufficient and accurate information. However, how to trade off better between protocol efficiency and implementation complexity is an issue to be further addressed.

10.6 Summary

UWAN MAC protocols that are modified from the popular protocols of RWNs can better reuse existing technologies to minimize development cost and maximize interoperability with existing networks. Many modifications take into account long propagation delay in MAC design by setting relevant parameters to be aware of the delay. However, collision avoidance by simply using guard time or guard bands may significantly reduce bandwidth utilization. Thus several MAC proposals try to leverage long propagation delay for more concurrent transmissions, such as parallel handshaking, packet train and acting during waiting period. However, precise transmission plan is still necessary for a better exploiting of large space-time volume offered by long propagation delay. Furthermore, this kind of protocol cannot thoroughly take into account the peculiar features of UWANs in MAC protocol design. A more comprehensive survey can be found in [1].

Several interesting MAC protocol approaches have been proposed to handle the peculiar features of UWANs, including leveraging of long propagation delay, signal-based reservation, scheduling-based MAC and receiver-initiated/centric protocols, while none of them have been tested adequately in practice since underwater field trial is costly. In this case, computer simulation becomes a major validation method, and this situation may last. To make simulation results more convincible, it is necessary to take into account adequately the effect of underwater acoustic channel characteristics on channel capacity, accuracy of time synchronization and localization as well as overheads for physical transmission and information collection. All the above requirements invoke professional communication and network simulation packages developed for UWANs. Actually, there are several simulation packages with specific functions developed for UWANs as reported in the literature, such as WOSS [18] and Aqua-Sim [77]. These packages themselves also need sufficient validation against real environments. It is useful for MAC research to provide verified simulation configurations for some typical underwater networking scenarios.

References

1. Jiang, S.M.: State-of-the-art medium access control (MAC) protocols for underwater acoustic networks: a survey based on a MAC reference model. IEEE Commun. Surv. Tutor **20**(1), 1st Quarter (2018)
2. Syed, A.A., Ye, W., Krishnamachari, B., Heidemann, J.: Understanding spatio-temporal uncertainty in medium access with ALOHA protocols. In Proceedings of the ACM International WS, Underwater Networks (WUWNet), Montreal, Canada (2007)
3. Syed, A.A., Heidemann, J.: Contention analysis of MAC protocols that count. In: Proceedings of the ACM International Conferences Underwater Networks and Systems (WUWNet), Woods Hole, USA (2010)
4. Preisig, J.: Acoustic propagation considerations for underwater acoustic communications network development. In: Proceedings of the ACM International WS, Underwater Networks (WUWNet), Los Angeles, USA (2006)
5. Stojanovic, M.: Underwater acoustic communications: design considerations on the physical layer. In: Proceedings of the Annual Conferences Wireless on Demand Network Systems and Services (WONS), Garmisch-Partenkirchen (2008)
6. Lanzagorta, M.: Underwater communications. MORGAN and CLAYPOOL (2012)
7. Tomasi, B., Toso, G., Casari, P., Zorzi, M.: Impact of time-varying underwater acoustic channels on the performance of routing protocols. IEEE J. Ocean. Eng. **38**(4), 772–784 (2013)
8. Kodithuwakku, J., Letzepis, N., McKilliam, R., Grant, A.J.: Decoder-assisted timing synchronization in multiuser CDMA systems. IEEE Trans. Commun. **62**(6), 2061–2071 (2014)
9. Pompili, D., Akyildiz, I.F.: Overview of networking protocols for underwater wireless communications. IEEE Commun. Mag. 97–102 (2009)
10. Melodia, T., Kulhandjian, H., Kuo, L.-C., Demirors, E.: Advances in underwater acoustic networking. In: Basagni, S., Conti, M., Giordano, S., Stojmenovic, I. (eds.), Mobile Ad Hoc Networking: The Cutting Edge Directions, pp. 804 – 852. Wiley-IEEE Press (2013)
11. Freitag, L., Grund, M., Singh, S., Partan, J., Koski, P., Ball, K.: The WHOI micro-modem: an acoustic communications and navigation system for multiple platforms. Proceedings of the MTS/IEEE OCEANS, Washington, DC, USA **2**, 1086–1092 (2005)
12. Liao, W.-H., Huang, C.-C.: SF-MAC: a spatially fair MAC protocol for underwater acoustic sensor networks. IEEE Sens. J. **12**(6), 1686–1694 (2012)
13. Rice, J., Creber, B., Fletcher, C., Baxley, P., Rogers, K., McDonald, K., Rees, D., Wolf, M., Merriam, S., Mehio, R., Proakis, J., Scussel, K., Porta, D., Baker, J., Hardiman, J., Green, D.: Evolution of Seaweb underwater acoustic networking. Proceedings of the MTS/IEEE OCEANS, Providence, RI, USA **3**, 2007–2017 (2000)
14. Pompili, D., Melodia, T., Akyildiz, I.F.: A CDMA-based medium access control for underwater acoustic sensor networks. IEEE Trans. Wirel. Commun. **8**(4), 1899–1909 (2009)
15. Xie, G.G., Gibson, J.A.: A networking protocol for underwater acoustic networks. Technical Report OMB No. 0704-0188, CS Department, Naval Postgraduate School, 2000
16. Hong, L., Hong, F., Guo, Z.W., Yang, X.H.: A TDMA-based MAC protocol in underwater sensor networks. In: Proceedings of the International Conferences on Wireless Communications, Networking and Mobile Computing (WiCOM), Dalian, China, pp. 1–4 (2008)
17. Hayajneh, M., Khalil, I., Gadallah, Y.: An OFDMA-based MAC protocol for underwater acoustic wireless sensor networks. In Proceedings of the International Conference on Wireless Communications and Mobile Computing and Mobile Computing (IWCMC), pp. 810–814 (2009)
18. Guerra, F., Casari, P., Zorzi, M.: World ocean simulation system (WOSS): a simulation tool for underwater networks with realistic propagation modeling. In: Proceedings of the ACM International WS, Underwater Networks (WUWNet), Berkeley, USA (2009)

19. Parrish, N., Roy, S., Arabshahi, P., Fox, W.: System design considerations for undersea networks: link and multiple access protocols. IEEE J. Sel. Areas Commun. **26**(9), 1720–1730 (2008)

20. Chen, Y.-D., Liu, S.-S., Chang, C.M., Shih, K.-P.: CS-MAC: a channel stealing MAC protocol for improving bandwidth utilization in underwater wireless acoustic networks. In: Proceedings of the MTS/IEEE OCEANS, Waikoloa, Hawaii, USA (2011)

21. Favaro, F., Azad, S., Casari, P., Zorzi, M.: Extended abstract: on the performance of unsynchronized distributed MAC protocols in deep water acoustic networks. In: Proceedings of the ACM International WS, Underwater Networks (WUWNet), Seatle, USA (2011)

22. Peleato, B., Stojanovic, M.: Distance aware collision avoidance protocol for ad-hoc underwater acoustic sensor networks. IEEE Commun. Lett. **11**(12), 1025–1027 (2007)

23. Guo, X.X., Frater, M.R., Ryan, M.J.: A propagation-delay-tolerant collision avoidance protocol for underwater acoustic sensor networks. In: Proceedings of the OCEANS - Asia Pacific, Boston, MA, USA (2006)

24. Molins, M., Stojanovic, M.: Slotted FAMA: a MAC protocol for underwater acoustic networks. In: Proceedings of the MTS/IEEE OCEANS, Boston, MA, USA (2006)

25. Zhou, Z., Peng, Z., Cui, J.-H., Jiang, Z.: Handling triple hidden terminal problems for multichannel mac in long-delay underwater sensor networks. IEEE Trans. Mob. Comput. **11**(1), 139–154 (2012)

26. Hsu, C.-C., Lai, K., Chou, C.-F., Lin, K.C.: ST MAC: spatial-temporal MAC scheduling for underwater sensor networks. In: Proceedings of the IEEE INFOCOM Rio de Janeiro, Brasil, pp. 1827–1835 (2009)

27. Luo, Y., Pu, L., Peng, Z., Zhou, Z., Cui, J.-H.: CT-MAC: A MAC protocol for underwater MIMO based network uplink communications. In: Proceedings of the ACM International Conferences on Underwater Networks and Systems (WUWNet), Los Angeles, USA (2012)

28. Zhou, Z., Le, S., Cui, J.H.: An OFDM based MAC protocol for underwater acoustic networks. In: Proceedings of the ACM International Conference Underwater Networks and Systems (WUWNet), Woods Hole, USA (2010)

29. Sozer, E.M., Stojanovic, M., Proakis, J.G.: Underwater acoustic networks. IEEE J. Ocean. Eng. **25**(1), 72–83 (2000)

30. Catipovic, J., Brady, D., Etchenmendy, S.: Development of underwater acoustic modems and networks. Oceanography **6**(3), 112–119 (1993)

31. Gibson, J., Larraza, A., Rice, J., Smith, K., Xie, G.: On the impacts and benefits of implementing full-duplex communications links in an underwater acoustic network. In: Proceedings of the International Mine Symposium Monterey, CA, USA, pp. 204–213 (2002)

32. Khalil, I., Gadallah, Y., Khreishah, M.H.: An adaptive OFDMA based MAC protocol for underwater acoustic wireless sensor networks. Sensors **12**(7), 8782–8805 (2012)

33. So, J., Vaidya, N.: Multi-channel MAC for Ad Hoc networks: handling multi-channel hidden terminals using a single transceiver. In: Proceedings of the Annual ACM International Conferences on Mobile Computing and Network (MobiCom), Philadelphia, USA, pp. 222–233 (2004)

34. Syed, A.A., Ye, W., Heidemann, J.: T-Lohi: a new class of MAC protocols for underwater acoustic sensor networks. In: Proceedings of the IEEE INFOCOM, Phoenix, Arizona, USA, pp. 789–797 (2008)

35. Wei, X., Zhao, L., Li, X., Zou, C.R.: A distributed power control based MAC protocol for underwater acoustic sensor networks. In: Proceedings of the IEEE International Conferences on Circuits and Systems for Communications (ICCSC), pp. 688 – 692 (2008)

36. You, L.N., Jiang, S.M., Wei, G.: A multi-channel MAC using no dedicated control channels for wireless mesh networks. In: Prof. International Conference on Wireless Communication and Signaling Processing (WCSP), Nanjing, China, pp. 2996–3000 (2009)

37. Ahn, J., Krishnamachari, B.: Performance of propagation delay tolerant ALOHA protocol for underwater wireless networks. In: Nikoletseas, S., Chlebus, B.S., Johnson, D.B., Krishnamachari, B. (eds.), Distributed Computing in Sensor Systems (DCOSS). LNCS, vol. 5067, pp. 1–26. Springer, Berlin (2008)

38. Basagni, S., Petrioli, C., Petroccia, R., Stojanovic, M.: Choosing the packet size in multi-hop underwater networks. In: Proceedings of the MTS/IEEE OCEANS, Sydney, Australia (2010)
39. Stojanovic, M.: Optimization of a data link protocol for an underwater acoustic channel. In: Proceedings of the MTS/IEEE OCEANS, Washington, DC, USA (2005)
40. Vieira, L.F.M., Kong, J., Lee, U., Gerla, M.: Analysis of ALOHA protocols for underwater acoustic sensor networks. In: Proceedings of the ACM International WS, Underwater Networks (WUWNet), Los Angeles, USA (2006)
41. De, S., Mandal, P., Chakraborty, S.S.: On the characterization of ALOHA in underwater wireless networks. Math. Comput. Model. **53**(11–12), 2093–2107 (2011)
42. Su, R.Y., Venkatesan, R., Li, C.: Acoustic propagation properties of underwater communication channels and their influence on the medium access control protocols. In: Proceedings of the IEEE International Conferences on Communication (ICC), Ottawa, Canada, pp. 5015–5019 (2012)
43. Gibson, J.H., Xie, G.G., Xiao, Y., Chen, H.: Analyzing the performance of multi-hop underwater acoustic sensor networks. In: Proceedings of the OCEANS, Europe, Aberdeen, UK (2007)
44. Xiao, Y., Zhang, Y.P., Gibson, J.H., Xie, G.G.: Performance analysis of p-persistent aloha for multi-hop underwater acoustic sensor networks. In: Proceedings of the International Conferences on Embedded Software and System (ICESS), Zhejiang, China, pp. 305–311 (2009)
45. Zhang, Y.P.: Performance of p-persistent slotted Aloha for underwater sensor networks. In: International Conference on Computing, Networking and Communication (ICNC), Honolulu, USA, Feb. 2014, pp. 583–587
46. Xie, P., Cui, J.-H.: Exploring random access and handshaking techniques in large-scale underwater wireless acoustic sensor networks. In: Proceedings of the MTS/IEEE OCEANS, Boston, MA, USA (2006)
47. Climent, S., Sanchez, A., Capella, J.V., Serrano, J.J.: Simulating MAC protocols under real underwater sensor networks assumptions. In: Proceedings of the ACM International Conferences on Underwater Networks and Systems (WUWNet), Los Angeles, USA (2012)
48. Casari, P., Tomasi, B., Zorzi, M.: A Comparison between the Tonelohi and Slotted FAMA MAC protocols for Underwater Networks. In: Proceedings of the MTS/IEEE OCEANS, Quebec City, Canada, pp. 381–384 (2008)
49. Petrioli, C., Petroccia, R., Potter, J.: Performance evaluation of underwater MAC protocols: from simulation to at-sea testing. In: Proceedings of the MTS/IEEE OCEANS, Santander, Spain, pp. 1–10 (2011)
50. Shahabudeen, S., Motani, M.: Short paper: performance analysis of a MACA based Protocol for Adhoc underwater networks. In: Proceedings of the ACM Internatinal WS, Underwater Networks (WUWNet), Berkeley, USA (2009)
51. Shahabudeen, S., Motani, M., Chitre, M.: Analysis of a high-performance MAC protocol for underwater acoustic networks. IEEE J. Ocean. Eng. **39**(1), 74–89 (2014)
52. Yang, M., Gao, M.S., Foh, C.H., Cai, J.F.: DC-MAC: a data-centric multi-hop MAC protocol for underwater acoustic sensor networks. In: Proceedings of the IEEE International Symposium on Computers and Communications/ (ISCC)., Kerkyra, pp. 491–496 (2011)
53. Zhu, Y.B., Jiang, Z.H., Peng, Z., Zuba, M.: Toward practical MAC design for underwater acoustic networks. In: Proceedings of the IEEE INFOCOM, Turin, pp. 683–691 (2013)
54. Noh, Y., Wang, P., Lee, U., Torres, D., Gerla, M.: DOTS: a propagation delay-aware opportunistic MAC protocol for underwater sensor networks. In Proceedings of the IEEE International Conference on Network Protocols (ICNP), Kyoto, Japan, pp. 183–192 (2010)
55. Rodoplu, V., Park, M.K.: An energy-efficient MAC protocol for underwater wireless acoustic networks. In: Proceedings of the MTS/IEEE OCEANS, Washington, DC, USA (2005)
56. Zheng, L., Cai, L.: Flipped diversity aloha in wireless networks with long and varying delay. In: Proceedings of the IEEE Global Tele. Conferences (GLOBOCOM), Houston, TX, USA, pp. 1–5 (2011)
57. Zhu, Y.B., Peng, Z., Cui, J.H., Chen, H.: Toward practical MAC design for underwater acoustic networks. IEEE Trans. Mob. Comput. **14**(4), 872–886 (2015)

58. Bandyopadhyay, S., Coyle, E.J.: An energy efficient hierarchical clustering algorithm for wireless sensor networks. In: Proceedings of the IEEE INFOCOM, San Francisco, USA, pp. 1713–1723 (2003)

59. Chirdchoo, N., Soh, W.-S., Chua, K.C.: RIPT: a receiver-initiated reservation-based protocol for underwater acoustic networks. IEEE J. Sel. Areas Commun. 26(9), 1744–1753 (2008)

60. Syed, A.A., Ye, W., Heidemann, J.: Comparison and evaluation of the T-Lohi MAC for underwater acoustic sensor networks. IEEE J. Sel. Areas Commun. 26(9), 1731–1743 (2008)

61. Hsu, C.-C., Kuo, M.S., Chou, C.F., Lin, K.C.: The elimination of spatial-temporal uncertainty in underwater sensor networks. ACM/IEEE Trans. Netw. 21(4), 1229–1242 (2013)

62. Noh, Y., Lee, U., Han, S.V., Wang, P., Torres, D., Kim, J., Gerla, M.: DOTS: a propagation delay-aware opportunistic MAC protocol for mobile underwater networks. IEEE Trans. Mob. Comput. 13(4), 766–782 (2014)

63. Fullmer, C.L., Garcia-Luna-Aceves, J.J.: Solutions to hidden terminal problems in wireless networks. In: Proceedings of the ACM SIGCOMM, Cannes, France (1997)

64. Rodoplu, V., Park, M.K.: UWAN-MAC: an energy-efficient MAC protocol for underwater acoustic wireless sensor networks. IEEE J. Ocean. Eng. 32(2), 710–720 (2007)

65. Casari, P., Lapiccirella, F.E., Zorzi, M.: A detailed simulation study of the UWAN-MAC protocol for underwater acoustic networks. In: Proceedings of the MTS/IEEE OCEANS, Aberdeen, UK (2007)

66. Gollakota, S., Katabi, D.: Zigzag decoding: combating hidden terminals in wireless networks. In: Proceedings of the ACM SIGCOMM, New York, NY, USA, pp. 159–170 (2008)

67. Kilfoyle, D.B., Preisig, J.C., Baggeroer, A.B.: Spatial modulation experiments in underwater acoustic channel. IEEE J. Ocean. Eng. 30(2), 406–415 (2005)

68. Li, Z., Guo, Z., Qu, H., Hong, F., Chen, P., Yang, M.: UD-TDMA: a distributed TDMA protocol for underwater acoustic sensor networks. In: Proceedings of the IEEE International Conference on Mobile Adhoc and Sensor Systems (MASS), Macau, China, pp. 918–923 (2009)

69. Yu, Y., Giannakis, G.B.: High-throughput random access using successive interference cancellation in a tree algorithm. IEEE Trans. Inform. Theory 53(12), 4628–4639 (2007)

70. Zheng, L., Cai, L.: AFDA: asynchronous flipped diversity ALOHA for emerging wireless networks with long and heterogeneous delay. IEEE Trans. Emerg. Top. Comput. 3(1), 64–73 (2015)

71. Jiang, S.M.: A logical MIMO MAC approach for uplink access control in centralized wireless. In: Proceedings of the IEEE International Conferences on Communication Systems (ICCS), Guangzhou, China (2008)

72. Jiang, S.M.: Future Wireless and Optical Networks: Networking Modes and Cross-Layer Design. Springer, London (2012)

73. Li, B., Zhou, S., Freitag, L., Stojanovic, M., Willett, P.: Multicarrier communication over underwater acoustic channels with nonuniform doppler shifts. IEEE J. Ocean. Eng. 33(2), 198–209 (2008)

74. Sklar, B.: Digital Communications: Fundamentals and Applications 2nd edn. Prentice-Hall (2002). ISBN 7-5053-7870-8

75. Diamant, R., Shirazi, G., Lampe, L.: Robust spatial reuse scheduling in underwater acoustic communication networks. In: Proceedings of the. IEEE Vehicular Technology Conference (VTC) - Fall, San Francisco, CA, pp. 1–5 (2011)

76. Kredo, K., II, Djukic, P., Mohapatra, P.: STUMP: exploiting position diversity in the staggered TDMA underwater MAC protocol. In: Proceedings of the IEEE INFOCOM, Rio de Janeiro, Brasil (2009)

77. Martin, R., Zhu, Y.B., Pu, L., Dou, F., Peng, Z., Cui, J.H., Rajasekaran, S.: Aqua-sim next generation: a NS-3 based simulator for underwater sensor networks. In: Proceedings of the ACM International Conferences on Underwater Networks and Systems (WUWNet), Arlington, VA, USA (2015)

Chapter 11
Routing in UWANs

Abstract This chapter reviews some typical UWAN routing protocols in terms of routing strategies, key issues addressed and basic idea of the proposed schemes as well as their feasibility in UWANs, according to the typical application scenarios of underwater networks.

11.1 Overview

This section discusses briefly the major challenges facing UWAN routing protocol design and category of the proposed protocols.

11.1.1 Primary Challenges

UWANs have some unique characteristics as mentioned in Chap. 9, which will affect routing protocol design. They also cause many existing routing protocols developed for terrestrial RWNs unsuitable for UWANs directly as explained below.

- The very small bandwidth and high power consumption for both transmission and reception make the following operations and schemes undesirable: frequent broadcast-based path search, redundant packet forwarding and schemes with large packet overhead. Therefore, geographic routing protocols, which exploit relative position information such as depth or pressure without exchanging link state information or requiring route maintenances, become a favourable option [1].
- Many routing protocols assume symmetric links, with which a reverse path will be automatically available in parallel with the forwarding path. However, the asymmetric feature of underwater links causes such kind of protocol unable to work properly in UWANs. In this case, a routing protocol designed for UWANs cannot only use handshake schemes for path search because the source node may not receive routing reply only with handshaking. An asymmetric link scenario may be caused by a directional beam with less than 360°-width rather than an omni-directional coverage (see Fig. 11.6), or by an upslope bathymetric profile,

© Springer Nature Singapore Pte Ltd. 2018

S. Jiang, *Wireless Networking Principles: From Terrestrial to Underwater Acoustic*,

https://doi.org/10.1007/978-981-10-7775-3_11

in which, the channel may experience much higher bit error rates (BER) in the direction toward the upslope than in the opposite direction [2].

- The dynamics and unreliability of a underwater acoustic channel due to multipath fading and interference as well as random terminal mobility caused by current and wind may happen frequently along a long latency path due to the slow acoustic propagation speed. In this case, a proactive routing protocol to secure an end-to-end path before data transmission may not work well because network connect intermittency may occur during the search process. Furthermore, in a rapidly changing channel, a channel status probed by short control packets may not be suitable for transmitting long packets since a long packet can be more easily affected by errors [3].
- The bandwidth of an underwater acoustic channel depends on the frequency as well as the distance to the receiver [4, 5], i.e., the shorter the communication distance, the higher the channel bandwidth. Thus, a routing protocol cannot use a long-distance link in constructing a path to satisfy the bandwidth requirement of applications. Although a multi-hop routing protocol can make sense in this case, the power consumption due to increased transmission and reception operations should be also taken into account to prolong UWAN lifetime.
- Since radio signal cannot propagate long enough in underwater environments, GPS cannot be used to locate underwater nodes. In this case, a routing protocol for UWANs cannot assume the availability of the information on the precise location of underwater nodes. It is possible for a node to have the information on the coarse relative position of a node through measuring the strength and angle of arrival (AOA) of the received signal. However, it is difficult for a an underwater node to know its orientation without help of other location services.

11.1.2   Protocol Category

To minimize the consumption of bandwidth and energy caused by either message broadcast for path search or flooding-like routing, geographic routing [6] is often adopted in UWANs because it can exploit some (relative) position information of underwater nodes to guide path search and packet forwarding. To handle intermittent connectivity caused by the mobility in sparse UWANs and dynamics of acoustic link quality, opportunistic routing is also often adopted to maximize packet delivery based on "store-carry-forward" schemes. Therefore, many UWAN routing protocols are based on these two routing approaches.

There are many routing protocols proposed for UWANs so far along with several survey papers available in the literature such as [7]. Although several routing strategies are investigated for UWANs, one fact is that it is difficult to design a couple of routing protocols that can support all underwater application scenarios with the expected performance [8]. This results because different applications may require different network topologies, which largely affect routing protocol design. Actually, the major applications of UWANs are underwater data collection, by which, the data

Table 11.1 Typical application scenarios of UWAN routing protocols and proposals

Nodal mobility	One sink		Multiple sinks		Zero sink
	Stationary	Mobile	Stationary	Mobile	
Stationary	[9]	Mobicast [10]	VBF [11]	[12]	[13]
Mobile	[14]	SEA [15]	DBR [16]		[17]

collected by underwater nodes are transferred to the surface nodes for processing. The surface node is also called sink, and can be either stationary or mobile. One or multiple sinks may be used in one UWAN. A group of AUVs can also form a UWAN for cooperative operation in an ad hoc networking mode without using fixed sinks.

The routing protocols under review is organized according to the network topologies corresponding to the above application scenarios listed in Table 11.1 by considering the mobility of both sink and underwater nodes. From the following discussion, we will find that only the following scenario has not been considered by UWAN routing protocols: multiple mobile sinks with mobile underwater nodes. Actually, compared to anchored underwater nodes, those floating in water do show small-scale mobility with current because their motion regions are under control. This application scenario has been considered by many routing protocols. What is really missing is large-scale mobility, with which, a underwater node can move anywhere without limitation, like wave gliders. Usually, a group of such nodes may form a Mobile Ad Hoc Network (MANET), and there is no need to have a long-term sink in this case. This scenario is similar to mobile UWANs without specified sinks.

11.1.2.1 Geographic Routing

One major advantage of the geographic routing [6] is that the routing operations are localized and do not require route discovery or route maintenance (i.e., stateless routing), which allows simple greedy routing to be used to forward packets to the destination according to the location information [2, 15]. However, the complexity and cost of a geographic routing protocol depends on the type of location information that the protocol relies on. According to the reference point used to measure the location, the location information can be classified into absolute location and relative position. With the former, every node in the network adopts a common reference point to measure their locations, such as those provided by the GPS. This type of location information is meaningful to every node. With the latter, a node measures its neighbor's location with only reference to itself. The primary location information includes the relative distance and the angle of arrival (AOA) of signals from this node to the measured node, and is meaningful only to the measuring node.

As mentioned earlier, the GPS signal cannot propagate long enough in UWANs, and an underwater node cannot have the information on absolute locations for either itself or other underwater nodes. The water surface node can get its absolute position information through GPS, and update it to the underwater nodes by sending messages to them. This node can be used by other nodes as a reference to localize roughly themselves by jointly using a method to obtain the relative position information mentioned above. This method can work well in a small UWAN, in which the surface node can reach every underwater node, whereas routing in this case is not difficult. In a large UWAN, the above method cannot work well alone, and message exchange between underwater nodes is needed, which will consume more bandwidth and energy. If every node is stationary, the position information of each node can be pre-measured and pre-configured in each node during the network deployment. However, this method is not scalable and does not work for mobile UWANs. To solve this problem, location service protocols have been studied, through which, a node can get the information on the absolute position of a requested node.

11.1.2.2   Location Service

Inspired by the pheromone trail of ants, a bio-inspired Phero-Trail location service protocol is studied for the Sensor Equipped Aquatic (SEA) Swarm [15]. As illustrated in Fig. 11.1a, the protocol uses a 2D trajectory projection of the mobile sink to point to its location. To this end, the mobile sink forwards its current location vertically upwards to the upper hull, which is a collection of the nodes deployed to store the 2D trajectory projection reported by the mobile sink. The hull provides location service as requested, and a hull at zero depth is just the water surface. The sequence of projections like a pheromone trail will be slowly diffused in the water current through controlling its length with a pheromone expiration timer. As illustrated in

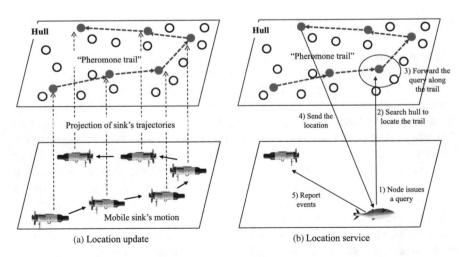

(a) Location update   (b) Location service

Fig. 11.1 Phero-Trail location service protocol for SEA swarm [15]: **a** As the mobile sink moves, it sends location updates to the mobile nodes on the hull; **b** A query is forwarded to the node holding the current location of the mobile sink along the trail

Fig. 11.1b, a querying node issues a search packet to the hull for a location query, and the receiving node on the hull searches the trail and then forwards the search packet to the node that has the current location of the mobile sink along the trail. Then this node updates this location information to the querying node, which then reports the event to the mobile sink according to its current position information.

The provisioning of location service in UWANs is not trivial because there is not a common positioning facility available to every underwater node. In the terrestrial environment, the same GPS service is available everywhere, and can be used by all kinds of wireless network nodes equipped with a GPS signal receiver. However, in underwater environments, each UWAN will require an independent location service system probably consisting of a set of nodes floating on the water surface, which will significantly increase the implementation cost of a UWAN. Once the absolute position information is available, some geographic routing protocols proposed for terrestrial wireless networks can be used with some enhancement to handle asymmetries of underwater acoustic links. On the other hand, it is relatively easy for a node to have the information on the relative position of its neighbors. Most data transmissions in UWANs are vertical upward to the water surface, while underwater depth and water pressure have a deterministic relationship. Thus, there are many UWAN routing protocols that rely on the relative position information.

11.1.2.3 Opportunistic Routing

In UWANs, limited communication distances and sparse node distributions due to high cost of underwater nodes will easily cause connection intermittence. When all nodes are stationary, either being mounted on the bottom or fixed to pillars, connection intermittence is mainly caused by abnormal operations of nodes. In this case, there is no opportunity to resume the network connection until some artificial interventions are invoked. If some nodes are mobile, such as AUVs used as data mule, these nodes can cruise the area and replace the mis-operational nodes temporarily, and even to collect the data directly without further routing. In the latter case, opportunistic routing is unnecessary. If all nodes are mobile, such as a team of AUVs, opportunistic routing can be exploited to forward packets in case of connection intermittence, which will become easier by jointly using the GPS and satellite communication if available [18].

11.1.2.4 Centralized Routing

A stationary sink is not only a common destination of the underwater nodes, but sometimes can function as a central unit to coordinate routing. For example, with one of the earliest UWAN routing protocols proposed in [9], a master node on the surface is used to establish a network topology by broadcasting a probe to its neighbors. Each receiver appends its identity to this probe and re-broadcasts the probe to its neighbors. As this process continues, a connected tree rooted from the master is created, over which routing is performed. A similar process has also been proposed for Seaweb in [19]. Actually, the principle of this kind of routing has no much difference from that used in terrestrial (wireless) local area networks so that its detail is ignored here. However, this kind of protocol usually assumes the availability of a reverse path in the opposite direction of the forward path used by the central unit to collect topology information. It will not work with asymmetric acoustic links, and cannot work properly with mobile nodes, which may cause frequent changes in network topology.

11.2   Stationary Sinks

A stationary sink can be installed either in a pillar or in a stationary buoy. A pillar can be fixed, while a buoy may move or shift following current or wind within a predictable area. In the following, we introduce some geographic UWAN routing protocols, and those taking into account asymmetric acoustic links and exploiting multipath features in protocol design, as well as cross-layer designed protocols.

11.2.1   Relative Position-Based Routing

Many UWAN routing protocols are based on relative position, and are among the earliest routing protocols proposed for UWANs.

11.2.1.1   Vector-based-Forwarding (VBF)

The Vector-based-Forwarding (VBF) routing protocol [11] is a hop-by-hop routing protocol. It allows a node to make a routing decision for a received packet based on its position relative to the routing vector set between the source and the sink, e.g., $\overrightarrow{S_1 S_0}$ as illustrated in Fig. 11.2. Each node has to measure the distance relative to the packet relaying node (called forwarder) and the angle of arrival (AOA) of the signal from the forwarder. Each packet needs to carry the information on the positions of the source, the sink and the forwarder as well as the radius of routing pipe, i.e., W in the figure. Whether or not a receiving node will continue the forwarding of a received packet depends on how much it is close to the routing vector. Once receiving a packet, the node computes its distance relative to the routing vector. If the estimated distance is less than a predefined threshold, the node is qualified to forward this packet; otherwise, it simply discards the packet.

To reduce unnecessary duplicated forwarding of the same packet, each qualified node that has received a packet needs to hold it for a time period so that the most optimized node can be determined to forward it. This time period is mainly determined

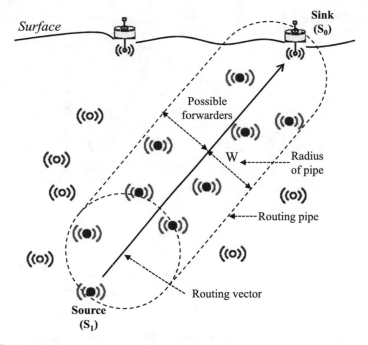

Fig. 11.2 Principle of vector-based-forwarding (VBF) [11]

by the relative distance between the current node under consideration and the forwarder of the packet, W and transmission range. The forwarded packet carries the position of the current forwarder. Furthermore, a localized and distributed self-adaptation algorithm is developed to allow nodes to weight the benefit of forwarding packets to avoid flooding of every packet by discarding low benefit packets to reduce energy consumption.

The Hop-by-Hop VBF (HH-VBF) [20] aims to overcome the following weaknesses of VBF. The availability of only single end-to-end routing pipe with the same radius affects routing performance. HH-VBF tries to set a pipe hop-by-hop. For both VBF and HH-VBF, it is necessary to estimate the positions of underwater nodes [2] because it is difficult to use only the locally available information on the relative position to calculate the routing vector.

11.2.1.2   Depth-Based Routing (DBR)

A stateless Depth-Based Routing (DBR) is investigated in [16] for a mobile UWAN with multiple stationary sinks randomly deployed on the water surface. All nodes forward data packets greedily to the sinks. Its basic idea is that a node will continue to forward a received packet only if its depth is smaller than that carried by the packet. Thus, each data packet carries the information on the depth of its recent forwarder,

Fig. 11.3 Depth-based
routing (DBR) [16]

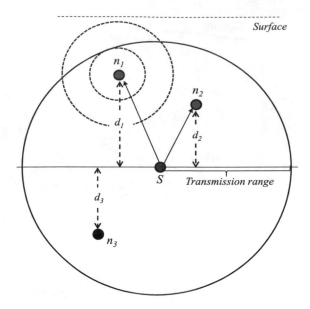

and is updated at every hop. As illustrated in Fig. 11.3, node S is the sender, and
nodes $n_1 \sim n_3$ are within its transmission range indicated by the solid line circle.
When node S sends a packet, n_1, n_2 and n_3 all will receive it. Node n_3 will discard the
packet since it is below S. Among n_1 and n_2, obviously, n_1 should continue the packet
forwarding, while n_2 should be prevented. To this end, a received packet should not
be relayed immediately and will be held for some time. The packet holding time is
calculated based on the depth of the receiving node and the difference between the
forwarder's and the receiver's depths. The difference between the holding times of
two neighboring nodes should be large enough to allow receivers to hear timely the
forwarding of a receiver with the smaller depth to avoid redundant relay.

The major problem of DBR is a void region probably encountered by a non-sink
forwarding node, which cannot find any neighbors with a lower pressure than its own
pressure (called local maximum), resulting in a dead-end. This problem is handled in
[21] by allowing every local maximum node to secure a node whose depth is lower
than its current depth, and has a recovery path to reroute the packets to another node.
This node may be either another local maximum node having a new recovery path,
or a point able to greedily forward packets. As illustrated in Fig. 11.4, there are two
local maximum nodes: LM_1 and LM_2. A packet can be re-routed from LM_1 to LM_2
to the sink, and will be delivered to a node on the ocean surface.

The void region problem is eliminated by the following enhancement, called
Void-Aware Pressure Routing (VAPR) [1, 21]. It employs the information on surface
reachability to set up the next-hop direction toward the surface for each node, so that
local opportunistic directional forwarding can always be used for data packet delivery
even in the presence of a void. To this end, a partial view of the network topology

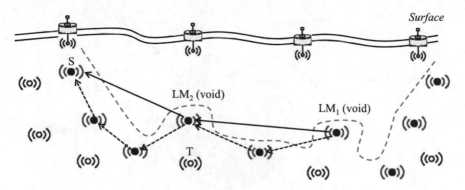

Fig. 11.4 Void-aware pressure routing (VAPR): recovery mode of the local maximum node [1]

is provided to each node, and the surface sonobuoy uses periodic beacon to build directional trails toward the surface, along which nodes can adopt a geo-opportunistic forwarding scheme.

11.2.2 Priority Routing

The routing protocol proposed in [14] routes packets adaptively based on the types of messages and application requirements for a mobile UWAN with a stationary sink. It exploits message redundancy and resource reallocation to allow more important packets to obtain more resources so that they can be delivered earlier. To this end, each packet in transit is assigned a priority based on its status (e.g., emergency level and residual lifetime) and the node status (e.g., residual battery energy and neighbor density). The higher priority of a packet in transit, the more copies of its duplicate will be sent. The underwater nodes are deployed at different depths and controlled by buoyancy, and nodes located at the same depth are belong to the same layer. A sink deployed in the center of a pre-defined area on the water surface. It assumes that each node know its 3D position through some localization service.

For example, the whole routing spectrum can be divided into 4 routing states each corresponding to one of the following priority scales: [0, 25], (25, 50], (50, 75] and (75, 100]. These four routing states are indexed sequentially from 1 to 4. When a node choosing routing state i ($i = 1 \ldots 4$), it can send i copies of the corresponding packet. In Fig. 11.5, node S is the sink, and node A, whose neighbors include nodes B, C, D and E, needs to make a routing decision based on the routing states of its packets in transit. Here R is the transmission range of a sensor node. The forwarding area i corresponding to a packet at routing state i is a sphere tangent to A's transmission sphere (ATS) at the intersection point of AS and the ATS. A higher priority with a larger routing state has a larger forwarding area. Here, node A with a packet at priority of 45 corresponding to routing state 2 first forwards the packet to one node

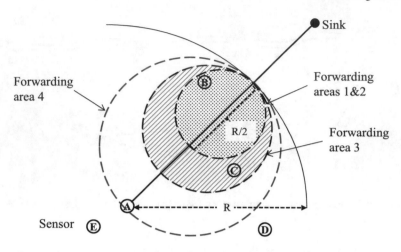

Fig. 11.5 An example of priority routing [14]

in the forwarding area 2 (e.g., node B). Then it waits until another node moving into forwarding area 2, and will forward another copy of this packet to this node. Thus, total two duplicates of this packet will be forwarded corresponding to priority 2. Neighboring nodes at the upper layer are closer to the sink, and will be selected first for packet forwarding.

11.2.3 Asymmetric Link Connectivity

Consider the following asymmetric link connectivity scenario, which is caused by directional signal with a less-than-360° beam width emitted from a practical modem rather than a uniform omni-directional signal assumed by many routing protocols. The beam width can be 120° for a wide beam and 210° for an omni-directional transmission, both of which are less than 360°. As illustrated in Fig. 11.6, although nodes A, B and C are in their mutual transmission ranges, due to their less-than-360° beam widths, only A can be heard by B and C, while there is no connectivity in other directions. In this case, routing protocols using handshaking between a pair of adjacent nodes cannot work well.

A link-state-based Adaptive Feedback Routing (LAFR) protocol discussed in [22, 23] for single sink UWAN takes into account the impact of the above asymmetric link connectivity in routing protocol design. LAFR is a table-driven routing protocol, consisting of route discovery and routing maintenance schemes. The routing discovery process contains a mechanism called link detection, which is specifically designed to detect the symmetry of link connectivity. To this end, a node needs to periodically broadcast a link detection message, which carries its identity and the list of the senders that have sent out link detection messages successfully received by the

Fig. 11.6 Scenarios for asymmetric link connectivity [22–24]

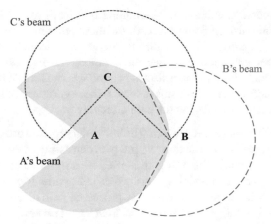

Nodes B and C can hear node A, not vice versa; B cannot hear C.

node. Upon receiving a link detection message, the receiver can identify whether the connectivity of the corresponding link is symmetric by checking whether its identity is in the list of the senders carried by this message.

11.2.4 Multipath Routing

Multipath routing employs more than one path between a source and the destination to improve routing performance in terms of reliability, throughput and delay for packet delivery, transmitting packet replicas along these paths simultaneously. This kind of routing protocol is particularly useful in a harsh network environment, where the quality of an end-to-end path is not stable with possible intermittent connectivity.

A multipath routing protocol with multiple sinks first discovers link- or node-disjoint routes, which are used to support multiple transmissions based on the network topology [25, 26] established during the route discovery phase. Many routing protocols have a broadcast-based route discovery phase, during which all possible paths can be found. A link- or node-disjoint route means that the related link or node is not part of other simultaneously active routes. As discussed in [27], most multipath routing protocols proposed for terrestrial wireless networks assume interference-free transmissions available at the data link layer using MAC protocols [28]. However, the design of MAC protocols for UWANs is a challenging issue as discussed earlier, and it is difficult to have a cost-effective MAC protocol that can assure collision-free reception. In this case, most multipath routing protocols proposed for terrestrial wireless networks cannot perform as expected.

There are several multipath routing protocols proposed for UWANs as summarized in [27]. These protocols either assume bidirectional links (i.e., symmetrical links) [29], which may not exist in UWANs, or only consider link- or node-disjointness in determining multipath route [30, 31], which is insufficient to minimize collision especially without efficient MAC protocols in place.

A proposal in [27] aims to directly choose multipath routes that cause little cross-path interference to other routes among the discovered multipath routes. It adopted a source-initiated path discovery similar to those of AODV and DSR, with the following additional mechanisms to handle the asymmetry of link connectivity and cross-path interference during the path discovery process.

- To facilitate discovering link-disjoint routes, a broadcast path discovery packet carries the information of a full path consisting of the address of each node composing the path. This full path is used by an immediate node to discard path discovery packets describing non-link-disjoint paths (up to the current intermediate node) with respect to the already processed path discovery requests.
- The full path is used to construct a path reversely back to the source, which is used to send a path reply for each discovered path. This procedure can avoid selecting a path with unidirectional links, over which a reply along this kind of path cannot reach the source node.
- To minimize cross-path interference, before forwarding a path reply to the source, every intermediate node checks whether its neighbors have already transmitted a path reply for the same path discovery request, and flags the path reply accordingly if any. The source node discards flagged ones, and uses all other routes described in the collected path reply packets. When all routes carried in the reply packets are flagged, the path with the lowest end-to-end delay is selected.

11.2.5 Cross-Layer Design

Some UWAN routing protocols jointly uses lower layer functions to improve routing performance, such as the power control of the physical layer for network topology control.

11.2.5.1 Focused Beam Routing (FBR)

The Focused Beam Routing (FBR) protocol [32, 33] is cross-designed with the physical layer for a mobile UWAN with multiple stationary sinks. It assumes that each node knows its own position and those of the destinations (i.e., sinks). The initial transmission is performed at the lowest power level, and each node can adjust its transmission power following a set of different power levels when necessary. Before a node transmits a data packet, a short control packet (RTS) is broadcast for neighbor search at the initial power level, and each node that receives this RTS should reply to the sender with another short control packet (CTS). If the sender fails in receiving any CTS corresponding to the RTS, it increases its transmission power recursively, and tries again subject to the maximum power level. If a transmitting

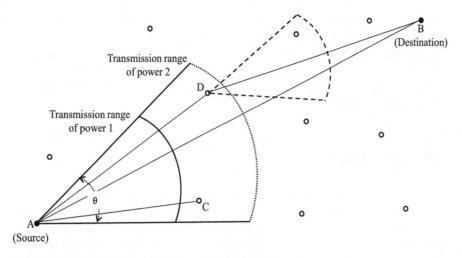

Fig. 11.7 An example of focused beam routing (FBR) [32]

node cannot receive any CTS corresponding to its RTS at the maximal transmission power level, it will shift its cone, and try again to look for neighbors from the left to the right of the main cone. If there are multiple responses, the one closer to the sink is chosen as the next hop for relay. Note that, here RTS/CTS handshaking is regarded as the MAC layer mechanism, but it cannot completely avoid reception collision at the node. It is also used by the routing protocol for neighbor discovery.

As illustrated in Fig. 11.7, where source A wants to transmit to destination B, node A issues an RTS to its neighbors, which contains the locations of nodes A and B. Upon receiving the RTS, the node first calculates its location relative to the AB line to determine whether it is suitable for packet relaying. If the node lies within a cone of angle $+\frac{\theta}{2}$ emanating from node A towards node B, then it is suitable for relaying, and responds node A with a CTS. However, if there are no suitable nodes covered by the transmission range at transmission power level 1 (P_1), then node A increases the transmission power to P_2 and sends a new RTS, which can reach nodes D and C. In this case, both nodes D and C send a CTS to A. Since C is closer to A than D, A forwards data packet to D, which will follow the same procedure to decide the next hop for packet forwarding.

11.2.5.2 Channel-Aware Routing

A data packet can be orders of magnitude longer than a control packet so that a data packet can be more easily affected by transmission errors, and channel quality measured by experience of short packets may not be certainly suitable for transmitting long packets. Thus, in [3], a channel-aware routing protocol (CARP) is proposed to route data packets along a multi-hop path to the sink. It explicitly takes into account link quality in the relay selection, i.e., selecting a relaying node according to the history of successful transmissions received by the sink. CARP also tries to take advantage of modem power control when available to select proper transmission power so that shorter control packets will experience packet error rate similar to that experienced by longer data packets in order to select robust and reliable links for data packet forwarding.

11.3 Mobile Sinks

A mobile sink is a mobile agent such as an AUV, which can cruise a large area to collect data stored in underwater nodes. It is usually equipped with more sophisticated communication and computing facilities with more power supply. It can save communication energy by avoiding multiple-link communication, and achieve high data rates with short link communication. For example, a Sensor Equipped Aquatic (SEA) swarm system reported in [15] is a group of mobile underwater sensors moving with water current. Each sensor monitors local underwater activities, and reports critical data or events in realtime to a distant mobile sink through multi-hop routing. A location service is needed to provide the dynamic position of the mobile sink to active underwater source nodes.

11.3.1 Routing in 3D UWANs

Here the mobile sink moves according to a pre-defined path to collect data from sensors developed in a 3D underwater environment. A distinctive feature of data delivery in this case is that the engagement between the mobile sink and a set of sensor nodes for data collection is temporary in short-term. That is, a set of nodes can only happen to be in the space region covered by the mobile agent at a particular time point. Such encounter can last for some time, which is called encountering interval denoted by $[t_{start}, t_{end}]$), and depends on the relative speed between the mobile sink and the sensor nodes. So, the data delivery between them have to be completed during such encountering period without missing any nodes. If some nodes fail in contact with the mobile agent during this period, we say there are holes.

A mobile geocast scheme (Mobicast) is investigated in [34] to overcome the above hole problem and minimize energy consumption of sensor nodes, while maximizing data collection in a UWAN with one mobile sink to collect data. The underwater nodes are randomly distributed and may be drifted by ocean current. This scheme adopts a geographic zone prescribed by an AUV, called 3D zone of relevance (ZOR^3). It defines an area, in which the mobile sink will collect data from all sensor nodes during the encountering interval. Assume that an AUV travel along a circle in a given observed area. The AUV constructs a series of $ZOR^3 s$ crossing over different encountering intervals. Only sensor nodes located in the ZOR^3 at the actual time

Fig. 11.8 3D ZOR (ZOR^3) and 3D ZOF (ZOF^3) for Mobicast protocol [34]

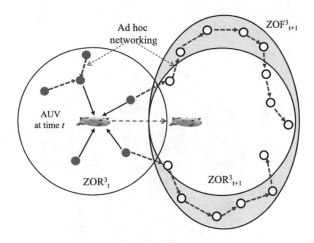

interval will wake up to send data to the AUV. Nodes outside of a ZOR^3 will still keep connected to nodes within the ZOR^3 through ad hoc networking, and this extended geographic zone is called 3D zone of forwarding (ZOF^3). Such an extended ZOR^3 can cover all sensor nodes to be present in the upcoming encountering interval. The AUV delivers a mobicast message when it is in the ZOR^3 at time t (ZOR_t^3) to wake up all sensor nodes to be present in the ZOR_{t+1}^3 as illustrated in Fig. 11.8.

11.3.2 Sector-Based Routing

The Sector-based Routing with Destination Location Prediction (SBR-DLP) [35] adopts a sector approach to reduce collision caused by reply messages in searching a mobile destination node. The destination node is assumed to move with pre-planned movements known to all other nodes. It also periodically broadcasts a message to notify its one-hop neighbors if it deviates from its schedule significantly so that its location can be predicted by other nodes. SBR-DLP routes a packet to the destination node hop-by-hop without need of an end-to-end path to the destination.

As illustrated in Fig. 11.9, when node S wishes to send or relay a packet to destination D, it needs to find a next relaying node by broadcasting a search packet. A neighbor responds to node S if it is nearer to node D than the distance between nodes S and D, using the predicted location of node D. To reduce reply collisions at node S, each neighbor first determines the sector of node S that it is located in, which is used to schedule the transmission of its reply. To this end, the node locates the first sector by ensuring that the sector is bisected by line SD, and the subsequent sectors are determined according to their angular differences from SD.

All sectors are labeled according to their priorities for packet relay. After determining the sector, the neighbor writes its sector number (j) and its estimated distance from the predicted destination location in the reply to node S. It then schedules the

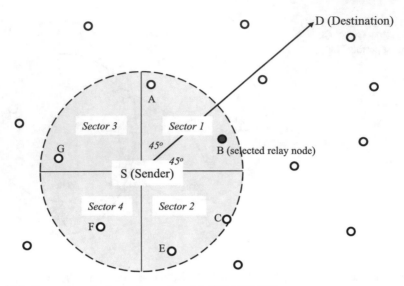

Fig. 11.9 Forwarder selection at the sender with SBR-DLP [35]

transmission to occur after an offset given by $\alpha(j-1)\tau_{max}$, where $0 \le \alpha \le 1$ and τ_{max} is the maximum propagation delay. α is determined according to the sector number, i.e., the larger this number, the smaller α should be. Then, node B is selected as the relaying node.

11.3.3   Multiple Mobile Sinks

Several proposals suggest utilizing multiple mobile agents (i.e., either a relay or a sink) in large stationary UWANs. In [12], a 2D UWAN consists of sensor nodes anchored to the seabed, and is divided into several clusters. Each cluster has a sink, and the sinks are connected by fibre optics. Cluster creation is independent of mobile agents, which are used in the following scenarios. It can be used to fill up a gap between nodes to enhance the connectivity of the UWAN. A gap may be temporal due to occasional changes in channel conditions. Once the gap disappears, it leaves. If a gap is long-term due to the failure of some nodes, it is used as a replacement. When the network is partitioned, a mobile agent can be used as a mobile sink to restore the connectivity. Thus, mobile agents can improve routing performance if they can be deployed rapidly in the right site when necessary.

In [36], multiple mobile sinks are used as "virtual cluster heads" for data collection from underwater sensors in a 2D UWAN. When a mobile sink approaches to underwater nodes, they will organize themselves to form a temporary cluster headed by the mobile sink, and the nodes send data to the head. A routing protocol using four mobile sinks for a 3D UWAN, MobiSink, is investigated in [37]. A 3D area is divided into four regions each at different depths, in which nodes are deployed randomly. A mobile sink cruises horizontally in its region to collect data, and forward the collected data to the base station on the water surface. However, in both cases, routing is not a difficult issue.

11.4 No Specified Sinks

In this case, routing between any pairs of nodes in a UWAN is possible. Here no sink is specified, and a sink if any is treated identically as other nodes in terms of routing function, which is similar to terrestrial wireless ad hoc networks. In the following, we introduce four UWAN routing protocols, which either consider some special features of underwater acoustic channels in routing protocol designs or try to make routing protocols adaptive to underwater environments.

11.4.1 Non Line-of-Sight (NLOS)

With a greedy forwarding protocol, each packet is forwarded to the next hop that is the closest to the destination node. However, such a node may have no link connecting the destination due to a blocked line-of-sight (LOS) path. In [38], a non-LOS (NLOS) link based on surface reflection, which is the interface between water and air, is used to overcome the obstacle and noisy LOS links. Such a link can also boost simultaneous transmissions and minimize interferences between different paths. As illustrated in Fig. 11.10, where BC, CF, DE, DC and DF each is the corresponding

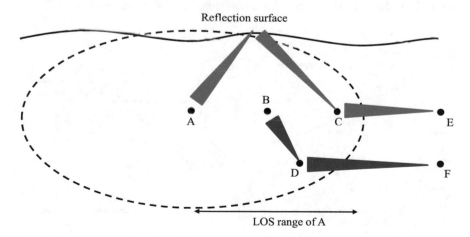

Fig. 11.10 Non line-of-sight routing via surface reflection [38]

LOS transmission range of each other, simultaneous transmissions cannot occur with omni-directional antennas. However, NLOS links with directional antenna can handle the obstacle along the LOS path by bypassing node B, which is just the destination node of another path FDB. A NLOS link can also enable simultaneous multi-hop transmission, i.e., ECA via the surface reflection and FDB as illustrated in the figure.

11.4.2   Routing for Wave Gliders

A underwater glider is a battery-powered AUV, mainly relying on local intelligence with minimal onshore operator dependence. Data and configuration information may be relayed between gliders before they can reach a surface station for data collection and analysis. A glider's motion usually follows sawtooth trajectories, which can be used to predict their positions. However, drifting and self-localization errors will impact the position estimate. Thus, the routing protocol discussed in [17] adopts a statistical approach to predict a glider's positions by using a confidence region instead of a single point. With the proposed protocol, a glider sends packets to a region rather than to a point.

As illustrated in Fig. 11.11, where glider 1 has a packet to forward to glider 7, its estimated position is P_7 within the corresponding uncertain region of C_{17}. Glider 1 estimates a path to glider 7 as follow: $1 \rightarrow C_{12} \rightarrow C_{14} \rightarrow C_{17}$. The figure also shows a path with a deterministic geographic routing, i.e., $1 \rightarrow 3 \rightarrow 5 \rightarrow 8$ (since glider 8 is closer to P_7 than glider 7), and a possible path with a shortest path routing: $1 \rightarrow 6 \rightarrow 7$, which however will not work when glider 7 moves out of the range of glider 6.

A glider needs to estimate first the regions of its neighbors and that of the destination glider when it has a packet to forward. Then, it forwards the packet to the region of the best neighbor. The selected neighboring glider broadcasts periodically

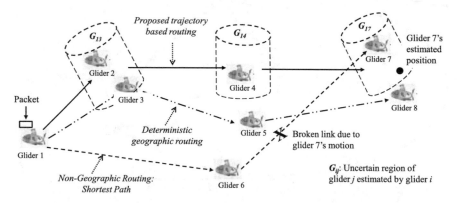

Fig. 11.11   Regional routing versus point routing [17]

the information on its velocity, next turning points and destined location to the source glider for region estimation. Accordingly, each glider has also to broadcast the location information sent by its neighbors.

11.4.3 Season-Adaptive Routing

A season-adaptive routing protocol investigated in [13] is based on the fact that different acoustic refraction effects are observed in different seasons. For example, in summer, acoustic waves in a shallow water are refracted downward because warm surface layers tend to increase acoustic speeds near the surface. In winter, the acoustic speed is more uniform, and typically slower close to the surface while faster near the sea bottom. This feature causes sound waves close to the surface to be slightly bent upward to probably insonify the upper water layers.

The proposed routing protocol takes into account the above effect, suggesting a season-adaptive next hop selection scheme for packet forwarding. That is, in summer, a node located near the surface beyond a given maximum distance should not be considered as a relaying node for the transmission from a deeper location. As illustrated in Fig. 11.12, where a source node (S) is placed on the left, and two destination nodes (D_1 and D_2) are placed on the right at different depths. A suggested summer routing path consists of the links connecting nodes at increasingly greater depth until the proximity of the receivers. Then, a direct delivery via a vertical channel is carried out.

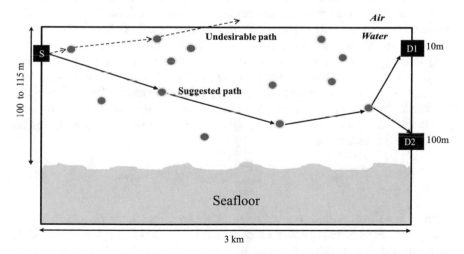

Fig. 11.12 Example of a season-adaptive routing [13]

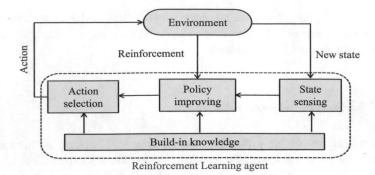

Fig. 11.13 Framework of reinforcement learning (RL) [41]

11.4.4   Smart Routing Protocols

A smart routing protocol based on the Reinforcement Learning (RL) technique called (QELAR) is discussed in [39] to optimize the distribution of energy consumption across network nodes to maximize the network lifetime.

11.4.4.1   Reinforcement Learning (RL)

As illustrated in Fig. 11.13, RL technique provides a learning framework to achieve certain objective based on experiences, i.e., the current state of a system and the reinforcement received from the environment are used to determine actions. The decision process is described by a tuple of [S,A,P,R], where S, A, P and R respectively denote the set of states, actions, state transition probabilities and rewards [40].

Q-learning is an estimation method that can yield near-optimal policies without much computations based on an off-policy temporal difference approach [41]. With this approach, the agent approximates the Q-values iteratively. Particularly for a Markovian decision process, it uses the value of state-action pair $Q(s, a)$, and aims to find the maximum expected reward $Q^*(s, a)$ that can be received by taking an action a_t at the state s_t, i.e.,

$$Q^*(s_t, a_t) = r_t + \gamma \sum_{s \in S} P^{a_t}_{s_t, s_{t+1}} \max_a Q^*(s_{t+1}, a), \qquad (11.1)$$

where r_t is a direct reward received after taking an action in the current state s_t at time t, and the summation indicates the maximum reward that can be received by all possible actions in the future state s_{t+1}. γ is a discount factor ranging in [0,1), which is used to discount the rewards in the future: $\gamma = 0$ means that the system only considers the current reward, while $\gamma = 1$ means that the system will more take into account a long-term reward. $P^a_{s,s'}$ is the probability of going to the next state $s' \in S$

from state $s \in S$ with a given action $a \in A$, where A and S denote the set of actions and states, respectively.

11.4.4.2 Application for Routing

With QELAR [39, 42], the whole network is treated as a system, whose states are related to each individual packet. The S and A are made of all the states and the actions related to each node in the network, and the state and action in the system are all related to each packet. Particularly, when node s_1 has a packet, the system state related to this packet is just denoted by s_1, and a_s denotes the action taken by a node to forward the packet to node s. Accordingly, a state transition occurs only when a packet is forwarded from one node (e.g., s_n) to another (e.g., s_m), and each forwarding action may either succeed with probability $P_{s_n s_m}^{a_m}$ or fail with probability $P_{s_n s_m}^{a_m}$. For unicast transmission, $P_{s_n s_n}^{a_m} = 1 - P_{s_n s_m}^{a_m}$. Particularly if $m = n$, $P_{s_n s_m}^{a_m} = 1$. Note that in this case, the item s_{t+1} in (11.1) indicates the next hop of packet forwarding.

To prolong the network lifetime, Q-learning is applied to forward each packet to the destination with a maximum reward and balanced traffic load in the network to avoid a situation that some sensor nodes are over-used in order to prolong the network lifetime. To this end, the following reward functions are defined.

$$r_t = P_{s_n s_m}^{a_m} R_{s_n s_m}^{a_m} + P_{s_n s_n}^{a_m} R_{s_n s_n}^{a_m}, \tag{11.2}$$

where $R_{s_n s_m}^{a_m}$ and $R_{s_n s_n}^{a_m}$ indicate the reward function for a successful and failed forwarding operation, respectively, and are defined as follows:

$$\begin{aligned}
R_{s_n s_m}^{a_m} &= -g - \alpha_1[c(s_n) + c(s_m)] + \alpha_2[d(s_n) + d(s_m)], \\
R_{s_n s_n}^{a_m} &= -g - \beta_1 c(s_n) + \beta_2 d(s_n).
\end{aligned} \tag{11.3}$$

Here g denotes a constant punishment (cost) for packet forwarding since this operation consumes power and bandwidth, and causes delay to other transmitted packets. Both α and β are weights that can be tuned. $c(s_n)$ is the cost function of residual energy of node n, and $d(s_n)$ is the reward function of energy distribution of all direct neighbors of node n in its transmission range. They are defined as follows:

$$\begin{aligned}
c(s_n) &= 1 - \frac{E_{res}(s_n)}{E_{init}(s_n)}, \\
d(s_n) &= \frac{2}{\pi} \arctan[E_{res}(s_n) - \bar{E}_{res}(s_n)],
\end{aligned} \tag{11.4}$$

where $E_{res}(s_n)$ and $E_{init}(s_n)$ denote the residual energy and initial energy of node n, and $\bar{E}_{res}(s_n)$ is the average residual energy in the direct neighbors of node n.

Figure 11.14 illustrates a simple example on how to calculate $Q(s_t, a_t)$ for routing decision. There are five network nodes (i.e., $1 \sim 5$), and node 1 is the source node

Fig. 11.14 An example of Q-learning for routing [39, 42]

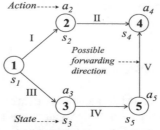

$V(s_n)$	Value	Step
$V(s_1)$	-1.0	I, III
$V(s_2)$	-1.0	II
$V(s_3)$	-1.0	IV
$V(s_4)$	0.0	Dest.
$V(s_5)$	-1.0	V

$$V(s_n)=max_a Q(s_n,a)$$

(a) Node, state and action   (b) $V(s_n)$ and $Q(s_n,a)$

that has a packet to forward to the destination of node 4. The corresponding states and actions for packet forwarding are depicted in Fig. 11.14a. Here the initial values of Q and V are 0, where $V(s_n) = max_a Q(s_n, a)$, which means that the maximum reward from all possible actions. Let $\gamma = -1$, $g = 1$ and $R^{a_m}_{s_n s_m} = 1$, which means that each packet forwarding is successful so that $P^{a_m}_{s_n s_n} = 0$. For simplicity, $c(s_n)$ and $d(s_n)$ are not considered so that $R^{a_m}_{s_n s_m} = R^{a_m}_{s_n s_n} = -g$. Thus from (11.2), we have $r_t = -g$.

Obviously, node 1 at state s_1 has two possible actions for packet forwarding: forwarding to node 2 (a_2) and forwarding to node 3 (a_3). For the first packet forwarding, it calculates the rewards for each action with initial $V(s_2) = 0$ as follows:

$$\begin{aligned}
Q(s_1, a_2) &= r_t + \gamma[P^{a_2}_{s_1 s_2} V(s_2) + P^{a_2}_{s_1 s_1} V(s_1)] \\
&= -g + \gamma V(s_2) \\
&= -1.0, \\
Q(s_1, a_3) &= r_t + \gamma[P^{a_3}_{s_1 s_2} V(s_2) + P^{a_3}_{s_1 s_1} V(s_1)] \\
&= -g + \gamma V(s_2) \\
&= -1.0.
\end{aligned} \tag{11.5}$$

Since $Q(s_1, a_2) = Q(s_1, a_3) = -1$, $V(s_1) = max_a Q(s_1, a) = -1$ (from a_2 and a_3), and both node 2 and node 3 can be used as the next hop. Suppose that node 2 is randomly selected for the packet forwarding (step I). Since node 4 is the destination, the packet should be forwarded directly to it without considering other possible directions (step III), and the following calculation is carried out to update $V(s_2)$:

$$\begin{aligned}
V(s_2) = Q(s_2, a_4) &= r_t + \gamma[P^{a_4}_{s_2 s_4} V(s_4) + P^{a_4}_{s_2 s_2} V(s_2)] \\
&= -g + \gamma V(s_4) \\
&= -1.0.
\end{aligned} \tag{11.6}$$

For the second packet forwarding, the above calculation is repeated with the only difference that now $V(s_2) = -1.0$ rather than the initial $V(s_2) = 0$. Thus, now $Q(s_1, a_2) = -1 + 0.5 \times (-1.0) = -1.5 < Q(s_1, a_3)$, which means that the second packet should be forwarded to node 3 (step III), and $V(s_1) = -1.0$. For node 3, it can send the packet either back to node 1 or toward node 5, depending on the reward calculation as follows:

$$
\begin{aligned}
Q(s_3, a_1) &= r_t + \gamma[P^{a_5}_{s_3 s_1} V(s_1) + P^{a_1}_{s_3 s_3} V(s_3)] \\
&= -g + \gamma V(s_1) \\
&= -1.5, \\
Q(s_3, a_5) &= r_t + \gamma[P^{a_5}_{s_3 s_5} V(s_5) + P^{a_5}_{s_3 s_3} V(s_3)] \\
&= -g + \gamma V(s_4) \\
&= -1.0.
\end{aligned}
\tag{11.7}
$$

Thus, the packet should be forwarded to node 5 (step IV), which then will directly forward the packet to node 4 (step V), and $V(s_3) = -1.0$. Similarly, we can have $V(s_5) = Q(s_5, a_4) = -1.0$.

To fully implement the proposed scheme, the routing protocol needs to monitor the environment closely and learn to extract needed information. When a node hears a packet, either control packet or data packet, no matter whether or not it is designated as the next forwarder, it extracts the sender neighbor's information such as the network topology and residual energy of surrounding nodes. The Q-learning technique has been applied by the same authors to design other adaptive routing protocols in [43, 44].

11.5 Discussion

The above discussion shows that it is impossible for only a few routing protocols to support cost-effectively all underwater applications, while an optimal routing protocol design should leverage different favorable features present in different application scenarios [8]. The major application of UWANs is underwater monitoring and surveillance, which results in many routing protocols proposed for UWANs that have sinks with the majority traffic flowing upward to the water surface. In this case, relative position that can be measured by water pressure can be used to improve path search efficiency. If underwater nodes are stationary, routing is relative simple because once a path is found, it can be used for long time. This can reduce routing overhead and energy consumption for re-search and maintenance of paths especially with long propagation delay.

The major challenges source from routing in large 3D UWANs, where some sensor nodes also acting as relays float in the water and may drift with water current, resulting in frequent topological changes that may break links. Typical solutions proposed for this problem include multipath routing and leverage of mobile agents. More challenges arise in mobile UWANs that have no fixed sinks because they are typically sparse networks with node mobility. A regional routing protocol based on track prediction is investigated to handle this problem. The routing becomes much more challenging in wide and deep water areas, where UWANs become more sparse, and it is much more difficult to obtain support from powerful units on water surface to control and relay data from deep underwater nodes. In this case, a fibre optical network deployed on the seafloor can help solving the problem, but the cost for the deployment and maintenance of such a network is very high so that it cannot be available everywhere.

Different from MAC protocol design, a long propagation delay does not affect the design strategy for UWAN routing protocols so much although it may considerably affect routing performance. Many UWAN routing protocols still utilize broadcast route search and unicast route reply which are adopted extensively in RWNs. Such kind of routing strategies cause very large end-to-end latency due to not only long propagation delay but also half-duplex underwater communication caused by technological limitation. Unicast forwarding is vulnerable to asymmetric acoustic channels, and one possible solution is utilizing multipath forwarding instead of single path forwarding.

Another difficult routing issue in UWANs is to handle large scale shadow zones, in which no acoustic signal can be received. In this case, only mobile agents can help, while they can also be used to improve connectivity in cases of connection gaps or asymmetric links. However, a mobile agent is usually much more expensive than a fixed underwater node, and there is a risk of losing it when it is out of control. A hard operational issue is how to deploy mobile agents to right places rapidly and to call them back successfully when necessary especially in remote deep water areas. The solution to this problem depends on techniques of underwater communication and localization as well as energy supply of mobile agents.

Similar to UWAN MAC protocols, whose performance largely depends on underwater acoustic communication, the performance of a UWAN routing protocol is largely affected by underlying MAC protocols. However, the majority of the proposed UWAN routing protocols are validated by computer simulation without taking into account adequately the impact of UWAN MAC protocols, similar to UWAN MAC protocols as discussed in Sect. 10.3. Therefore, a question remains about the performance of a routing protocol validated only by simulation in the actual underwater environments.

A large UWAN behaves more like RF-based opportunistic mobile networks with following differences. (i) Connection intermittence in mobile RWNs is mainly due to node mobility and power-off operations of nodes, while in UWANs, it is mainly caused by asymmetric links and shadow zones, which depend on propagation environments. (ii) Opportunities in mobile RWNs are mainly created by node mobility, while in UWANs mainly depending on changes in propagation environments. This is due to the fact that acoustic signal can propagate over long distances at reduced channel rates in good propagation environments. Although mobile agents can bring opportunities, they are expensive with risk of loss as mentioned above, and it may take long time to complete deployment in large UWANs due to their low motion speeds. These differences make the store-then-forward routing strategy extensively used by RF-based opportunistic networks inefficient in UWANs.

11.6 Summary

There are also other ways to categorize UWAN routing protocols, which classify routing protocols into clustering routing, AUV-assisted routing, geographic-based routing and special routing assuming that underwater nodes have some special features (e.g., buoys, ropes and anchors). No matter what kind of protocol categories are used, routing protocol design is limited by narrow bandwidth, large communication energy consumption and sparse node distribution in UWANs. Routing performance is mainly affected by the following factors in UWANs: large end-to-end latency, asymmetric acoustic communication links and shadow zones, and mobility in 3D UWANs as well as MAC protocols. This chapter briefly reviews some design approaches that try to handle these problems for different application scenarios, which are useful to design better UWAN routing protocols. A more comprehensive survey can be found in [8].

References

1. Noh, Y., Lee, U., Wang, P., Choi, B.S.C., Gerla, M.: VAPR: void-aware pressure routing for underwater sensor networks. IEEE Trans. Mob. Comput. **12**(5), 895–908 (2013)
2. Otnes, R., Asterjadhi, A., Casari, P., Goetz, M., Husøy, T., Nissen, I., Rimstad, K., van Walree, P., Zorzi, M.: Underwater Acoustic Networking Techniques. Springer, Germany (2012)
3. Basagni, S., Petrioli, C., Petroccia, R., Spaccini, D.: Channel-aware routing for underwater wireless networks. In: Proceedings of the MTS/IEEE OCEANS, Yeosu, Korea (2012)
4. Stojanovic, M.: On the relationship between capacity and distance in an underwater acoustic communication channel. In: Proceedings of the ACM International WS, Underwater Networks (WUWNet), Los Angeles, USA (2006)
5. Zorzi, M., Casari, P., Baldo, N., Harris, A.F.: Energy-efficient routing schemes for underwater acoustic networks. IEEE J. Sel. Areas Commun. **26**(9), 1754–1766 (2008)
6. Souiki, S., Feham, M., Feham, M., Labraoui, N.: Geographic routing protocols underwater wireless sensor networks: surveys. Int. J. Mob. Netw. (IJWMN) **6**(1), 69–87 (2014)
7. Li, N., Martínez, J.-F., Chaus, J.M.M., Eckert, M.: A survey on underwater acoustic sensor network routing protocols. Sensors **16**(414), 1–28 (2016)
8. Lu, Q., Liu, F., Zhang, Y., Jiang, S.M.: Routing protocols for underwater acoustic sensor networks: a survey from an application perspective. In: Zak, A. (ed.), Advances in Underwater Acoustics. INTECH (2017). ISBN 978-953-51-3609-5
9. Xie, G.G., Gibson, J.H.: A network layer protocol for UANs to address propagation delay induced performance limitations. In: Proceedings of the MTS/IEEE OCEANS, Honolulu, HI, USA, pp. 2087–2094 (2001)
10. Chen, Y., Zhang, S.Q., Xu, S.G., Li, G.Y.: Fundamental trade-offs on green wireless networks. IEEE Commun. Mag. **49**(6), 30–37 (2011)
11. Xie, P., Cui, J.-H.: VBF: vector-based forwarding protocol for underwater sensor networks. In Proceedings of the IFIP Networking Conferences on Coimbra, Portugal, pp. 1216–1221 (2006)
12. Seah, W.K.G., Tan, H.X., Liu, Z., Ang, M.H.: Multiple-UUV approach for enhancing connectivity in underwater ad-hoc sensor networks. In: Proceedings of the MTS/IEEE OCEANS, Washington, DC, USA **2**, 2263–2268 (2005)
13. Casari, P., Asterjadhi, A., Zorzi, M.: On channel aware routing policies in shallow water acoustic networks. In: Proceedings of the MTS/IEEE OCEANS, Waikoloa, Hawaii, USA (2011)

14. Guo, Z., Colombit, G., Wang, B., Cui, J.-H., Maggiorinit, D., Rossit, G.P.: Adaptive routing in underwater delay/disruption tolerant sensor networks. In: Proceedings of the Annual Conferences on Wireless on Demand Network Systems and Services (WONS), Garmisch-Partenkirchen, pp. 31–39 (2008)
15. Vieira, L.F.M., Lee, U., Gerla, M.: Phero-trail: a bio-inspired location service for mobile underwater sensor networks. IEEE J. Sel. Areas Commun. **28**(4), 553–563 (2010)
16. Yan, H., Shi, Z., Cui, J.-H.: DBR: depth-based routing for underwater sensor networks. In: Proceedings of the IFIP Networking Conference, Singapore pp. 72–86 (2008)
17. Chen, B.Z., Hickey, P.C., Pompili, D.: Trajectory-aware communication solution for underwater gliders using WHOI micro-modems. In: Proceedings of the Annual IEEE Communications Society Conference on Sensor, Mesh and Ad Hoc Communications and Networks (SECON), Boston, USA, 2010
18. Lindgren, A., Doria, A., Davies, E., Grasic, S.: Probabilistic routing protocol for intermittently connected networks. Internet Research Task Force (IRTF) (2012)
19. Rice, J.A., Ong, C.W.: A discovery process for initializing underwater acoustic networks. In: Proceedings of the International Conference on Sensor Device Technologies and Applications (SENSORCOMM), Venice, pp. 408–415 (2010)
20. Nicolaou, N., See, A., Cui, J.-H., Maggiorini, D.: Improving the robustness of location-based routing for underwater sensor networks. In: Proceedings of the MTS/IEEE OCEANS, Aberdeen, UK, pp. 1–6 (2007)
21. Lee, U., Wang, P., Noh, Y., Vieira, L.F.M., Gerla, M., Cui, J.-H.: Pressure routing for underwater sensor networks. In: Proceedings of the IEEE INFOCOM, San Diego, CA, USA, pp. 1–9 (2010)
22. Zhang, S., Li, D.S.: A beam width and direction concerned routing for underwater acoustic sensor networks. In: Proceedings of the IEEE International Conference on Mobile Ad-hoc and Sensor Networks, Dalian, China, pp. 17–24 (2013)
23. Zhang, S., Li, D.S., Chen, J.: A link-state based adaptive feedback routing for underwater acoustic sensor networks. IEEE SENSORS J. **13**(11), 4402–4412 (2013)
24. Jiang, S.M. (2018) On reliable data transfer in underwater acoustic networks: a survey from networking perspective. IEEE Commun. Surv. Tutor. **PP**, 99 (2018)
25. Ogier, R.G., Rutenburg, V., Shacham, N.: Distributed algorithms for computing shortest pairs of disjoint paths. IEEE Trans. Inf. Theory **39**(2), 443–455 (1993)
26. Lal, C., Laxmi, V., Gaur, M.S.: A node-disjoint multipath routing method based on AODV protocol for MANETs. In: Proceedings of the IEEE International Conference on Advanced Information Networking and Applications (AINA), Fukuoka, Japan, pp. 399–405 (2012)
27. Azad, S., Casari, P., Zorzi, M.: Multipath routing with limited cross-path interference in underwater networks. IEEE Wireless Commun. Lett. **3**(5), 465–468 (2014)
28. Marina, M.K., Das, S.R.: Ad hoc on-demand multipath distance vector routing. Wireless Commun. Mob. Comput. **6**, 969–988 (2006)
29. Goetz, M., Azad, S., Casari, P., Nissen, I., Zorzi, M.: Jamming-resistant multi-path routing for reliable intruder detection in underwater networks. In: Proceedings of the ACM International Conference on Underwater Networks and Systems (WUWNet), Seattle, WA, USA (2011)
30. Chen, Y.-S., Juang, T.-Y., Lin, Y.-W., Tsai, I.-C.: A low propagation delay multi-path routing protocol for underwater sensor networks. J. Internet Tech. **11**(2), 153–165 (2010)
31. Zhou, Z., Peng, Z., Cui, J.-H., Shi, Z.: Efficient multipath communication for time-critical applications in underwater acoustic sensor networks. ACM/IEEE Trans. Netw. **19**(1), 28–41 (2011)
32. Jornet, J.M., Stojanovic, M., Zorzi, M.: Focused beam routing protocol for underwater acoustic networks. In: Proceedings of the ACM International WS. Underwater Networks (WUWNet), San Francisco, USA, pp. 75–82 (2008)
33. Jornet, J.M., Stojanovic, M., Zorzi, M.: On Joint frequency and power allocation in a cross-layer protocol for underwater acoustic networks. IEEE J. Ocean. Eng. **35**(4), 936–947 (2010)
34. Chen, Y.-S., Lin, Y.-W.: Mobicast routing protocol for underwater sensor networks. IEEE SENSORS J. **13**(2), 737–749 (2013)

35. Chirdchoo, N., Soh, W.S., Chua, K.C.: Sector-based routing with destination location prediction for underwater mobile networks. In: Proceedings of the International Conference on Advanced Information Networking and Applications Workshops (WAINA), Bradford, UK, pp. 1148–1153 (2009)

36. Wang, J.C., Li, D.S., Zhou, M., Ghosal, D.: Data collection with multiple mobile actors in underwater sensor networks. In: International Conference on Distributed Computing Systems (Workshop), Beijing, China, pp. 216–221 (2008)

37. Shah, P.M., Ullah, I., Khan, T., Hussain, M.S., Khan, Z.A., Qasim, U., Javaid, N.: MobiSink: cooperative routing protocol for underwater sensor networks with sink mobility. In: Proceedings of the IEEE International Conference on Advanced Information Networking and Applications (AINA), Crans-Montana, Switzerland (2016)

38. Emokpae, L., Younis, M., Signal reflection-enabled geographical routing for underwater sensor networks. In: Proceedings of the IEEE International Conference on Communication (ICC), Ottawa, Canada, pp. 147–151 (2012)

39. Hu, T.S., Fei, Y.S.: QELAR: a machine-learning-based adaptive routing protocol for energy-efficient and lifetime-extended underwater sensor networks. IEEE Trans. Mob. Comput. 9(6), 796–809 (2010)

40. Sutton, R.S., Barto, A.G.: Reinforcement Learning: An Introduction, pp. 688–698. The MIT Press (1998)

41. Wang, P., Wang, T.: Adaptive routing for sensor networks using reinforcement learning. In: Proceedings of the IEEE International Conference on Computer and Information Technology, Seoul, Korea (2006)

42. Hu, T.S., Fei, Y.S.: QELAR: a q-learning-based energy-efficient and lifetime-aware routing protocol for underwater sensor networks. In: Proceedings of the IEEE International Performance, Computing, and Communications Conference (IPCCC), Austin, Texas, USA, pp. 247–255 (2008)

43. Hu, T.S., Fei, Y.S.: MURAO: a multi-level routing protocol for acoustic-optical hybrid underwater wireless sensor network. In: Proceedings of the Annual IEEE Communications Society Conference on Sensor, Mesh and Ad Hoc Communications and Networks (SECON), Seoul, Korea, pp. 218–226 (2012)

44. Hu, T.S., Fei, Y.S.: An adaptive routing protocol based on connectivity prediction for underwater disruption tolerant networks. In: Proceedings of the IEEE Global Telecommunications Conference (GLOBOCOM), Atlanta, USA, pp. 65–71 (2013)

Chapter 12
Transfer Reliability Control in UWANs

Abstract Reliable data transfer attempts to guarantee a destination node to receive successfully what has been sent to it. What leading to reception failures include poor channel quality due to interference and fading, congestion causing data loss, and collision leading to data corruption as well as network attacks deviating to data forwarding. This chapter will introduce some typical reliable transfer schemes proposed for the data link layer, the network layer and the transport layer. Network attacks will be discussed with network security in Chap. 13.

12.1 Overview

As discussed in Chap. 2, in RWNs, the main methods to provide reliable data transfer over an unreliable path include redundancy and retransmission. The typical redundancy scheme is FEC, while the typical retransmission scheme is ARQ. It is difficult for FEC to guarantee transfer reliability. When the BER is too large for the receiver to correct each error, retransmitting the impacted data with ARQ should be invoked for successful reception.

The main challenges facing UWANs for reliable data transfer stem from the following aspects. (i) Underwater acoustic channels are characterized by poor quality and high dynamics due to time-varying and multipath propagation, fading and motion-induced Doppler distortion, which result in high and changing BER [1]. (ii) The small capacity of underwater acoustic channels makes it undesirable to apply widely redundancy schemes. (iii) The long propagation delay makes ARQ schemes very inefficient. Furthermore, underwater acoustic links may be asymmetric, which affects establishment of a feedback channel required by ARQ. Meanwhile, the following technological limitations make reliable transfer protocol design more difficult. Most available underwater acoustic modems can operate only in half-duplex mode, and consume large energy for both transmission and reception, while underwater network nodes are often battery-operated [2].

To deal with the peculiar features of UWANs mentioned above, many proposed schemes using different design strategies such as cross-layer design and hybrid design with various mechanisms. Typically, these schemes attempt to improve the performance of ARQ by making it suitable for half-duplex links such as packet train [3],

© Springer Nature Singapore Pte Ltd. 2018   315
S. Jiang, *Wireless Networking Principles: From Terrestrial to Underwater Acoustic*,
https://doi.org/10.1007/978-981-10-7775-3_12

or creating full-duplex links. Coding schemes such as erasure codes [4] and network coding [5, 6] are also applied to improve reliable transfer performance. Meanwhile, multipath transmission, power control and cooperative transmission etc. have also been jointly used with either FEC or ARQ for transfer reliability control.

The following sections review some typical approaches and schemes proposed for transfer reliability control in UWANs. The main reference for this part is [7].

12.2   Reliable Transfer Architecture

This section briefly reviews the architecture and basic mechanisms proposed for reliable data transfer, which include FEC and ARQ as well as network coding. Note that network coding itself does not have capability for transfer reliability control. It can allow more redundancy transmission due to its higher transfer efficiency than that without network coding, and its multipath transmission nature can also be used to provide redundancy, hence improving reliability control performance [7].

The major functionalities for reliable data transfer are usually distributed into the four lower layers of the Open System Interconnection (OSI) model, and can be grouped into two levels: link and path, as illustrated in Fig. 12.1. Link-level functions attempt to provide reliable transmission over links between neighboring nodes, mainly combating transmission errors caused by interferences. Path-level functions aim to provide end-to-end reliable transfer across a network, mainly handling packet loss during a packet's journey to the destinations [7].

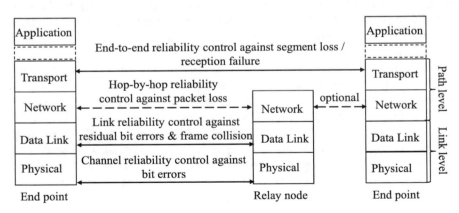

Fig. 12.1 A general architecture for end-to-end transfer reliability control [7]

12.2.1   Link-Level Functions

They include those implemented on the physical layer and data link layer. As summarized in [7], physical layer functions are used to improve channel quality and control errors. The first part attempts to improve channel robustness and capacity against various factors deteriorating channels. This can be achieved by using modulation and interference alignment [8] as well as transmit diversity with antenna technologies, such as smart antenna (e.g., [9]) and MIMO [10]). The second part mainly applies channel coding to enable error correction to minimize reception errors, typically with FEC. However, it is difficult and not cost-effective to guarantee error-free reception at the physical layer. The logical link control of the data link layer using ARQ should be implemented for error-free reception via retransmission. ARQ can also be used to recover collided frames due to MAC failure [7].

12.2.2   Path-Level Functions

They compose of hop-by-hop reliability control on the network layer and end-to-end reliability control on the transport layer. As discussed in [7], when an error-free packet travels across a network, it may experience congestion, leading to packet loss. In this case, a hop-by-hop retransmission of each lost packet can be performed with an ARQ-like scheme, i.e., the upstream node of the link retransmits the packets lost at the downstream node. However, such implementation complicates network nodes. In practice, hop-by-hop reliability control may not be adopted if end-to-end reliability control is in place. This structure actually is adopted by TCP/IP network. For end-to-end transmission reliability control, an ARQ-like (e.g., TCP) protocol is performed between a source-destination pair without involving any intermediate relaying nodes (e.g., routers). Only the source node keeps retransmitting a packet until it has been positively acknowledged by the destination node, subject to a maximum number of retransmissions. Note that, an end-to-end reliability control scheme has not to assume the availability of any lower-layer reliability control. For example, in TCP/IP network, TCP performs both transmission reliability control and congestion control, leading to very simple and robust IP network.

12.2.3   Typical Error Control Schemes

Figure 12.2 summarizes the typical error control schemes adopted by reliable transfer protocols in UWANs except convolutional codes due to their computational complexity that are undesirable for battery-operated underwater nodes. Both the bit-level FEC and the link-level ARQ have been discussed in error control for RWNs (Chap. 2). The typical path-level ARQ for reliable transfer control in RWNs is performed end-to-end between source and destination nodes, such as TCP discussed in Chap. 6.

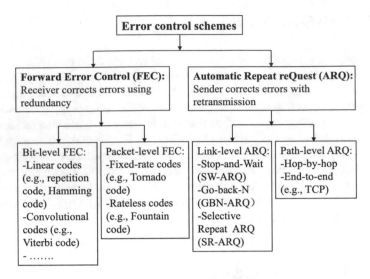

Fig. 12.2   Typical error control schemes [7]

For UWANs, besides these error control schemes mentioned above, the packet-level FEC and the hop-by-hop reliable transfer control of the path-level ARQ are also used jointly. The link-level ARQ may be performed jointly with FEC by each link of a network connection hop-by-hop to provide path-level reliable control. The packet-level FEC encodes a message of k packets into a set of longer encoded messages each with n packets such that the original message can be recovered from a subset of the encoded messages. One type of such scheme requires a constant code rate $\frac{k}{n}$ (e.g., tornado codes [11]), which is not suitable for highly dynamic underwater acoustic channels. Another type of such scheme called rate-less codes releases such constraint [12]. With such codes, no theoretical limit is imposed to the number of encoding packets to be transmitted to the receiver for error recovery. The encoded packets can be sent as many as necessary until the packets are fully recovered, such as fountain codes [13, 14] to be discussed in Sect. 12.4.1.

12.3   Data Link Layer

Reliable transfer on the data link layer attempts to provide 1-hop reliable transmission between two neighboring nodes. For half-duplex communication with most commercially available underwater acoustic modems, SW-ARQ discussed in Chap. 2 can be directly applied. Its throughput is low because the channel is idle during the waiting period for ACK. Its efficiency depends on packet size, link delay and link quality, and an optimal packet size exists for the maximum efficiency [15]. As mentioned earlier, GBN-ARQ and SR-ARQ can outperform SW-AQR by allowing packet transmission and ACK reception to process simultaneously, but both require full-duplex communication. There are several proposals aiming to enable full-duplex underwater communication [7].

To realize full duplex communication, a channel is split into two subchannels: one for transmission and the other for reception, using either frequency-division duplex (FDD) or time-division duplex (TDD). Dividing a bandwidth-limited underwater acoustic channel with FDD reduces data rate due to guard bands necessary between subchannels. TDD requires time synchronization, which is another difficult issue in UWANs as mentioned in Chap. 9. Furthermore, the time used by a node to switch between transmission and reception states may affect channel utilization. For both, asymmetric traffic loads in two directions have to be considered by both FDD and TDD to optimize channel utilization [7].

12.3.1 Enhanced SW-ARQ

As discussed in [7], the following schemes are proposed to enhance SW-ARQ against the peculiar features of UWANs. Typically, with group transmission or packet train, a group of packets are transmitted without waiting for acknowledgement on each packet one by one. Actually, such enhancement is similar to modifying SR-ARQ to run over half-duplex underwater acoustic links to be discussed in Sect. 12.3.2.

One earlier packet train scheme is discussed in [3], using accumulative ACK. A transmitter sends M frames and then stops to wait for the ACK, while the receiver waits an M-frame duration, and then sends an ACK for all the received successfully frames. Once positively acknowledged, the transmitter will send another group of M frames, which include the unacknowledged frames. As analyzed in [1], its throughput efficiency can be maximized by selecting optimal frame sizes as a function of transmission ranges, transmission rates and error probability. However, to deliver ACKs, it is necessary to reserve a feedback channel, which excludes the possibility to send urgent data from the receiver during an RTT period when necessary.

Similarly, the continuous SW-ARQ or juggling-like SW (JSW) scheme [16] further leverages long propagation delays by allowing a sender to transmit maximal M data frames continuously at the beginning of transmission without waiting for the corresponding ACK. This ACK corresponds to an earlier transmitted data frame rather than the most recently transmitted one. The setting of M increases with RTTs so that the long propagation delay can be leveraged.

Figure 12.3 shows an example with $M = 3$, i.e., the sender continuously sends 3 data frames, and then stops further transmission and waits for the ACKs for previously sent frames. Assume that the ACKs are received error-free. After the sender successfully receives the ACK for frame 1, it switches to the transmission state to transmit frame 4, and then switches back to the listening state to wait until it has received an ACK before transmitting the next data frame. This scheme can work without explicit time slot synchronization between the sender and the receiver [16].

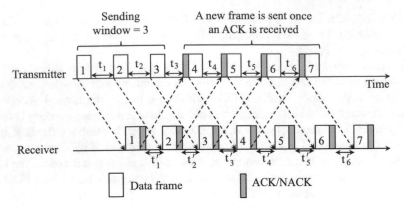

Fig. 12.3 Transmission flow with JSW [16]

To run JSW, a node should be able to finish its transmission and switch to listening state before the ACK arrives. This issue is related to determine M for optimal protocol performance. Analytical and numerical studies in [17] show that JSW can provide good data streaming throughput but not for small data transfers. Thus, a rateless code JSW is proposed to overcome this limitation, by allowing the frames encoded for a file to be transmitted without waiting for acknowledgement. The sender is informed only when enough frames have been received to recover the file [18].

12.3.2   Enabling of SR-ARQ

To make JSW adaptive to various communication distances with different destination nodes, the Underwater Selective Repeat (USR) scheme [19, 20] establishes a TDD based full-duplex channel to run SR-ARQ to leverage long propagation delays. It packs several data frame transmissions within the same RTT, while keeping the receiver silent when it is receiving ACKs. The RTT is defined as the time interval between a data frame transmission and the reception of the corresponding ACK, and is estimated accordingly. To this end, at the beginning, a node adopts SW-ARQ to send only one data frame, and then waits for the corresponding ACK. If no ACK is received, the node will repeat the above procedure after a backoff. When the ACK is received, the node estimates the RTT, which is used by the node to estimate the maximum number of frames that can be transmitted before it enters to the state of waiting for ACKs. The transmission of data frames and the reception of ACKs are interlaced in time so that the same channel can be used for both data communication and feedback [7].

Let τ denote the propagation delay, and T_d and T_a the transmission time of a data frame and of an ACK frame, respectively. For a destination node, window size M can be computed as follows [20]:

$$M = \max\left(1, \left\lfloor \frac{k\tau}{T_d + T_a + \delta} \right\rfloor\right), \qquad (12.1)$$

where δ is a guard time used to prevent tight scheduling of data frames and ACKs, and k is an adaptation factor to limit the portion of the RTT that will be considered in computing M [20].

With $M > 1$, the sender can transmit multiple frames in the window. To prevent the sender from receiving an ACK while transmitting a data frame, it has to wait a fixed time W before sending the next frame in the window. W is determined such that the ACK reception takes place in the middle of the waiting time with a constant RTT. Then, a relationship between W and other parameters is established to sort out W as follows [20]:

$$W = \frac{T_a + 4\tau - 2(M - 1)T_d}{2M - 1}. \qquad (12.2)$$

As illustrated in Fig. 12.4, where $M = 4$, the sender first transmits a data frame D1 and then waits for ACK1 in the SW-ARQ mode to estimate the RTT, which is used to compute M and waiting time W. So, the sender sends 4 data frames and then stops to wait for ACKs. The remaining operation is similar to JSW.

12.3.3 Hybrid-ARQ (HARQ)

Such kind of scheme has also attracted lots of attention to improve ARQ performance in UWANs, jointly using the following erasure codes [21] with ARQ: cross-layer designed bit-level FEC and jointly designed packet-level FEC.

Two types of HARQ schemes for a half-duplex UWAN are discussed in [22], jointly using random binary linear code as bit-level FEC on the data link layer. With

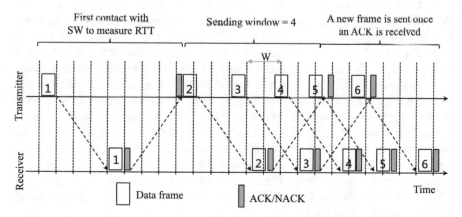

Fig. 12.4 Principle of the underwater selective repeat (USR) protocol with $M = 4$ ($\tau = 4T_d$, $\delta = 0.5T_d$, $W = 1.5T_d$, $T_a = 0.5T_d$) [20]

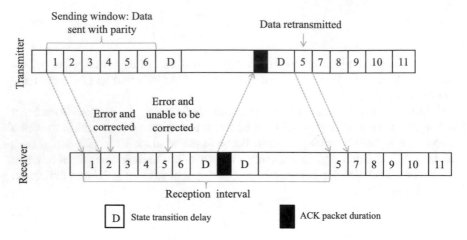

Fig. 12.5 Type-I HARQ in UWANs for sending window equal to 6 ($M = 6$) [7, 22]

Type-I HARQ, the sender transmits a bath of data frames subject to the size of sending window, and then waits for the ACK. If the received data frames are error-free or the errors can be corrected, an ACK is returned to the sender. After receiving the ACK, the sender can transmit new data frames if any. If no ACK is received for the expected data frames, retransmission is triggered. As illustrated in Fig. 12.5, after 6 data frames are transmitted during a sending window (i.e., $M = 6$), the sender switches to the listening state to receive the ACK. Suppose both frames 2 and 5 are received in error while frame 2 can be corrected by the receiver. Frame 5 will be retransmitted by the sender in the next sending window, along with new data frames $7 \sim 11$ [7].

The major difference between Type-I and Type-II HARQ is that with Type-II, the sender calculates a parity frame for each transmitted frame, but not transmit them together. The data frames are sent first. If some of them are received in error, they will not be dropped. The receiver asks the sender to transmit the parity frame for error correction, which will be successful if the parity frame is received error-free, and fail otherwise. In the latter case, a retransmission of the data frame is triggered, while the previously received parity frame is still used for frame recovery when necessary. Similarly, parity frames may also be retransmitted. As illustrated in Fig. 12.6, frames 2 and 5 are received in error. The receiver asks the sender to transmit their parity frames P2 and P5. However, only P2 is successfully used to correct frame 2, and frame 5 is asked for retransmission. These two retransmissions take place alternately until the reception is successful subject to a pre-defined maximum number of retransmissions [7]. In [23], the above two HARQ and USR [20] schemes mentioned earlier are analyzed for a star-topology 1-hop multiuser TDMA-based UWAN. There are also several HARQ schemes proposed for reliable multicast, which are discussed in Sect. 12.4.4.

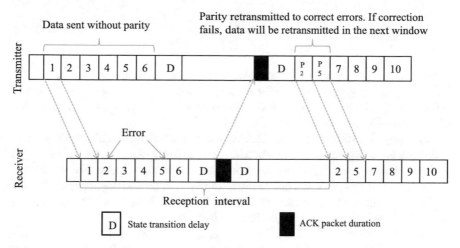

Fig. 12.6 Type-II HARQ in UWANs for $M = 6$ [7, 22]

12.4 Network Layer

As discussed in [7], reliable transfer schemes on the network layer attempt to provide end-to-end reliable transmission across the network hop-by-hop, jointly using various erasure codes (e.g., fountain codes), network coding and multiple path transmissions as well as cooperative transmissions.

12.4.1 Fountain Codes

Digital fountain codes [24] are more suitable for highly dynamic underwater acoustic channels because the amount of redundancy can be adjusted on the fly until the full recovery is achieved, making data dissemination process to adapt to diverse error rates. An efficient implementation of fountain codes can be realized with available encoding and decoding algorithms [13, 14] by means of simple XOR operations [25, 26].

12.4.1.1 Principle

The simplest fountain codes can be realized through random linear codes as described below [24]. Consider a set of original k packets: $S = \{s_1, s_2, \ldots, s_k\}$. At each clock cycle n ($n = 1, 2, \ldots k, \ldots$), the encoder generates randomly a k-bit binary vector $G_n = \{g_1, g_2, \ldots, g_k\}$, where g_i is set to either 0 or 1. The corresponding transmitted packet t'_n is set to the bitwise sum of the source packets (modulo 2) as follows:

$$t'_n = \sum_{i=1}^{k} g_i s_i. \tag{12.3}$$

After a node has successfully received enough encoded packets t_n, i.e., $T = \{t_1, t_2, \ldots, t_n\}$ ($n \geq k$) along with the corresponding binary vector G'_i, it can decode S if there are at least k independent G'_i, denoted by $G' = \{G'_1, G'_2, \ldots, G'_n\}$, through matrix inversion by solving $T = SG'$.

The Luby Transform (LT) codes [13] are the first practical fountain codes, with the following encoding steps. (i) The encoder first randomly chooses a number $d_n \leq k$ (called degree) from a degree distribution, which depends on the number of original packets k. (ii) It then chooses, uniformly at random, d_n distinct input packets, which are XORed together to generate one LT coded packet. The degree distribution is the key to achieve high successful decoding probability.

The major advantage of the LT coding is its simplicity of decoding operation. Figure 12.7 shows an example of decoding 3 original packets based on 4 received encoded packets. These packets are illustrated in Fig. 12.7a, which also depicts the coding relationship between the original and encoded packets. An original packet can be determined by the decoder when it has a connection with a decoding point that has no connection with other original packets. As illustrated in Fig. 12.7b, s_1 has a connection with t_1, while t_1 does not connect other packets, thus $s_1 = t_1$. After decoding of s_1, it is XORed to other encoded packets received by the decoder according to the coding relationship, and repeat the above decoding process until all the original packets have been decoded.

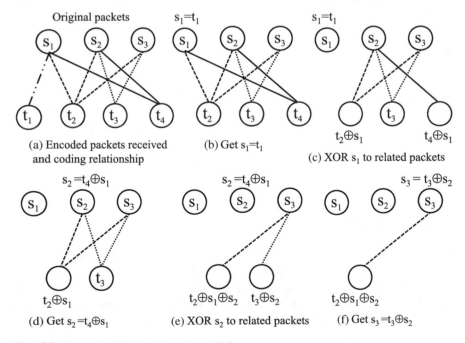

(a) Encoded packets received and coding relationship

(b) Get $s_1 = t_1$

(c) XOR s_1 to related packets

(d) Get $s_2 = t_4 \oplus s_1$

(e) XOR s_2 to related packets

(f) Get $s_3 = t_3 \oplus s_2$

Fig. 12.7 Example of LT decoding process [24]

12.4.1.2 Fountain Code Based Adaptive Multi-hop Reliable Data Transfer (FOCAR)

The FOCAR protocol for multi-hop reliable data transfer [26] integrates fountain codes with hop-by-hop retransmission-upon-failure, considering the impact of large state transition delays of the currently available half-duplex acoustic modems. With the proposed per-hop transfer scheme, after a source node receives a certain number of packets from the high layer, it switches to the transmission mode. It first groups m packets into one block, and encodes the packets in each block into M ($M \geq m$) packets with fountain codes. Then, it sends the M coded packets to its downstream node and switches to the reception mode. With fountain codes, if the receiver can correctly receive m of the M coded packets, it can correctly decode the original m data packets. In this case, it sends a positive ACK to the relaying node; otherwise, a negative ACK indicating the number of unrecovered packets is sent, and the source node will transmit more encoded packets to its downstream node for recovery [7].

The settings of m and M affect the performance of reliable transmission and the amount of the redundancy. For optimization, FOCAR first collects the information on packet error rates of each hop from the source to the destination, and then solves an approximate convex optimization problem to get optimal solutions, which will be distributed to every node. The following two components are used: per-hop reliable data transfer and multi-hop optimization. Per-hop reliable data transfer aims to reduce the number of retransmissions by making sure that every transmission of the coded packets within one hop can be successfully decoded with a target probability as follows:

$$\sum_{k=0}^{m-1} \binom{M}{k} P_{ER}^{M-k} (1 - P_{ER})^k \geq 1 - \delta, \tag{12.4}$$

where P_{ER} is the packet error rate of the acoustic channel, and is a system design parameter. Given M, the above relationship can be used by each node to determine m subject to δ.

For multi-hop optimization, the above-mentioned coding operation is also conducted by a relaying node along the path when there are more than m data packets in its FIFO queue. That is, when the relaying node is in the transmission mode and its next hop is in the receiving mode, it will try to code first m packets into M packets, and send them to its next hop. The optimization problem to minimize average end-to-end delay subject to m and M as well as traffic load is further studied in [26].

12.4.2 Network Coding

The basic ideal of network coding is to allow networking units to perform packet-level coding by summarizing several packets for transmission to increase network capacity [5].

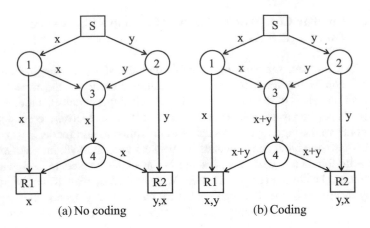

Fig. 12.8 Coding versus no coding in the network: an edge can transmit only one packet once [5]

12.4.2.1   Principle

As illustrated in Fig. 12.8a, when no coding is performed at node 3 after it has received packets x and y, it can either transmit x or y. If x is transmitted, only R2 can receive both x and y. However, if node 3 linearly summarizes x and y to transmit x+y, both R1 and R2 can receive x and y as illustrated in Fig. 12.8b. Compared to no-coding schemes, network coding can be used to transmit more redundancy to reduce the cost of such transmission.

A random linear network coding scheme is described below [27]. Given a single source-destination pair, the following coding is performed by any node in the path except the destination node for n packets (denoted by $M_1, \ldots, M_n$) travelling along this path. The node linearly combines these packets to compute $n'(n' \geq n)$ outgoing coded packets as follows:

$$X = \sum_{j=1}^{n} g_j M_j, \qquad (12.5)$$

where g_j is a binary coefficient (i.e., either 0 or 1), and set $(g_1, \ldots, g_n)$ is called encoding vector, which is forwarded along with packet X. The above encoding can be performed recursively with coded packets [27].

The original packet can be decoded through matrix inversion by solving the following system [27]:

$$\{X_i'\} = \left\{ \sum_{j=1}^{n} g_{i,j}' M_j \right\}, \quad i = 1, \ldots K, \qquad (12.6)$$

where X_i' is a coded packet received by the node, and set $g_{i,j}'$ $(j = 1, \ldots n)$ is the corresponding encoding vector carried by this packet. To solve the above system for decoding n packets, the node needs to receive $K \geq n$ coded packets.

12.4.2.2 Performance Study

The efficiency of the above network coding scheme for error control is demonstrated by simulation in [28] for a 6×6 grid network for the following two scenarios. (i) Network coding is performed only by the relaying nodes. (ii) The coding is also performed by the source nodes. The following forwarding schemes are compared: (i) single-path forwarding, (ii) multipath forwarding, and (iii) multipath coded forwarding. A routing pipe is determined by the vector-based forwarding routing protocol [29] as illustrated in Fig. 12.9. The simulation results show that network coding with source coding performs best. This work is further deepened in [30, 31] by providing theoretical study to set configurations for both efficient error recovery and energy consumption, with a proposal working as follows. The original packets from the source are divided into generations, each of which contains K packets. These packets are linearly combined into a generation using randomly generated coefficients. A relaying node in the forwarding path first linearly combines packets from different paths but belonging to the same generation, and then relays them.

12.4.2.3 Multipath Transmission

Packet-level redundancy can also be provided by multipath transmission, which can make use of bandwidth available in the network to send more copies of the same packet simultaneously. Several such schemes jointly using power control and FEC have been investigated.

Long propagation delays cause the RTT in UWANs significantly larger than that in terrestrial networks. For example, RTTs in UWANs are often in the order of tens of seconds, while few milliseconds in terrestrial networks [32]. Long RTTs severely degrade TCP performance when a congestion is inferred because it takes much time for the source node to restore its TCP congestion window ($cwnd$) to the normal level. In [32], a routing protocol called Linear Coded Digraph Routing (LCDR) is proposed to maximize TCP throughput in underwater acoustic mesh networks. The basic idea

Fig. 12.9 Routing pipe for network coding [28]

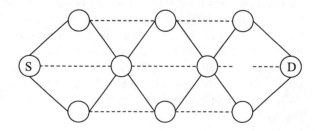

of LCDR is to maximize packet redundancy for lost packet recovery through making use of available bandwidth with multipath transmissions from both the source node and relaying nodes to the destination node and using network coding for packets [7]. It consists of the following components:

- Multipath routing: LCDR finds a directed acyclic graphic (DAG), which is rooted at node A and ends at node G. Packets from nodes A to node G are routed from A to G over the links in this DAG.
- Local sequencing: Each node buffers and locally sequences the packets received from multipath before forwarding to reduce the number of out-order packets experienced by the underlying TCP flows, because these packets may arrive at the buffer out of order due to different delays over different paths, which will be treated as a congestion indication by the TCP module.
- FEC: An ingress node creates forward redundant packets so that a sufficient number of packets can reach the gateway node to recover the dropped packets. This is realized by using network coding through random linear combination of queued packets before forwarding them. The same operation is also suggested for relaying node, which will first linearly combine appropriate packets in its queue before forwarding.

12.4.3   Cooperative Transmissions

Cooperative transmission based reliable transmission for an 1-hop UWAN is investigated in [33], aiming to reduce inefficiency of error control caused by long RTTs. A node is allowed to participate in transmission to other nodes by providing another copy of the same packet to the destination node [7]. As illustrated in Fig. 12.10a, the cooperation region of the destination node is the shadow area, which is the intersection between the transmission areas of the source node and the destination node. Once the destination node receives an erroneous packet, it first asks a node closest to it to retransmit the packet. If not successful, other nodes will be asked one-by-one until the packet is received successfully.

This scheme is extended to a multi-hop UWAN in [34], allowing relaying nodes in a specific source-to-destination route to provide alternative paths. Neighbors or other relaying nodes closer to the destination node may opportunistically take part in retransmission, and these nodes are called cooperators. Whenever a retransmission is required, the destination node requests it from the cooperators first by sending a NACK, and waits for the data. After the destination node successfully receives the data, it sends an ACK to the source node, which also indicates the cooperators that the cooperation is completed. Here the additional cooperation region of a relaying node is the intersection between the transmission areas of the relaying node and the source node as well as the destination node, as illustrated in Fig. 12.10b. It also adopts implicit acknowledgement using a backward overheard packet as an ACK, i.e., the ACK signal is replaced with the overheard data packet returning back from the next hop [35].

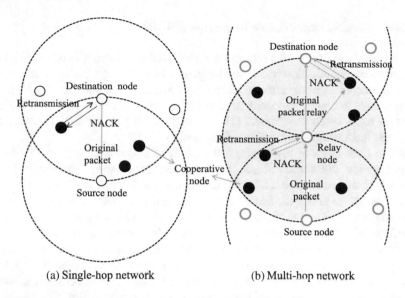

(a) Single-hop network (b) Multi-hop network

Fig. 12.10 Cooperative regions for one-hop and multi-hop UWANs [34]

12.4.4 Reliable Broadcast

As discussed in [7], a reliable broadcast protocol has to control errors of the messages received by different nodes efficiently without causing retransmission storms. Two hybrid FEC/ARQ reliable broadcast protocols are investigated in [36]. They leverage the relationship between communication distance and bandwidth available for an underwater acoustic link since the frequency band available for communication is reduced and shifted toward lower frequencies as distance increases [37]. The power control of typical underwater devices allows communication range to vary on the order of tens of kilometers [36], as shown by empirical formulas (9.7) in Chap. 9.

The proposed two schemes are based on a Simple Reliable Broadcast (SRB) protocol. With SRB, every node re-broadcasts a received broadcast message to its neighbors. If a node has not received all the packets of a message, it waits a predefined time and broadcasts a retransmission request to its neighbors. The neighbors will broadcast the requested packets after winning medium contention through MAC such as CSMA [7]. For high BER channels, a FEC-based SRB (FSRB) is more suitable by encoding each message before transmitting it so that the receiver can correct errors, while retransmission is necessary only if FEC fails. For both, every node repeats the broadcast, which may cause more redundant transmissions in dense networks. To minimize computational cost for message forwarding with FSRB, the same encoding mechanism can be used by each node to avoid decoding/re-encoding of each received broadcast message before repeating the broadcast.

12.4.4.1 Single-Band Reliable Broadcast (SBRB)

As discussed in [7], the SBRB protocol expands SRB by allowing a node to employe long range communication in a lower-frequency band but at a high-power level to notify all neighbors of the initiation of a broadcast. Then the node uses short-range transmissions in a high-frequency band but at a low-power level to send broadcast messages to its neighbors (Fig. 12.11). This signal will not collide with these long range notification signals at different frequencies.

Once a node successfully receives all packets of a broadcast message, it tries to contend for the channel to re-broadcast the received message at short-range bandwidth and transmission power if it wins the contention. Then, it sends the first packet of the message to notify the nearest neighbors, and then sends a notification to all other nodes with the long-range bandwidth and transmission power, telling them an undergoing broadcast process. Once this message is sent, the node completes the broadcast transmission on the short-range band. If the node fails in contention, it will give up re-broadcasting if it receives the same broadcast from one of its neighbors to avoid unnecessary broadcasting; otherwise, it contends for the channel again. If any node does not receive all packets of the broadcast message after a certain time, it asks its neighbors to retransmit them.

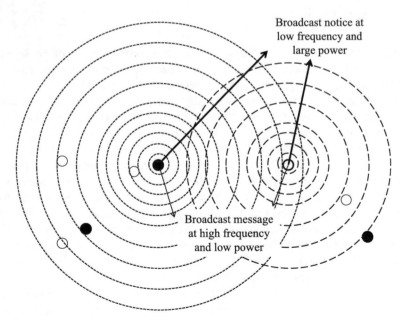

Fig. 12.11 Principle of reliable broadcast based on SRB

12.4.4.2 Dual-Band Reliable Broadcast (DBRB)

Here the main difference from SBRB is that low-frequency band for long distances at high-power level is used to send some FEC data for error correction. Every node sends some redundant data in long distance band to repeat broadcast so that some nodes may have sufficient redundancy to reconstruct the message. The simulation study shows that both SBRB and DBRB much outperform other protocols in terms of energy consumption and completion time, while DBRB performs best in the case of high BER [7].

12.5 Transport Layer

Only a few schemes have been proposed for the transport layer mainly due to the following reasons. First, long propagation delays in UWANs significantly increase end-to-end latency, so that transport-layer schemes alone cannot perform well without support of lower-layer schemes. Second, if lower-layer schemes is in place, some transport layer protocols developed for RWNs can be used here (e.g., Semi-TCP) since this layer is less aware of networking difference.

To handle time-varying characteristics of underwater acoustic channels, an Adaptive RTT-driven transport layer flow and error control (ARTFEC) scheme is investigated for a multi-hop UWAN in [38]. It combines an adaptive congestion window (ACWND) control and Q-learning timeout selection. ACWND is an RTT-based data flow control scheme, adjusting congestion window according to TCP throughput during a congestion duration, which is called congestion throughput and calculated based on varying underwater acoustic communication environments estimated through RTT feedbacks from data and ACK packets as described below [7].

During a particular time epoch, after a number of TCP segments have been transmitted following the current window size, the sender will re-calculate congestion throughput and $cwnd$ when the number of acknowledged TCP segments is less than the number of sent segments after timeout. Then, all unacknowledged segments will be retransmitted. When all sent segments have been acknowledged before timeout, only $cwnd$ is re-calculated. When a particular segment is indicated by the receiver to the sender as "being lost", its retransmission will be triggered after timeout to avoid unnecessary retransmission. However, when a particular segment is indicated as "being error", its retransmission will be triggered immediately, and $cwnd$ is also re-calculated.

As summarized in [7], an optimal timeout selection is implemented with a Q-learning algorithm. It is a model-free reinforcement learning technique, and can be used to yield near-optimal policies without much computations, using an off-policy temporal difference approach [39, 40], in which the agent approximates the Q-values iteratively. More description can be found in Sect. 11.4.4.1.

12.6   Discussion

Tables 12.1 and 12.2 summarize the reliable transfer schemes reviewed in this book for link-level and path-level, respectively, highlighting the major design points. For the link-level schemes, the duplex mode of underwater acoustic links assumed by each scheme is noted. Since most path-level schemes jointly use coding, the adopted coding schemes are highlighted.

The above discussion shows that end-to-end transfer reliability control involves the physical layer, the data link layer and the network layer as well as the transport layer. This feature is unique and different from other major networking technologies such as MAC and routing protocols, but similar to network security to be discussed in Chap. 13. To optimize reliable transfer performance in harsh UWAN environments with less consumption of channel resource and energy, most schemes are jointly designed or cross-layered.

An optimal design for end-to-end transfer reliability control in UWANs requires a systematical plan for function distribution and cooperation from the physical layer up to the transport layer, in order to minimize transmission redundancy, computation loads, buffering operation and consumption of channel resource and energy. Such plan needs a throughout investigation by further considering asymmetric features in UWANs that somewhat have been ignored by most schemes. Typically, these features include asymmetric underwater acoustic links, asymmetric capacity and capability of nodes in UWANs. For example, an asymmetric link scenario may be caused by a directional beam with less than 360°-width rather than an omni-directional coverage

Table 12.1   Typical schemes for link-level reliability control

Schemes/Ref.	Duplex	Major design points
Ref. [3]	Half	Packet train
JSW [16]	Half	Packet train
USR [19]	Full	Time division duplex
Type-I,II HARQ [22]	Half	Bit-level FEC

Table 12.2   Typical schemes for path-level reliability control

Schemes/Ref.	Codes	Major design points
FOCAR [26]	Fountain codes	Hop-by-hop control Feedback channel (Half duplex)
Ref. [28]	Network coding (NC)	Joint design
Ref. [33]	(NA)	Cooperative transmission
SBRB [36]	(NA)	Power control
DBRB [36]	(NA)	Power control, bit-level FEC dual bands
ARTFEC [38]	(NA)	Q-learning

E2E=End-to-end control, HBH=Hop-by-hop control

as illustrated in Fig. 11.6, or by an upslope bathymetric profile, in which, the channel may experience much higher BERs in the direction toward the upslope than in the opposite direction [41]. Regarding asymmetric capacity and capability, a sink node on water surface is usually much powerful than an underwater node in terms of communication capacity, energy supply and buffer capacity etc.

12.7 Summary

Different from other networking technologies such as medium access control (MAC) protocols and routing protocols, the performance of end-to-end reliable transfer depend on the performance of the physical to the transport layer. An optimal design needs a systematical plan for distribution and cooperation of the relevant functions on these layers, and a throughout investigation taking into account major asymmetric features in UWANs is still necessary. Furthermore, both bit-level and packet-level coding is helpful to further improve reliable transfer performance, and fully adaptive coding is especially important for very resource-constrained and time-varying UWANs. A more comprehensive survey can be found in [7].

References

1. Stojanovic, M.: Optimization of a data link protocol for an underwater acoustic channel. In: Proceedings of the MTS/IEEE OCEANS, Washington, DC, USA (2005)
2. Partan, J., Kurose, J., Levine, B.N.: A survey of practical issues in underwater networks. In: Proceedings of the ACM International WS, Underwater Networks (WUWNet), Los Angeles, California, USA (2006)
3. Proakis, J.G., Sozer, E.M., Rice, J.A., Stojanovic, M.: Shallow water acoustic networks. IEEE Commun. Mag. **39**(11), 114–119 (2001)
4. Luby, M.G., Mitzenmacher, M., Shokrollahi, M.A., Spielman, D.A.: Efficient erasure correcting codes. IEEE Trans. Inform. Theory **47**(2), 569–584 (2001)
5. Bassoli, R., Marques, H., Rodriguez, J., Shum, K.W., Tafazolli, R.: Network coding theory: a survey. IEEE Commun. Surv. Tutor. **15**(4), 1950–1978, Fourth Quarter (2013)
6. Fong, S.L., Yeung, R.W.: Variable-rate linear network coding. IEEE Trans. Inform. Theory **56**(6), 2618–2625 (2010)
7. Jiang, S.M.: On reliable data transfer in underwater acoustic networks: a survey from networking perspective. IEEE Commun. Surv. Tutor. **PP**, 99 (2018)
8. Zhao, N., Richard Yu, F., Jin, M.L., Yan, Q., Leung, V.C.M.: Interference alignment and its applications: a survey, research issues, and challenges. IEEE Commun. Surv. Tutor. **18**(3), 1779–1803, Third Quarter (2016)
9. Boukalov, A.O., Häggman, S.-G.: System aspects of smart-antenna technology in cellular wireless communications an overview. IEEE Trans. Microwave Theory Tech. **48**(6), 919–928 (2000)
10. Alamouti, S.M.: A simple transmit diversity technique for wireless communications. IEEE J. Sel. Areas Commun. **16**(8), 1451–1458 (1998)
11. Luby, M., Mitzenmacher, M., Shokrollahi, A., Spielman, D., Stemann, V.: Practical loss-resilient codes. In: Proceedings of the annual ACM symposium theory of computing (STOC), El Paso, TX, USA, pp. 150–159 (1997)

12. Maymounkov, P., Mazieres, D.: Rateless codes and big downloads. In: Proceedings of the International WS Peer-to-Peer Systems in Berkley, USA (2003)
13. Luby, M.: LT codes. In: Proceedings of the Annual IEEE Symposium on Foundations of Computer Science (FOCS), Washington, DC, USA, pp. 271–280 (2002)
14. Shokrollahi, A.: Raptor codes. IEEE Trans. Inform. Theory **52**(6), 2551–2567 (2006)
15. Schwartz, M.: Telecommunication Networks, Protocols. Modeling and Analysis. Addison-Wesley Publishing Company, Massachusetts, USA (1987)
16. Gao, M.S., Soh, W.-S., Tao, M.X.: A transmission scheme for continuous ARQ protocols over underwater acoustic channels. In: Proceedings of the IEEE International Conference on Communications (ICC), Dresden, Germany (2009)
17. Chitre, M., Soh, W.-S.: Reliable point-to-point underwater acoustic data transfer: to juggle or not to juggle? IEEE J. Ocean. Eng. **40**(1), 93–103 (2015)
18. Jiang, J.F., Han, G.J., Zhu, C.S., Chan, S., Rodrigues, J.J.P.C.: A trust cloud model for underwater wireless sensor networks. IEEE Commun. Mag. 110–116 (2017)
19. Azad, S., Casari, P., Guerra, F., Zorzi, M.: On ARQ strategies over random access protocols in underwater acoustic networks. In: Proceedings of the MTS/IEEE OCEANS, Santander, Spain (2011)
20. Azad, S., Casari, P., Zorzi, M.: The underwater selective repeat error control protocol for multiuser acoustic networks: design and parameter optimization. IEEE Trans. Wireless Commun. **12**(10), 4866–4877 (2013)
21. Rizzo, L.: Effective erasure codes for reliable computer communication protocols. ACM SIGCOMM Comput. Commun. Rev. (CCR) **27**(2), 24–36 (1997)
22. Yu, J., Chen, H., Xie, L., Cui, J.-H.: Performance analysis of hybrid ARQ schemes in underwater acoustic networks. In: Proceedings of the St. John's, MTS/IEEE OCEANS (2014)
23. Lin, A.J., Chen, H.F., Xie, L.: Performance analysis of ARQ protocols in multiuser underwater acoustic networks. In: Proceedings of the MTS/IEEE OCEANS, Washington, USA (2015)
24. MacKay, D.J.C.: Fountain codes. IEE Proc. Commun. **152**(6), 1062–1068 (2005)
25. Casari, P., Tomasi, P., Zorzi, M.: Towards optimal broadcasting policies for HARQ based on Fountain codes in underwater networks. In: Proceedings of the IEEE/IFIP Wireless On-demand Network System and Services (WONS), Garmisch-Partenkirchen, Germany (2008)
26. Zhou, Z., Mo, H., Zhu, Y., Peng, Z.: Fountain code based adaptive multi-hop reliable data transfer for underwater acoustic networks. In: Proceedings of the IEEE International Conference on Communications (ICC), Ottawa, Canada, pp. 6396–6400 (2012)
27. Fragouli, C., Boudec, J.-Y.L., Widmer, J.: Network coding: an instant primer. ACM SIGCOMM Computer Communication Review (CCR) (2006)
28. Guo, Z., Xie, P., Cui, J.-H., Wang, B.: On applying network coding to underwater sensor networks. In: Proceedings of the ACM Internatinal WS. Underwater Networks (WUWNet), Los Angeles, USA, pp. 109–112 (2006)
29. Xie, P., Cui, J.-H.: VBF: vector-based forwarding protocol for underwater sensor networks. In: Proceedings of the IFIP Networking Conferences on Coimbra, Portugal, pp. 1216–1221 (2006)
30. Guo, Z., Wang, B., Cui, J.-H.: Efficient error recovery using network coding in underwater sensor networks. In: Proceedings of the IFIP Networking Conferences Atlanta, Georgia, USA, pp. 109–112 (2006)
31. Guo, Z., Wang, B., Xie, P., Zeng, W., Cui, J.-H.: Efficient error recovery with network coding in underwater sensor networks. Ad Hoc Netw. **7**(4), 791–802 (2009)
32. Huang, C.-Y., Ramanathan, P., Saluja, K.: Routing TCP flows in underwater mesh networks. IEEE J. Sel. Areas Commun. **29**(10), 2022–2032 (2011)
33. Lee, J.W., Cheon, J.Y., Cho, H.S.: A cooperative ARQ scheme in underwater acoustic sensor networks. In: Proceedings of the MTS/IEEE OCEANS, Sydney, Australia (2010)
34. Lee, J.W., Cho, H.S.: A cooperative ARQ scheme for multi-hop underwater acoustic sensor networks. In: Proceedings of IEEE Symposium on Underwater Technology and Workshop on Scientific Use of Submarine Cables and Related Tech, Tokyo, Japan (2011)

35. Lee, J.W., Kim, J.P., Shen, J.H., Jiang, Y.S., Cheol, K., Son, K., Cho, H.S.: An improved ARQ scheme in underwater acoustic sensor networks. In: Proceedings of the MTS/IEEE OCEANS, Kobe, Japan (2008)
36. Casari, P., Harris, A.F., III: Energy-efficient reliable broadcast in underwater acoustic networks. In: Proceedings of the ACM International WS. Underwater Networks (WUWNet), Montreal, Canada, pp. 49–56 (2007)
37. Stojanovic, M.: On the relationship between capacity and distance in an underwater acoustic communication channel. In: Proceedings of the ACM International WS, Underwater Networks (WUWNet), Los Angeles, USA (2006)
38. Wang, P., Li, J.H., Zhang, X.: Adaptive RTT-driven transport-layer flow and error control protocol for QoS guaranteed image transmission over multi-hop underwater wireless networks: design, implementation, and analysis. In:Proceedings of the IEEE International Conferences on Communications (ICC), Sydney, Australia, pp. 5142–5147 (2014)
39. Wang, P., Wang, T.: Adaptive routing for sensor networks using reinforcement learning. In: Proceedings of the IEEE International Conference of Computer and Information Technology, Seoul, Korea (2006)
40. Hu, T.S., Fei, Y.S.: QELAR: a machine-learning-based adaptive routing protocol for energy-efficient and lifetime-extended underwater sensor networks. IEEE Trans. Mob. Comput. 9(6), 796–809 (2010)
41. Otnes, R., Asterjadhi, A., Casari, P., Goetz, M., Husøy, T., Nissen, I., Rimstad, K., van Walree, P., Zorzi, M.: Underwater Acoustic Networking Techniques. Springer, Heidelberg (2012)

Chapter 13
Security in UWANs

Abstract Weak security is Achilles' heel of many wireless networks because of the severe security situation caused by the broadcast nature of wireless media, mobility and heterogenous user profiles. Communication capacity, computation capability and energy supply are constrained for countering against security threats in wireless networks. This situation becomes even worse in UWANs because the resources are much more constrained while security situation is more server due to special networking environments. The peculiar features of UWANs cause existing solutions proposed for radio wireless networks (RWNs) cannot be used directly in UWANs (Stojanovic, ACM Mob Comput Commun Rev 11(4): 34–43, 2007, [1], Jiang and Xu, Adv Mater Res 317–319:1002–1006, 2011, [2], Dini and Duca, Sensors 12(11):15133–15158, 2012, [3]). This chapter discusses some typical proposals for UWAN security.

As mentioned in Chap. 8, the fundamental for confidentiality, integrity, authentication and non-repudiation is cryptography, particularly symmetric and asymmetric/public key cryptography. However, they cause ciphertext expansion due to padding and additional fields to be added in encryption as to be discussed in Sect. 13.1.1. Furthermore, high-layer cryptographic mechanisms suffer from heavy computational complexity, especially in resource-constrained UWANs [4–8]. Particularly, public key cryptography widely used for digital signature and authentication (e.g., Ron Rivest, Adi Shamir and Leonard Adlemanv (RSA) introduced in Appendix B) are almost inapplicable in UWANs [9, 10]. Due to specific networking environments, UWANs suffer from some potential attacks that are difficult to defend against as to be discussed in Sect. 13.2.

The above issues are discussed in this chapter with a brief survey on some typical approaches and schemes proposed to address them. The main reference for this part is [11].

13.1 Cryptographic Primitives for UWANs

Some peculiar characteristics of UWANs impose challenges on securing UWANs, especially on the applicability of popular cryptographic primitives and suitability of existing countermeasures in harsh underwater environments. This section discusses

© Springer Nature Singapore Pte Ltd. 2018 337
S. Jiang, *Wireless Networking Principles: From Terrestrial to Underwater Acoustic*,
https://doi.org/10.1007/978-981-10-7775-3_13

cryptographic primitives investigated to protect confidentiality and integrity and to provide authentication and non-repudiation in UWANs.

13.1.1   Challenges to Popular Cryptographic Primitives

The major problems of the existing cryptographic primitives include ciphertext expansion and computational complexity for both asymmetric/public and symmetric key cryptography, which consume much bandwidth and energy, and make these primitives unsuitable for very resource-constrained UWANs, as discussed below [11].

13.1.1.1   Ciphertext Expansion

Message padding and codes to identify message modification and provide message authentication make the message length to increase after applying cryptography [12], resulting in an increased transmission and energy consumption. For example, with the standardized AES encryption, the block size is 128 bits, and the message expansion due to padding is around 18% for a typical UWAN message of 720 bits [3]. For message authentication using digital signature, a digest is usually appended to an authenticated message, also causing expansion and communication overhead. For instance, the size of a digest produced by SHA-256 is 256 bits, which causes an overhead up to around 35% of an average UWAN message.

13.1.1.2   Public Key

It is adopted widely in RWNs for symmetric key distributions and digital signature as well as securing the transfer of credential information. RSA is a widely used public-key system, but it is computationally complex with thousands or even millions of multiplication instructions for single cryptographic operation. This causes a resource-constrained device to take an order of tens of seconds or even up to minutes to perform encryption and decryption [13]. Furthermore, it usually takes a microprocessor thousands of nanojoules to do a simple multiplication for a 128-bit result [14]. For real applications, the public modulus should be more than 1024 bits to guarantee security. These features make RSA not suitable for UWANs [15].

13.1.1.3   Symmetric Key

As discussed in [11], a cipher using symmetric keys is often used to protect confidentiality due to its super performance over asymmetric ones. However, key distribution in resource-constrained UWANs is difficult because it is almost impossible to implement an online distribution center in oceans to allocate secret keys. Combining pseudo-random key generators and key pre-distribution have no true randomness, probably causing cryptanalytic break. With pre-installing keys on nodes, there is a risk that a single compromised node may make a number of nodes sharing the common key unsafe [10].

13.1.2 Symmetric Key for Reciprocal Channels

As discussed in [11], the randomness of the input parameters for a key generation function affects the security level of generated keys since the algorithms for key generation is usually well-known. The basic idea behind key generation for reciprocal channels is reciprocity theorem, i.e., the secret key between two legitimate nodes is generated through sharing a common source of randomness. This is possible even when the quality of the communication channels between them is worse than that with eavesdroppers [16].

13.1.2.1 Principle

The typical sources of shared randomness include impulse response of a reciprocal channel, frequency selectivity and received signal strength (RSS) [17]. As illustrated in Fig. 13.1, the shared randomness of the communication channels between Alice and Bob can be captured by themselves but not by Eve due to different channel correlations. Such randomness can be obtained through measuring the same probing signals sent by Alice and Bob to each other, such as a pure tone signal over a time period synchronously. The signals received by each of them are the random sources to generate a shared key [11].

Fig. 13.1 Principle of symmetric key generation in reciprocal channels [11]

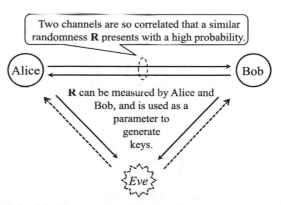

Two channels are so correlated that a similar randomness **R** presents with a high probability.

Alice ⟷ Bob

R can be measured by Alice and Bob, and is used as a parameter to generate keys.

Eve

It is hard for Eve to capture **R** because the channels associated with her are uncorrelated with those between Alice and Bob.

Such key generation can eliminate the deployment of a key distribution center, allowing each pair of nodes to update secret keys at any time. This is due to that the randomness of the generated key depends on the entropy naturally available in the environment. For example, a pair of nodes of a reciprocal channel can produce a shared key through the locally measured RSS [18]. An adversary can hardly guess the secret key generated by them if it is not physically near them, and the spatial diversity of a wireless channel can ensure the secret in the key generation [10].

However, the discrepancy in the measurements of a pair of nodes of a channel increases with the time interval between when they receive the corresponding probing signals sent to each other, especially in time-varying communication channels, and affects successful key generation rates. Particularly for time-varying underwater acoustic channels, long propagation delays and low transmission speeds increase this time interval. How to overcome this problem to improve key generation performance is still a research issue, and some research results reported in the literature are discussed below.

13.1.2.2   Key Generation for OFDM Channels

In an OFDM system, channel frequency response and BCH codes are exploited for information reconciliation in the key generation [19]. The proposed scheme as discussed below is validated by a lake test, which shows that the coherence of reciprocal channels leads to a high probability of successful key generation.

Key Generation

Figure 13.2 depicts the secret key generation process used in the test, which is described below [11].

- Alice sends two OFDM blocks to Bob. The first one contains the frame number, and second one carries a probing signal. Once Bob has successfully received them, he replies immediately to Alice the same OFDM probing signal to minimize the

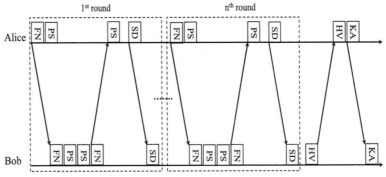

FN=Frame number, PS=Probing signal, SD=Syndrome, HV=Hash value, KA=Key acknowledgement

Fig. 13.2 Symmetric key generation for reciprocal channels with OFDM

interval between the probing signals in order to keep higher channel correlation. Then the decoded frame number is sent to Alice for her to pair the OFDM probing signals.

- Alice first quantizes the observation of the underwater acoustic channel in the frequency domain, and uses it to generate the key. Then she sends the syndrome to Bob based on the predefined error correction code, which is used to help Bob to recover the sequences observed by Alice. Bob extracts the key generated by Alice from the received syndrome with the help of the quantized channel frequency response observed by himself. These two steps repeat until a desired key length is reached.
- Bob generates a hash value of the generated key, and sends it to Alice. She does the same with the same hash function using her generated key as the input. If the two hash values are equal, they have successfully generated a secret key, and Alice sends a key acknowledgement to Bob.

Key Reconciliation

Due to noise and channel time variation, the two binary codewords observed by Alice and Bob (y_A and y_B) may not be identical, and a reconciliation process is proposed. It is carried out by an error correction scheme following Slepian-Wolf coding [20]. Let n denote the length of the codeword and m the length of the information word. Alice and Bob use BCH(n, m), and share the same 2^m information words of length m, $\{b_0, b_1, \ldots, b_{2m-1}\}$, and the corresponding code words of length n, $\{c_0, c_1, \ldots, c_{2m-1}\}$, respectively, as well as the generation matrix $\mathbf{G}$ and the parity check matrix $\mathbf{P}$. The process is described below.

1. Alice calculates a syndrome s_A of length $n-m$ based on y_A through $s_A = y_A \mathbf{P}^T$. Her key is generated as follows:

$$\hat{b}_A = \arg \min_{\mathbf{b}} \| y_A - \mathbf{bG} \| .$$

2. Alice sends s_A to Bob as the help information through a public channel. This information can also be overheard by Eve s_A. But she cannot exploits this information because the two channels from Alice to Bob and Alice to Eve are highly uncorrelated.

3. Bob recovers the coset leader $\mathbf{e}_A$ based on the received syndrome s_A as follows: $\mathbf{e}_A = y_A - \hat{b}_A \mathbf{G}$, and decodes its own key as follows:

$$\hat{b}_B = \arg \min_{\mathbf{b}} \| y_B - (\mathbf{e}_A + \mathbf{bG}) \| .$$

The major problem of such key generation approach is the uncertainty of successful key generation, and a lot of energy and bandwidth may be consumed to generate a sufficiently long key. A variety of RSS-based key generation schemes originally designed for RWNs are evaluated in [10] for UWANs with the following major observations. (i) The long transmission time of a probe signal in UWANs results in a low key generation rate. (ii) The long propagation delay and large transmission time cause the asymmetry of RSS measurements between two communicating parties more significant in UWANs, which causes a high rate of bit mismatch on a shared key [11].

13.1.3   Public Key with Elliptic Curve Cryptography (ECC)

Several non-RSA public-key algorithms have been devised for resource-constrained devices [21, 22]. A typical one is Elliptic Curve Cryptography (ECC) [23] due to its shorter key sizes and higher computational efficiency than other large integer-based algorithms [9] for the same security strength.

ECC is based on elliptic curves defined by a set of parameters that are chosen in such a way that it is difficult to solve the Elliptic Curve Discrete Logarithm Problem (ECDLP) in reasonable time [23]. To match the security of RSA, smaller elliptic curve keys are needed. In this case, theoretically it is possible to make a practical attack feasible many years before such an attack is available on an equivalently secure RSA scheme [25]. Therefore, ECC is expected to be more vulnerable than RSA to attacks using Shor's algorithm [26], which runs on a quantum computer for integer factorization.

Table 13.1 compares cipher key lengths of typical cipher schemes for an equivalent-level security, showing that the key size of ECC is double the symmetric ones but with a large superior over those of DSA/RSA. Accordingly, Elliptic Curve Digital Signature Algorithm (ECDSA) and Elliptic Curve Authenticated Encryption Scheme (ECAES) have also been devised [24].

13.1.4   Digital Signature

As discussed in [11], three digital signature schemes: ECDSA, Zhang-Safavi-Naini-Susilo (ZSS) [27] and Boneh-Lynn-Shacham (BLS) [28], are evaluated for underwater acoustic sensor networks (UWSNs) in terms of energy efficiency in [29]. Both BLS and ZSS are short signatures with the signature sizes of about 160 bits for a security level of 280. The signature generation is computationally efficient. BLS also supports signature aggregation to accumulate signatures from different signers and

Table 13.1 Key lengths (bits) for equivalent security with typical cipher schemes [11, 24]

Symmetric cipher	ECC cipher	DSA/RSA cipher
80	160	1024
112	224	2048
128	256	3072
192	384	7680
256	512	15360

DSA=Digital Signature Algorithm, which is based on discrete logarithms computation

Table 13.2 Signature size, generation time and authentication overhead [11, 29]

Schemes	Generation time (ms)	Length (bytes)	Overhead
ECDSA	134	40	Largest
ZSS	229	21	Medium
BLS	302	21	Lowest

on distinct messages into a single short value. Signatures shorter than 160 bits have also been studied. Their two typical independent security parameters (i.e., extension degree and hidden polynomial degree) can make them more flexible than ECC-based schemes. One can set the first one small to achieve shorter signatures, and tune the other independently to achieve the desired security level [23]. The above features make these schemes possible candidates suitable for UWANs [29].

Table 13.2 summarizes the major characteristics of these schemes in terms of key generation time and key length. This evaluation targets 80-bit security, which is equivalent to the security strength of RSA-1024 [29]. It only considers an end-to-end authentication scenario, in which, digital signature will be verified by resource-rich end points. The results show that these schemes perform well in RWNs may not do equally well in UWSNs, and BLS presents the lowest overhead, while ECDSA yields the largest one [29].

Note that these digital signature schemes are not suitable for link-layer authentication due to very constrained resources of these nodes and hop by hop operation necessary for authentication. Instead, Message Authentication Codes (MACs) can be used because they are more computationally efficient with approximately the same amount of processing, and equivalent-size codes for a given security level [30].

13.1.5 Reputation-Based Authentication

As discussed in [11], authentication is usually carried out through cryptographic operations, which are usually computationally complex with high energy consumption. Another type of authentication approach is proposed for resource-constrained networks such as trust models. They are based on reputation aiming to measure abnormality according to statistics. In the following, such a scheme for UWANs, called Attack-Resistant Trust Model based on multidimensional trust Metrics (ARTMM) [31], is introduced, taking into account the characteristics of underwater acoustic channels and node mobility.

13.1.5.1 Trust Model

As illustrated in Fig. 13.3, this model consists of the following components:

- Link trust (T_{link}) is assessed with link quality (L_q) and link usage (L_u). Link quality is measured jointly by packet loss rates (P_l) and packet error rates (P_e), i.e., $L_q = (1 - P_l)(1 - P_e)$, where P_e is determined by channel BERs. Link usage

Fig. 13.3 Structure of the
ARTMM reputation model

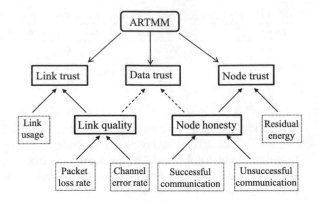

is the ratio of a link that has been used during a time window. Then,

$$T_{link} = \begin{cases} 0.5 + (L_q - 0.5) \times L_u \text{ if } L_q \geq 0.5, \\ L_q \times L_u, \qquad\qquad\quad \text{else.} \end{cases}$$

- Node trust is evaluated by node honesty (NH) and the residual energy available in the node (E_{res}). NH is measured according to the numbers of successful (s) and unsuccessful (f) communications via the node. Following the subjective logic framework [32], NH is calculated by

$$NH = \frac{2s + 1}{2(s + f + 1)}. \tag{13.1}$$

Then the direct node trust is calculated by

$$T_{node} = \begin{cases} 0.5 + (NH - 0.5) \times NC, \text{ if } NH \geq 0.5, \\ NH \times NC, \qquad\qquad\qquad \text{else,} \end{cases}$$

where,

$$NC = \begin{cases} 1 - p_{eng}, \text{ if } E_{res} > \theta, \\ 0, \qquad\quad \text{else,} \end{cases}$$

with θ denoting the energy threshold, while $p_{ene} \in [0, 1]$ being the energy consumption rate.

- Data trust depends on the link quality and the node honesty. To calculate T_{data}, it assumes that the mean (μ) is the most trusted one with the highest trust value for a set of data. Then,

$$T_{data} = 1 - 2 \int_{\mu}^{v_d} f(x) dx, \tag{13.2}$$

where $f(x)$ is the probability density function of the data items, and v_d is the numerical value of a data item.

A fuzzy membership function of trust value based on the inter-dependency property of three trust parameters is further defined by the following fuzzy sets:

- Completely untrust with trustvalue $\in$ [0, 0.25),
- Untrust with trustvalue $\in$ [0.25, 0.5),
- Uncertainty with trustvalue $= 0.5$,
- Trust with trustvalue $\in$ (0.5, 0.75),
- Completely trust with trustvalue $\in$ [0.75, 1].

13.1.5.2 Sliding Time Window

The sliding time window [33] is adopted to update trust values dynamically in order to adapt changing underwater environments. During a time window, the numbers of packets exchanged successfully and unsuccessfully between two neighbors as well as the residual energy level of nodes are recorded, which are used to calculate the current trust values. After a unit of time elapses, the window slides forward, and the historical trust values are used to update new trust values in the next time window.

13.1.5.3 Enhancement Proposals

The main weaknesses of ARTMM are summarized below. (i) The complexity of the algorithms may invoke lots of computation and communication overhead for information collection (e.g., residual energy). (ii) The trust evidence generation does not take into account the influences of malicious attacks [34]. (iii) The fuzzy set definitions are subjective, and cannot be adaptive to dynamic UWANs, while the fuzzy logic method cannot well describe the uncertainty of trust relationship with a definite real number [34]. Such uncertainty actually presents fuzziness and randomness among underwater nodes especially among strange nodes. Therefore, a trust model based on cloud theory is proposed in [34, 35] to solve the above problems, by assuming the availability of position information of each node in UWANs. The cloud model is based on the traditional fuzzy set theory and probability statistics theory [36]. It claims that this model is good for not only the combination of multiple trust attributes, but also the calculation of recommendation and indirect trust values.

13.2 Security Threats in UWANs

There are many security attacks in UWANs [37–40]. An attack can be passive or active according to actions taken by the attacker in order to enforce an attack. With a passive attack, the attacker does not transmit any signal, while sending signal necessary for an active attack, and it is very difficult to detect an ongoing passive attack.

Table 13.3 Network security with the layered network architecture [11, 37]

Layer	Typical threats	Defense strategy
Transport layer	Man-in-the-middle e.g., fake ACK	Authorization, encryption
Network layer	Routing attack, packet interception e.g., black hole, sinkhole	Routing process authentication
Data link layer	Attacks to location, neighbor (wormhole, Sybil), MAC, time synchronization	Abnormality monitoring
Physical layer	Channel jamming, signal eavesdropping	Spread spectrum communication

*Security issues of one layer may be addressed by higher layers

Fig. 13.4 A default security threat scenario [11]

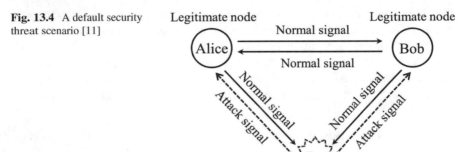

Malicious/Compromised node

In the following sections, we will summarize the typical attacks according to the layers from which an attack is initiated as summarized in Table. 13.3. A default scenario is illustrated in Fig. 13.4, where Alice and Bob are assumed to be two legitimate nodes, while Eve is a malicious or compromised node. Such settings will be used throughout the following discussion if unspecified otherwise.

Actually, the security situation in UWANs is largely affected by the special underwater networking environments, which are first discussed below.

13.2.1 Environmental Factors

Underwater nodes are usually deployed in unattended and even hostile environments [41]. It is possible for an adversary to compromise or even capture them because it is almost impossible to implement physical countermeasures to protect them all [42]. Identifying compromised security is often carried out through detecting anomalies in the expected communication and movement patterns [3]. However, this is a difficult issue especially for passive and carefully designed attacks. Long propagation delays make attack detection less efficient, while certain networking services such as localization and routing highly depend on the availability and freshness of the information [40]. Wormhole attacks also become much easier because either RF links or wired links can provide much faster data links than acoustic links.

To ensure security, removing compromised nodes are necessary but costly. A logical removal of a compromised node should be carried out through rekeying across the whole network because of possible leakage of the secret contained in compromised nodes [3]. However, this may not be sufficient because a compromised node can still jam the network. Very narrow bandwidth makes UWANs especially vulnerable to jamming attack, whilst batter-operated underwater nodes that are not easily recharged may be exhausted of energy by repeated attacks. Therefore, a physical removal is essential.

A threat source could be deployed anywhere and anytime, while an underwater acoustic link is open to any node within the communication range. In this case, an adversary can passively intercept acoustic signals for analysis, or actively disrupt network services such as localization, time synchronization and routing [42]. Sparse node density causes low availability of multipaths between nodes, which make multipath routing less efficient.

Different from radio communications, so far no standard model is available for underwater acoustic communication [43], and underwater channel conditions are affected by various properties of underwater environments such as depth, temperature and salinity. These make it difficult to distinguish between attack evidences and abnormal situations due to channel variations and differences in communication schemes.

13.2.2 Physical Layer Attacks

As discussed in [11], the typical active attack on this layer is channel jamming[1], and the passive attack is signal eavesdropping.

13.2.2.1 Channel Jamming

As illustrated in Fig. 13.5, one or several attackers emit signals to the channel to interfere the reception of legitimate nodes in order to paralyze the normal communication. Since reliable physical communication is the fundamental for all networking operations, the damage of such attack is large and even fatal to the network. Particularly, if a victim is certain special nodes such as a base station, access point or gateway or the root node, a successful jamming attack can paralyze the whole network [37].

---

[1] Some references categorize such attacks as Denial of Service (DoS). Conventionally, DoS usually refers to attacks targeting at some particular nodes such as servers attached to a network to disable their services by exhausting their resources. This paper follows this convention.

Fig. 13.5 Diagram of
jamming attacks

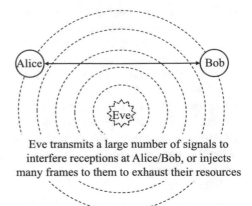

Eve transmits a large number of signals to
interfere receptions at Alice/Bob, or injects
many frames to them to exhaust their resources

Such attack consumes energy of the attacker, and it is relatively easy to identify
and locate the attacker. However, the attacker cannot obtain any useful information.
The most effective countermeasure is to eliminate attackers physically. To this end,
identification and localization of the attacker have to be carried out first, and is followed by demolishing operation. A legitimate node can also increase its transmission
power to improve signal-to-interference noise ratio (SNIR) at the receiver against
such attack but at cost of more power consumption. It is also possible to exploit
low-power transmission to tempt a smart attacker to keep attacking until it exhausts
its energy [11].

13.2.2.2 Signal Eavesdropping

An attacker just silently collects signals of communication between legitimate nodes
through listening. It is a easy, cheap and energy-efficient attack since wireless media
are broadcast by nature, and any node can receive the signal from others if it is located
in their communication ranges. The collected signals are the fundamental of many
attacks launched on higher layers [11].

13.2.3 Link Layer Attacks

Similar to the physical layer, there is also a link jamming attack, with which, an
attacker injects a large number of frames into the communication medium to block
legitimate nodes' access to the medium [44]. If the attacker does not follow the MAC
protocol to transmit frames, the effect of such attack is just equivalent to that caused
by channel jamming; otherwise, such an attack is a sort of MAC protocol attacks.
Actually, there are many other attacks that can be launched at this layer, and the
typical ones are discussed below [11].

Fig. 13.6 Diagram of replay
attack [38]

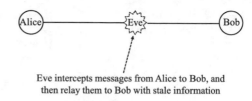

Eve intercepts messages from Alice to Bob, and
then relay them to Bob with stale information

13.2.3.1   Localization Attack

As discussed in [11], the information on a node's location is very important to construct network topologies to improve the performance of networking protocols such as MAC and routing as well as security. Such information usually includes positions, relative distances and the angle of arrival (AOA) of the received signal, which are usually obtained through measurement when GPS is not available. However, several attacks can influence successful information collection or the accuracy of the collected information to make a node to appear closer to or farther away from another. Some of these attacks are discussed below [45].

- Replay attack: As illustrated in Fig. 13.6, Eve intercepts a message sent by Alice to Bob, and then re-sends it to Bob, who gets imprecise locations of Alice. This result is caused by imprecise propagation time and signal strength because the distance is often estimated according to the signal arrival time, AoA or signal strengths. Eve can even delay a relay to make the real sender to appear farther away from the sender by jamming the normal reception of Alice's signal at Bob if they are close to each other. Eve may also send the response on behavior of Bob before it receives a request sent by Alice to Bob to make them seemingly closer to each other. Similar effect can also be achieved by making transmission power different from the pre-agreed level. All these attacks cause fake changes in the network topology [11].
- Non-Cooperation: A minimal number of anchor nodes are often required by a localization scheme for location estimation (e.g., three anchor nodes in [46]), while a distributed localization scheme further requires unknown nodes to cooperate for localization (e.g., [47]). If some nodes are compromised or destroyed, which causes the number of functional nodes to fall below a threshold, location estimation will fail. A similar attack can also be carried out by providing false information on positions of the anchor nodes or unknown nodes [11].

The performance of localization schemes can be also affected by the above-mentioned replay attack, as well as the wormhole attack discussed below [38].

13.2.3.2   Neighborship Attacks

As discussed in [11], such attack aims to establish fake neighbor relationship in order to deviate normal traffic to malicious nodes. For example, Eve can pretend herself as a

Fig. 13.7 Diagram of wormhole attack [11, 38]

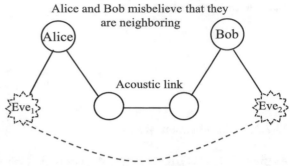

Alice and Bob misbelieve that they are neighboring

Acoustic link

A wormhole link faster than the acoustic link

legitimate node with stale link information (e.g., delay and cost) so that Alice or Bob mistakes Eve as a legitimate neighbor, and selects her as the next hop for routing. In this case, Eve can obtain illegally the information routed through her. Identification of malicious nodes through abnormality detection is not effective because sometimes an abnormality is caused by channel variations rather than attacks [38]. Typical such attacks include wormhole attack and Sybil attack as discussed below.

Wormhole Attack

As illustrated in Fig. 13.7, two malicious nodes use an out-band low-latency wormhole link (e.g. RF or wired links) between them to create a fake neighbor relationship between legitimate nodes [38, 45, 48]. With a wormhole link, one end (e.g., Eve_1) records frames and forwards them to its colluding end in another part of the network (e.g., Eve_2), which then replays the frames. This attack causes fake change in the network topology by either enlarging the neighborhood (e.g., [49]) or reducing the shortest routing path between two legitimate nodes (e.g., [50]) [45]. A wormhole link is most likely selected by legitimate nodes to set network connections for routing so that the malicious node can have all messages transmitted along this path to monitor the network, and inject messages or replay received messages with stale information to all connected nodes [37, 38].

Sybil Attack

With such an attack as illustrated in Fig. 13.8, an attacker can forge many identities of legitimate nodes, which are pretended in many places at once [38, 45, 48] by broadcasting fake messages to mislead the receivers to believe that they have some neighbors which actually do not exist at all. The misled nodes will forward frames to these forged neighbors so that the attacker can intercept them for further attack.

13.2.3.3   Attack to MAC Operations

As discussed in [11], an attacker aims to disturb/disrupt normal MAC operations or make nodes to consumer more energy for the normal MAC operation. For example, for the RTS/CTS handshake of IEEE 802.11, any node overhearing either an RTS or

Fig. 13.8 Diagram of Sybil attack [11, 38]

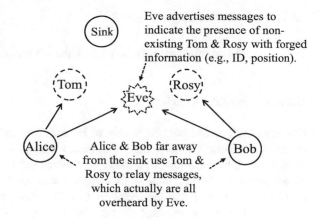

Eve advertises messages to indicate the presence of non-existing Tom & Rosy with forged information (e.g., ID, position).

Alice & Bob far away from the sink use Tom & Rosy to relay messages, which actually are all overheard by Eve.

a CTS should not transmit during the time period indicated by the RTS and the CTS. However, a malicious node may violate this rule, and transmits even after overhearing a CTS destined for a legitimate node, causing collision at the receiver. Alternatively, an attacker may repeat sending RTS frames to one legitimate node to exhaust its energy due to returning many CTS frames for each received RTS frame [37]. Such attack is more energy-efficient than channel jamming to paralyse network operations by avoiding blind broadcasting following MAC protocols, and it is more difficult to be detected. For MAC ACKs, a malicious node overhearing frames destined for a legitimate node with a weak link or located in a shadow zone can send a fake ACK to the sender [38] so that it will continue its transmission, which actually cannot be received by the real destination node over these links.

13.2.3.4   Time Synchronization Attacks

Precise time synchronization is important to schedule nodes' activities such as transmission, reception, sleeping or wake-up. Attacks to time synchronization process will affect the accuracy of the synchronized clocks, which further affects the efficiency of scheduled operations. Actually, many above-mentioned attacks can affect this accuracy. For example, Sybil, wormhole and replay attacks cause fake measurement results on ranges or round-trip time (RTT) between legitimate nodes. These results are the important parameters for time reference alignment. Message authentication based on cryptographic schemes can be used to prevent these attacks if attackers have no knowledge of the cipher keys or security protocol procedure [51]. However, it is possible for an attacker to compromise a legitimate node to obtain the essential information to impersonate a neighbor of a legitimate node under synchronizing with its neighbors to launch an insider attack [51].

13.2.4   Network Layer Attacks

Typical attacks on this layer include routing attacks and packet interception.

13.2.4.1   Routing Attack

Routing attack causes packets unable to be delivered to the destination node, and even worse forwards them to malicious nodes, using fake path information such as delay or cost. For example, in a black hole attack (or sinkhole attack), an attacker broadcasts a forged path with a seemingly lowest cost or shortest path toward a destination node to make the receivers to select this path. Actually, the selected path goes through the attacker so that it can analyze or even drop packets at its will [37]. Such attack becomes easier in wireless ad hoc networks [38] due to the broadcast nature of communication media and loose topology control. More smartly, an attacker may drop packets during certain time period or at certain percentage to make detection difficult. When multiple nodes are compromised, it is possible to launch collaborative attacks such as distributed DoS (DDoS), which is more difficult to prevent [11].

As discussed in [11], geographic routing protocols are more popular in UWANs because packets are forwarded according to the location information of nodes (e.g., depth) without using a dedicated route discovery process. However, such kind of protocol is especially vulnerable to location/neighbor spoofing attacks because broadcast information exchange process can be easily attacked. Cryptographic techniques are often used to secure routing protocols for protecting integrity and confidentiality as well as authentication and internal attack defense. However, the use of encryption increases not only the size of communication messages but also energy consumption due to high computational complexity. Network-wide security key distribution and maintenance are also two challenging issues in UWANs [52].

13.2.4.2   Packet Interception

A compromised node (e.g., a router) can intercept incoming packets for analyzing or selectively dropping them. For example, an attacker may send ACKs to the source node as if it is the destination node. Actually, it drops the received packets so that the destination node cannot receive what are sent to it by the source node. Moreover, a compromised node can even inject other packets to the destination node on behavior of the source node [37]. Intercepted packets can be used to launch other attacks on higher layers such as man-in-the-middle attack to be discussed below. By listening to the network and localizing nodes, an attacker can improve attack performance by attacking key nodes such as the root node [53].

Fig. 13.9 Diagram of
man-in-the-middle attacks

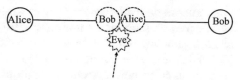

Eve impersonates Bob to Alice, and Alice to Bob so
that they believe that they are communicating to each
other directly

13.2.5   Transport Layer Attacks

As discussed in [11], typical attacks on this layer are related to TCP, such as SYN
attack for TCP connection setup, which is one kind of DoS attacks, and session attack
like man-in-the-middle. Countermeasures proposed for RWNs against such attacks
may be used here because one end of a TCP connection in UWANs is a sink or a
gateway, which is usually a powerful server and the potential target of such attacks.

13.2.5.1   Denial of Service (DoS)

In a TCP SYN attack, many malicious or compromised nodes synchronizes to flood
a large number of requests (i.e., SYN segments) to establish TCP connections to one
destination node, aiming to exhaust its memory so that it cannot accept other TCP
normal connection requests from legitimate nodes.

13.2.5.2   Man-in-The Middle

There are several TCP session attacks. With man-in-the-middle, an attacker secretly
relays the communication messages between a pair of nodes with possibly modification, and makes them to believe that they are directly communicating with each
others. As illustrated in Fig. 13.9, an attacker forges TCP ACKs with fake sequence
numbers of TCP segments in order to trigger unnecessary retransmission or block
necessary retransmissions [37]. An attacker can also split a TCP connection into two
portions so that it can intercept all TCP segments transferred through this connection.

13.3   Countermeasures Against Typical Threats

This section reviews some security schemes proposed to defend UWANs against
the major attacks from the physical layer to the network layer. These attacks include
signal eavesdropping, wormhole attack and routing attacks. Countermeasures against

other attacks on the transport layer can refer to those proposed for RWNs. Some countermeasures against the following UWAN attacks can be found in the literature: channel jamming [44, 54], location spoofing [42], Sybil attack [55], MAC attack [56] and time synchronization attack [51]. A cryptographic suite jointly considering cryptographic primitives, security functions and countermeasures against attacks is introduced at the end of this section.

13.3.1 Against Signal Eavesdropping

CDMA is a promising communication technology that is widely used in wireless networks. However, DS-CDMA is vulnerable to attacks because it is possible for attackers to identify blindly the spreading code used by legitimate nodes even when neither CSI nor training sequence is available [57]. A CDMA-based cooperative jammer with analog network coding (ANC) [58] is proposed in [8] to counter against signal eavesdropping. Such a jammer transmits the information, which is known a priori to Bob but not to Eve, using the same spreading code as used for the legitimate Alice-Bob link. Although the jammer's frame will interfere the reception at Bob, Bob can suppress the interference to decode Alice's frames by estimating the two multipath affected channels, while Eve cannot as described below.

13.3.1.1 Analog Network Coding (ANC)

The basic idea of the ANC is to allow concurrent signal transmissions over a wireless medium so that they intentionally interfere with each other to provide covert communications in underwater acoustic channels. Covert communication can be used to secure communication with the following properties. An input signal is much weaker than the ambient noise (e.g., -10 dB SNR in the signal band) so that it is difficult for a listener who has no prior knowledge about the signal to detect it. If the signals are like noise, it is difficult to decode them without a prior knowledge of the structure of the signal [11].

13.3.1.2 Jamming-through-ANC (J-ANC)

The J-ANC protocol considers a DS-CDMA link between Alice and Bob. Here, Eve has a better channel quality than Bob because she is closer to Alice as illustrated in Fig. 13.10. Eve also has perfect knowledge of all CSI and of the spreading code utilized by Alice and the friend cooperative jammer. The jammer is selected to transmit information modulated with the same spreading code assigned to the Alice-Bob link. Both Bob and Eve have to remove the jamming signal in order to retrieve Alice's frames. However, Bob knows the information bits transmitted by the cooperative

Fig. 13.10 System model for J-ANC [8]

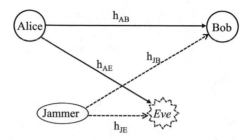

jammer a priori, but not Eve. So Bob can suppress the interference to retrieve Alice's frames, but Eve will fail to do so with a high probability [11]. More details are given below.

For conciseness, let $k \in \{A, J\}$ indicate the transmitters Alice or the jammer, and $k' \in \{B, E\}$ the receivers Bob or Eve. The baseband supervised and the information bits transmitted by Alice or the jammer all are expressed in the following form:

$$\mathbf{x}_k(i) = b_k(i)\sqrt{E_k}\mathbf{s}, \ i = 1, 2, \ldots,$$

where, $\mathbf{s} \in \frac{1}{L}\{\pm 1\}^L$ denotes the normalized spreading code of length L, E_k is the transmission energy per bit, $b_k(i) \in \{-1, 1\}$ is the ith bit modulated with the binary phase-shift-keying (BPSK) scheme.

Consider that the frames transmitted by the two transmitters (k) experience the same number of resolvable multipaths (M), and arrive at the two receivers (k') simultaneously. The received signals at Bob and Eve are respectively denoted by:

$$\mathbf{r}_B(i) = \sqrt{E_A}\mathbf{S}_A(i)\mathbf{h}_{AB} + \sqrt{E_J}\mathbf{S}_J(i)\mathbf{h}_{JB} + \mathbf{n}_B(i),$$

$$\mathbf{r}_E(i) = \sqrt{E_A}\mathbf{S}_A(i)\mathbf{h}_{AE} + \sqrt{E_J}\mathbf{S}_J(i)\mathbf{h}_{JE} + \mathbf{n}_E(i),$$

where,

$$\mathbf{h}_{kk'} = [h_{kk'}(1), h_{kk'}(2), \ldots, h_{kk'}(M)]^H$$

is the path channel coefficients (PCCs) from transmitter k to receiver k' for M paths, and $h_{kk'}(q)$ is the qth resolvable PCC modeled as quasi-static Rayleigh-distributed random variables that remain constant during a coherence time period of the channel (T_{CT}), $\mathbf{n}_{k'}$ is ambient noise, and

$$\mathbf{S}_k(i) \triangleq \mathbf{S}_k^0(i) + \mathbf{S}_k^+(i) + \mathbf{S}_k^-(i),$$

where $\mathbf{S}_k^0(i)$, $\mathbf{S}_k^+(i)$ and $\mathbf{S}_k^-(i)$ correspond to the spreading code matrices generated by transmitter k for transmitting bits $b(i)$, $b(i+1)$ and $b(i-1)$, respectively.

A joint channel estimation enabling Bob to estimate the CSI $(\hat{\mathbf{h}}_{kB})$ for channel kB uses a set of N_p pilot bits periodically repeated and inserted in each frame distanced less than T_{CT}, yielding the following estimation:

$$\hat{\mathbf{h}}_{AB} = [\mathbf{I}_M \ \mathbf{0}_M] \, \hat{\mathbf{h}}_{AB,JB},$$

$$\hat{\mathbf{h}}_{JB} = [\mathbf{0}_M \ \mathbf{I}_M] \, \hat{\mathbf{h}}_{AB,JB},$$

where,

$$\hat{\mathbf{h}}_{AB,JB} = \frac{1}{N_p} \sum_{i=1}^{N_p} \mathbf{S}_{AJ}^{\dagger}(i) \mathbf{r}_B(i),$$

with

$$\mathbf{S}_{AJ}^{\dagger}(i) \triangleq [\mathbf{S}_{AJ}(i)^H \mathbf{S}_{AJ}(i)]^{-1} \mathbf{S}_{AJ}(i)^H,$$

where

$$\mathbf{S}_{AJ}(i) \triangleq [\sqrt{E_A} \mathbf{S}_A(i), \sqrt{E_J} \mathbf{S}_J(i)]_{(L+M-1) \times 2M}.$$

13.3.2   Countermeasures Against Wormhole Attack

To defend UWANs against wormhole attacks, the Distributed-Visualisation of Wormhole (Dis-VoW) protocol [59] uses distances calculated according to the signal propagation delay to construct a local network topology within two hops. A Multi-Dimensional Scaling (MDS) scheme tries to visualize distortions in edge lengths and angles by requiring, each node to collect the distance estimations from its neighbors. However, Dis-VoW depends on secure distance estimation, which itself is vulnerable to attacks in UWANs, and the following scheme tries to solve this problem.

13.3.2.1   Direction of Arrival (DoA)

It is relatively easy for an attacker to manipulate signal power and transmission time but not the signal's DoA of a true neighbor. Therefore, DoA is used in the neighbor discovery protocol (NDP) against wormhole attacks in UWANs by a countermeasure proposed in [60], which does not rely on secure and accurate time synchronization, localization or high node density.

It consists of the following four protocols. (i) B-NDP involves two nodes for neighbor discovery. (ii) DV-NDP requires three nodes, and dramatically improves the wormhole resilience of B-NDP but decreases the probability for two true neighbors to successfully discover each other. (iii) SDV-NDP turns DV-NDP into a deterministic wormhole-resilient protocol, and (iv) MA-NDP accommodates node mobility during the execution of the above mentioned protocols [60]. Since the B-NDP is the base of the proposed scheme, it is introduced more in the following.

Fig. 13.11 Wormhole detection with DoA estimation [60]

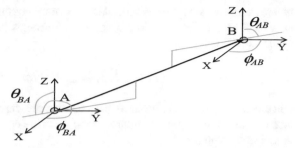

θ_{AB} θ_{BA}: the inclination angles of direction AB and BA
ϕ_{AB} ϕ_{BA}: the azimuth angles of direction AB and BA

13.3.2.2 DoA-Based Neighbor Discovery Protocol (NDP)

Assume that nodes A and B are true neighbors as illustrated in Fig. 13.11, where θ_{AB} and ϕ_{AB} indicate the inclination and azimuth angles of direction $\overrightarrow{AB}$, respectively, while $\overrightarrow{BA}$ is the opposite of $\overrightarrow{AB}$. For a pair of true neighbors, the following relationship holds:

$$\begin{cases} \theta_{AB} + \theta_{BA} = \pi, \\ \phi_{AB} - \phi_{BA} = \pm\pi, \end{cases}$$

which intuitively can be used by nodes A and B to verify the authenticity of their neighborship through comparing the respective DoA of their incoming signals. To this end, the following protocol is defined.

Now node A tries to discover its neighbors by broadcasting a request as follows:

$$A \longrightarrow * : ID_A, n_A, <prior-data>_{K_A^{-1}},$$

where n_A is a random nonce chosen by node A to thwart message replay attacks, $<.>_K$ denotes a digital signature operation over the prior-data with the private key K, which can be verified by using ID_A as the public key [61].

Once receiving the request, node B first estimates θ_{AB} and ϕ_{AB} as $\widehat{\theta_{AB}}$ and $\widehat{\phi_{AB}}$, respectively, and then uses ID_A to verify the authenticity of node A's public/private keys. If succeeds, node B unicasts to node A the following message:

$$B \longrightarrow A : ID_B, \widehat{\theta_{AB}}, \widehat{\phi_{AB}}, n_B, <prior-data>_{K_B^{-1}}.$$

Once receiving the reply, node A also estimates θ_{BA} and ϕ_{BA} as $\widehat{\theta_{BA}}$ and $\widehat{\phi_{BA}}$, respectively, uses ID_B as the public key to verify the authenticity of node B's public/private keys. If succeeds, node A checks the following conditions:

$$\begin{cases} |\widehat{\theta_{AB}} + \widehat{\theta_{BA}} - \pi| \le 2\sigma_\theta, \\ |\widehat{\phi_{AB}} - \widehat{\phi_{BA}} \pm \pi| \le 2\sigma_\phi, \end{cases}$$

where σ_θ and σ_ϕ are the predetermined maximum possible estimation errors of θ and ϕ, respectively. If the conditions hold, node A accepts node B as a true neighbor and sends the following reply:

$$A \longrightarrow B : ID_B, ID_A, \widehat{\theta_{BA}}, \widehat{\phi_{BA}}, < prior - data | n_A | n_B >_{K_A^{-1}} .$$

Once receiving the reply, node B first verifies the digital signature. If succeeds, it also checks the condition, similar to node A. If the conditions hold, it accepts node A as a true neighbor.

Note that, public-key based digital signature causes much computation and extra transmission for ciphertext, and estimation of the DoA of incoming acoustic signals may further burden underwater nodes. In UWANs consisting of unanchored nodes, water current will cause these nodes to change their positions frequently, which makes it difficult to determine proper σ_θ and σ_ϕ for a true pair of neighboring nodes.

13.3.3 Attack-Resilient Routing

Here we introduce two attack-resilient routing protocols adopted by two security suites, namely, SeFLOOD for the suite discussed in Sect. 13.3.4 and CARP for that discussed in [62].

13.3.3.1 Secure FLOOD

FLOOD is a NDP proposed for a clustered UWAN based on flooding. It requires each node to report the information on the link quality, such as signal attenuation and link delay to the master node. The master runs the Dijkstra algorithm with the link quality information to build the routing table, which is distributed to all the nodes [63]. However, this protocol is vulnerable to several attacks, such as spoofing attack that injects false information to the report [37].

A Secure FLOOD (SeFLOOD) proposed in [64] protects every control message using a cryptographic suite described in Sect. 13.3.4. It assumes that a node cannot be compromised physically by an attacker but only being attacked through the network. The major components are described below [11].

Securing Unicast

To protect unicast messages between nodes, a link key has to be established for each pair of nodes before the routing protocol starts. This key is used to encrypt each unicast message transmitted between them. SeFLOOD adopts a link-key table (LKT) instead of well-known key agreement protocols (e.g., Elliptic curve Diffie-Hellman [65] or the Blundo [66]) to distribute the keys for sake of both simplicity and efficiency. Each node has an LKT to store all pairwise link keys.

This protocol has an $O(n)$ storage overhead, where n is the number of nodes in the network, and usually small so that each node can have enough memory to store the LKT. For example, the LKT memory is about 16 Kilobytes for a network composed of 1,024 nodes each with a 128-bit link key. A node is typically equipped with 2 GB memory, which is enough to store such an LKT [64]. However, an LKT has to be pre-installed in each node, and if one of the nodes is compromised, then the routing protocol is vulnerable to attacks.

Securing Broadcast

To protect broadcast messages within a cluster, each node creates a cluster key. It distributes this key to each member in its cluster secretly by encrypting the cluster key with the corresponding link key. This transmission repeats until all the members have received the cluster key, which however yields a large overhead especially with frequent updating of cluster keys. Similar to the LKT, each node uses a cluster key table to maintain the cluster key for each cluster member. The cluster key is used to authenticate the messages broadcast within the cluster [64].

13.3.3.2 Resilient Channel Aware Routing Protocol (R-CARP)

The CARP [62] is a cross-layer designed routing protocol, exploiting link quality information to achieve robust, energy-aware and adaptive data forwarding. However, it is vulnerable to insider attacks such as sinkhole attack [6 /]. If one node is malicious or compromised, it can be chosen as a relay with high probability by advertising to its neighbors a high value of the utility function, such as available buffer space, residual energy and quality of the link. Thus, a Resilient CARP (R-CARP) employs jointly a BLS-based digital signature discussed in Sect. 13.1.4 and reputation-based mechanism for confidentiality and authentication. It tries to secure routing process mainly following the two steps discussed below [11].

Information Collection

A set up phase of the protocol allows each node to acquire hop-distance information from the sink, with which each node shares the same group key and a unique secret key, similar to SeFOOD. When a node has data to forward, it broadcasts a request message (PING) in order to find the best relay. Once a node receives the PING, it replies with a PONG message containing estimated information on hop distance from the sink and link quality information, which is estimated according to the number of control and data packets correctly received recently.

Reputation Estimation

For each received PONG message, the receiver calculates the reputation of the PONG's sender. The reputation is measured by the ratio of the number of packets confirmed by the sink that the receiver has forwarded and the total number of packets that the receiver has forwarded, both through the sender, according to the recent history. The higher this ratio, the better is the reputation of the sender. PONG messages are encrypted and authenticated so that any alter on the carried information can be detected.

13.3.4   A Cryptographic Suite

Different from single security schemes or mechanisms discussed above, several security suites [23, 68, 69] have also been investigated to secure UWANs systematically by providing simultaneously basic security (e.g., confidentiality, integrity, authentication and non-repudiation) and countermeasures against some attacks. One of them validated by field test is described below.

13.3.4.1   Overview

The security suite for a clustered UWAN comprising both fixed and mobile nodes [3, 68] aims to allow a node to join the system for initiating a mission, and to leave it after completion. It supports secure reconfiguration for changes in the network due to node's mobility. The suite consists of a secure routing protocol and a set of cryptographic primitives (e.g., cipher, digest and re-keying), which are used to secure underwater communication in one-to-one and one-to-many modes via the gateway by considering the characteristics of underwater acoustic channels. It has been implemented and tested in the field for group communication of underwater vehicles, and the results show that the communication and power consumption overhead introduced for security is limited, and sometimes negligible [70].

13.3.4.2   Security Services

The following security services are provided by the suite [3, 68]:

- Confidentiality: The cipherText stealing (CTS) technique is used for encryption to avoid the ciphertext expansion problem. CTS alters the processing of only the last two blocks of a plaintext by "stealing" a portion of the ciphertext of the second-last block to pad the last plaintext block. Then, the padded final block is encrypted as usual without ciphertext expansion [71].
- Integrity: It uses a 4-byte digest through truncating the real hash value such that the overhead is around 4.4% of the average UWAN message without harm to security [72]. In this case, an adversary has 1 in 232 chances to blindly forge a digest, and maximally needs to repeat 231 trial transmissions to the legitimate receiver to break a digest. In a UWAN with a 500-bps channel, sending 231 trial messages each with 184 bits requires about 306 months to break the digest.
- Group key management: Once a node leaves the system, the gateway generates and distributes a new group key to prevent the leaving node from reading new messages. A compromised node is removed logically by redistributing a new group key to all members but not the removed one. A Secure and Scalable Rekeying Protocol (S2RP) is adopted as the rekeying protocol, employing an one-way hash function proposed for resource-constrained devices in [73]. S2RP aims to provide an efficient proof of key authenticity by computing a digest of the key without requiring

additional information. The number of rekeying messages logarithmic in the number of nodes makes the key distribution highly scalable [3] (See Sect. 13.3.4.3).

- Secure routing: SeFLOOD [64] discussed in Sect. 13.3.3.1 is exploited to distribute group keys. After revoking a group key, the gateway deletes the compromised node from the routing tables and re-calculates the related paths to secure network reconfiguration. It also tries to support mobile nodes without adding overhead in terms of number of messages by assuming nodes loosely time-synchronized. Actually these nodes need to emerge periodically to synchronize with GPS.

13.3.4.3 Group Key Management

As discussed in [11], it is based on a key-chain scheme as illustrated in Fig. 13.12. A key-chain is a set of symmetric keys, in which each key is the hash value of the previous one in the key generation process. That is, given a key $K^{(i)}$ in the key-chain, one can compute all the previous keys $K^{(j)}$ ($j < i$) as follows: $K^{(j-1)} = H(K^{(j)})$, but not reverse except the key-chain creator. Here $K^{(i)} = H^{n-i}(s)$, where s is a seed randomly selected by the key-chain creator, and $H^i(s)$ means applying i times $H()$ to s. In this case, the next key can be authenticated by simply applying hash function $H()$.

Group keys are distributed following a logical key tree maintained by the gateway. As illustrated in Fig. 13.13, each key-node in this tree is associated to a key-chain. Each leaf is associated with a secret node-key, which the network node shares only with the gateway. Keys in the hash chain (i.e., $K_1 \sim K_7$) are only used in the rekeying process, and each network node maintains a key ring containing every such key in a way that the subtree rooted at a key node contains the leaf associated with the node-key. For example, the key ring of network node 4 (a leaf) is $[K_1, K_2, K_5]$. Since K_1 is the root of the key tree and shared by all nodes, it can be used as the group key.

Once a network node has left the group, all keys in its key ring have to be changed for security, which is carried out by the gateway through broadcasting rekeying messages containing the respective next keys. These message are encrypted with either the node-keys, unaffected keys or new keys in the hash chain, and can only be decrypted by the corresponding keys. After a node successfully decrypts the message, it authenticates the new key following the key-chain scheme mentioned above.

Fig. 13.12 Key chain for group key management [3, 11]

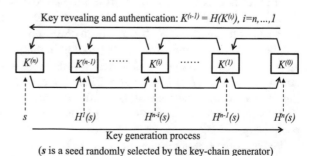

Key revealing and authentication: $K^{(i-1)} = H(K^{(i)})$, $i=n,...,1$

Key generation process

(s is a seed randomly selected by the key-chain generator)

Fig. 13.13 Key tree for
group key management [3]

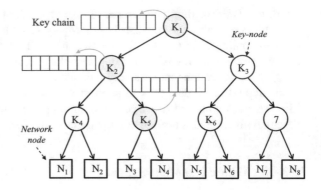

Now suppose N_4 has left the group, all keys in its key-ring (i.e., K_1, K_2, K_5) have been replaced by new ones: K_1^+, K_2^+, K_5^+, as follows:

1. Gateway $\rightarrow N_3 : E(K_5^+)$, encrypted with the node-key of N_3.
2. Gateway $\rightarrow N_3 : E(K_2^+)$, encrypted with K_5^+, which is a new authenticated key.
3. Gateway $\rightarrow \{N_1, N_2\} : E(K_2^+)$, encrypted with K_4, which is unaffected by the leaving of N_4.
4. Gateway $\rightarrow \{N_1, N_2, N_3\} : E(K_1^+)$, encrypted with K_2^+, which is a new authenticated key.
5. Gateway $\rightarrow \{N_5, N_6, N_7, N_8\} : E(K_1^+)$, encrypted with K_3, which is unaffected by the leaving of N_4.

This rekeying protocol requires $O(\log n)$ rekeying messages, where n is the number of nodes [3]. Compared with ECC-180 digital signature, which is as secure as RSA-1024, to authenticate 128-bit long group keys, a digital signature is 360 bits, resulting in a rekeying message of 488-bit length, which is 3.8125 times longer than this protocol [3].

13.4 Discussion

As discussed in [11], many security schemes have been investigated from the physical layer to the network layer in order to secure UWANs with a few of them being validated by field trials. Several interesting approaches specifically taking into account the peculiar features of UWANs have been proposed, such as energy-efficient cryptographic primitives, symmetric key generation in reciprocal channels, friend jammer against signal eavesdropping, DoA-based countermeasure against wormhole attack, and reputation-based authentication. There are also a couple of proposals addressing several security issues systematically. However, the research of UWAN security is still in its early stage, and several important issues have not been addressed adequately.

So far no clear strategic view on securing UWANs is available, taking into account thoroughly the peculiar features of UWANs. Network security is a complex issue that span layers from the physical to the application, and an optimal distribution of various security functions among these layers is paramountly important to minimize resource consumption without jeopardizing the security, especially for very resource-constrained UWANs. However, this issue almost has not been addressed in the literature [11].

Regarding the fundamental of network security (i.e., cryptographic primitives), although some proposals such as ECC can be more energy-efficient with less computation complexity than the popular primitives adopted in RWNs, the suitability of these cryptographic primitives need more validation and test in real UWANs against the peculiar features mentioned above. Due to difficulties in establishing a PKI in UWANs, symmetric key based security schemes become more important. However, the key generation and distribution in UWANs still need further study. Although properties of reciprocal channels can be leveraged to achieve the above purpose, successful key generation rates need to be further improved with less communication overhead. The performance of such approach for asymmetric underwater acoustic channels also requires study because of possible asymmetric responses to probing messages.

For authentication using reputation-based trust models, several issues need to be addressed before such kind of scheme can be applied in practice. Typical issues include securing the collected information to be used by the model, distinguishing between abnormality due to attack and that caused by malfunction of nodes, and setting thresholds widely used by such schemes adaptively according to dynamic networking environments.

Many schemes do not discuss how to re-configure security setting in UWANs, which is especially important to those security schemes that rely on pre-installed security configuration conducted during the deployment of a UWAN. As mentioned earlier, due to specific UWAN environments, it is difficult to protect each underwater node physically, which causes some of them to be compromised. On the other hand, it is also difficult for a UWAN to find out whether there are some nodes that have been compromised. In this case, periodically re-configuring security systems is necessary to remove the compromised nodes, while how to secure such re-configuration process is also an important issue [11].

Some schemes assumes the availability of precise time synchronization or location information of nodes, which is difficult to hold in UWANs as mentioned earlier, and these assumed schemes themselves may become security vulnerability.

13.5 Summary

This chapter discusses some security approaches and schemes specially designed for UWANs. As discussed above, it is not easy to secure UWANs due to harsh underwater networking environments and very constrained network resources as well as high cost for network deployment and maintenance. In this case, a strategic view of global cooperation of every layer is especially important to optimize resource utilization for cost-effective security protection. On the other hand, some features of UWANs also impose challenges to attackers, and should be leveraged in design security schemes. Meanwhile, applications may have different requirements on network security, how to take into account this feature in security scheme design is also an important issue. Please refer to [11] for more discussions on relevant issue.

References

1. Stojanovic, M.: On the relationship between capacity and distance in an underwater acoustic communication channel. ACM Mob. Comput. Commun. Rev. **11**(4), 34–43 (2007)
2. Jiang, H.F., Xu, Y.: Research advances on security problems of underwater sensor networks. Adv. Mater. Res. **317–319**, 1002–1006 (2011)
3. Dini, G., Duca, A.L.: A secure communication suite for underwater acoustic sensor networks. Sensors **12**(11), 15133–15158 (2012)
4. Akyildiz, I.F., Wang, X.D.: A survey on wireless mesh networks. IEEE Commun. Mag. **43**(9), S23–S30 (2005)
5. Krikidis, I., Thompson, J.S., McLaughlin, S.: Relay selection for secure cooperative networks with jamming. IEEE Trans. Wireless Commun. **8**(10), 5003–5011 (2009)
6. Melodia, T., Kulhandjian, H., Kuo, L.-C., Demirors, E.: Advances in underwater acoustic networking. In: Basagni, S., Conti, M., Giordano, S., Stojmenovic, I., (eds.), Mobile Ad Hoc Networking: The Cutting Edge Directions, pp. 804 – 852. Wiley-IEEE Press (2013)
7. Li, M., Kundu, S., Pados, D.A., Batalama, S.N.: Waveform design for secure SISO transmissions and multicasting. IEEE J. Sel. Areas Commun. **31**(9), 1864–1874 (2013)
8. Kulhandjian, K., Melodia, T., Koutsonikolas, D.: Securing underwater acoustic communications through analog network coding. In: Proceedings of the Annual IEEE Communications Society Conference on Sensor, Mesh and Ad Hoc Communications and Networks (SECON), Singapore, pp. 266–274 (2014)
9. Yan, H., Shi, Z.J., Fei, Y.: Efficient implementation of elliptic curve cryptography on DSP for underwater sensor networks. In: Proceedings of the Workshop on Optimization for DSP & Embedded System (ODES), Seattle, WA, USA, pp. 7–15 (2009)
10. Luo, Y., Pu, L.N., Peng, Z., Shi, Z.J.: RSS-based secret key generation in underwater acoustic networks: advantages, challenges, and performance improvements. IEEE Commun. Mag. **54**(2), 32–38 (2016)
11. Jiang, S.M.: Securing underwater acoustic networks: a survey. IEEE Commun. Surv. Tutorials (2017)
12. Menezes, A.J., van Oorschot, P.C., Vanstone, S.A.: Handbook of Applied Cryptography. CRC Press, Boca Raton (1996)
13. Brown, M., Cheung, D., Hankerson, D., Hernandez, J.L., Kirkup, M., Menezes, A.: PGP in constrained wireless devices. In: Proceedings of the USENIX Security Symposium (SSYM), Denver, Colorado, USA (2000)
14. Carman, D.W., Krus, P.S., Matt, B.J.: Constraints and approaches for distributed sensor network security. Technical Report 00–010. NAI Labs, Network Associates Inc, Glenwood, MD, USA (2000)
15. Yuan, C., Chen, W.P., Zhu, Y.Q., Li, D.Y., Tan, J.: A low computational complexity authentication scheme in underwater wireless sensor network. In: Proceedings of the IEEE International Conference on Mobile Ad-Hoc And Sensor Networks (MSN), Shenzhen China, pp. 116–123 (2015)
16. Maurer, U.: Secret key agreement by public discussion from common information. IEEE Trans. Inform. Theory **39**(3), 733–742 (1993)

17. Lal, C., Petroccia, R., Pelekanakis, K., Conti, M., Alves, J.: Toward the development of secure underwater acoustic networks. IEEE J. Ocean. Eng. **42**(4), 1075–1087 (2017)
18. Guillaud, M., Slock, D.T., Knopp, R.: A practical method for wireless channel reciprocity exploitation through relative calibration. In: Proceedings of the International Symposium on Signal Processing and Its Applications Sydney, Australia, pp. 403–406 (2005)
19. Huang, Y., Zhou, S.L., Shi, Z.J., Lai, L.F.: Experimental study of secret key generation in underwater acoustic channels. In: Proceedings of the Asilomar Conference on Signals, Systems, and Computers, Pacific Grove, CA, USA, pp. 323–327 (2014)
20. Lai, L.F., Liang, Y.B., Du, W.L.: Cooperative key generation in wireless networks. IEEE J. Sel. Areas Commun. **30**(8), 1578–1588 (2012)
21. Pompili, D., Akyildiz, I.F.: Overview of networking protocols for underwater wireless communications. IEEE Commun. Mag. 97–102 (2009)
22. Pompili, D., Melodia, T., Akyildiz, I.F.: A CDMA-based medium access control for underwater acoustic sensor networks. IEEE Trans. Wireless Commun. **8**(4), 1899–1909 (2009)
23. Ateniese, G., Capossele, A., Gjanci, P., Petrioli, C., Spaccini, D.: "SecFUN: Security framework for underwater acoustic sensor networks. In: Proceedings of the MTS/IEEE OCEANS, Genoa, Italy (2015)
24. López, J., Dahab, R.: An overview of elliptic curve cryptography. In: Technical Report, Institute of Computing, State University of Campina, Campinas, Brazil (2000)
25. Proos, J., Zalka, C.: Shor's discrete logarithm quantum algorithm for elliptic curves. Quantum Inf. Comput. **3**(4), 317–344 (2003)
26. Nielsen, M.A., Chuang, I.L.: Quantum Computation and Quantum Information (10th Anniversary Edition). Cambridge University Press, Cambridge (2010)
27. Zhang, F., Safavinaini, R., Susilo, W.: An efficient signature scheme from bilinear pairings and its applications. In: Proceedings of the International Workshop Practice & Theory in Public Key Cryptography (PKC), Singapore, pp. 277–290 (2004)
28. Dan, B., Lynn, B., Shacham, H.: Short signatures from the Weil pairing. In: Proceedings of the International Conference on the Theory and Application of Cryptology and Information Security, Gold Coast, Australia, pp. 514–532 (2001)
29. Souza, E., Wong, H.C., Cunha, I., Loureiro, A.A.F., Vieira, L.F.M., Oliveira, L.B.: End-to-end authentication in under-water sensor networks. In: Proceedings of the IEEE Symposium on Computers and Communications (ISCC), Split, Croatia, pp. 299–304 (2013)
30. Perrig, A., Szewczyk, R., Wen, V., Culler, D., Tygar, D.: SPINS: security protocols for sensor networks. In: Proc. Annual ACM International Conference on Mobile Computing and Networking (MobiCom), Rome, Italy (2001)
31. Han, G.J., Jiang, J.F., Shu, L., Guizani, M.: An attack-resistant trust model based on multidimensional trust metrics in underwater acoustic sensor network. IEEE Trans. Mob. Comput. **14**(12), 2447–2459 (2015)
32. Liu, Q., Liao, Y., Tang, B., Yu, L.: A trust model based on subjective logic for multi-domains in grids. In: Proceedings of the IEEE Pacific-Asia Workshop on Computational Intelligence and Industrial Application, Wuhan, China, pp. 882–886 (2008)
33. He, D.J., Chen, C., Chan, S., Bu, J.J., Vasilakos, A.V.: Retrust: attack-resistant and lightweight trust management for medical sensor networks. IEEE Trans. Inf. Tech. Biomed. **16**(4), 623–632 (2012)
34. Jiang, J.F., Han, G.J., Shu, L., Chan, S., Wang, K.: A trust model based on cloud theory in underwater acoustic sensor networks. IEEE Trans. Ind. Inf. **13**(1), 342–350 (2017)
35. Jiang, J.F., Han, G.J., Zhu, C.S., Chan, S., Rodrigues, J.J.P.C.: A trust cloud model for underwater wireless sensor networks. IEEE Commun. Mag. 110–116 (2017)
36. Li, D.Y., Meng, H.J., Shi, X.M.: Membership clouds and membership cloud generators. J. Comput. Res. Dev. **32**(6), 15–20 (1995)
37. Cong, Y.P., Yang, G., Wei, Z.Q., Zhou, W.: Security in underwater sensor network. In: Proceedings of the International Conference on Communication and Mobile Computing (CMC), Shenzhen, China, vol. 1, pp. 162–168 (2010)

38. Domingo, M.C.: Securing underwater wireless communication networks. IEEE Wireless Commun. Mag. 22–28 (2011)
39. Han, G.J., Jiang, J.F., Shu, L., Shu, L.: Secure communication for underwater acoustic sensor networks. IEEE Commun. Mag. 54–60 (2015)
40. Das, A.P., Thampi, S.M.: Secure communication in mobile underwater wireless sensor networks. In: Proceedings of the International Conference on Advances in Computing, Communications and Informatics (ICACCI), Kochi, India, pp. 2164–2173 (2015)
41. Patron, P., Petillot, Y.: The underwater environment: a challenge for planning. In: Proceedings of the The Workshop of the UK Planning and Scheduling Special Interest Group, Edinburgh, UK (2008)
42. Kong, J., Ji, Z., Wang, W., Gerla, M., Bagrodia, R., Bhargava, B.: Low-cost attacks against packet delivery, localization and time synchronization services in under-water sensor networks. In: Proceedings of the ACM Workshop on Wireless Security (WiSe), Cologne, Germany, pp. 87–96 (2005)
43. Urich, R.: Principles of Underwater Sound. McGraw-Hill, New York (1983)
44. Misra, S., Dash, S., Khatua, M., Vasilakos, A.V., Obaidat, M.S.: Jamming in underwater sensor networks: detection and mitigation. IET Commun. 6(14), 2178–2188 (2012)
45. Li, H., He, Y.H., Cheng, X.Z., Zhu, H.S., Sun, L.M.: Security and privacy in localization for underwater sensor networks. In: IEEE Commun. Mag. 56–62 (2015)
46. Bian, T., Venkatesan, R., Li, C.: Design and evaluation of a new localization scheme for underwater acoustic sensor networks. In: Proceedings of the IEEE Global Tele-communications Conference (GLOBOCOM), Hawaii, USA (2009)
47. Cheng, X.Z., Shu, H.N., Liang, Q.L., Du, D.H.-C.: Silent Positioning in Underwater Acoustic Sensor Networks. IEEE Trans. Veh. Tech. 57(3), 1756–1766 (2008)
48. Shahapur, S.S., Khanai, R.: Localization, routing and its security in UWSN - A survey. In: Proceedings of the International Conference on Electrical, Electronics and Optimization Techniques (ICEEOT), Bangkok, Thailand, pp. 1001–1006 (2016)
49. Chandrasekhar, V., Seah, W.: An area localization scheme for underwater sensor networks. In: Proceedings of the MTS/IEEE OCEANS, Singapore (2007)
50. Niculescu, D., Nath, B.: DV based positioning in ad hoc networks. Telecommun. Syst. 22(1–4), 267–280 (2003)
51. Hu, F., Malkawi, Y., Kumar, S., Xiao, Y.: Vertical and horizontal synchronization services with outlier detection in underwater acoustic networks. Wireless Commun. Mob. Comput. 8(9), 1165–1181 (2008)
52. Lal, C., Petroccia, R., Conti, M., Alves, J.: Secure underwater acoustic networks: current and future research directions. In: IEEE Underwater Communications and Networking Conference (UComms), Lerici, Italy (2016)
53. Lu, X.Y., Zuba, M., Cui, J.-H., Shi, Z.J.: Uncooperative localization improves attack performance in underwater acoustic networks. In: Proceedings of the IEEE Conference on Communication and Networks Security (CNS), San Francisco, CA, USA, pp. 454–462 (2014)
54. Zuba, M., Shi, Z.J., Peng, Z., Cui, J.H., Zhou, S.L.: Vulnerabilities of underwater acoustic networks to denial-of-service jamming attacks. Secur. Commun. Net. 8(16), 2635–2645 (2015)
55. Li, X., Han G.J., Qian, A.H., Rodrigues, J.: Detecting Sybil attack based on state information in underwater wireless sensor networks. In: Proceedings of the IEEE International Conference on Software, Telecommunication and Computer Networks (SoftCOM), Primosten, Croatia, (2013)
56. Ibragimov, M., Lee, J.-H.: Kalyani, M., il Namgung, J., Park, S.-H., Yi, O., Kim, C.H., Lim, Y.-K.: CCM-UW security modes for low-band underwater acoustic sensor networks. Wireless Pers. Commun. 89(2), 479–499 (2016)
57. Li, M., Batalama, S.N., Pados, D.A., Melodia, T., Medley, M.J.: Cognitive code-division links with blind primary-system identification. IEEE Trans. Wireless Commun. 10(11), 3743–3753 (2011)
58. Katti, S., Gollakota, S., Katabi, D.: Embracing wireless interference: analog network coding. In: Proceedings of the ACM SIGCOMM, Kyoto, Japan, pp. 397–408(2007)

59. Wang, W.C., Kong, J.J., Bhargava, B., Gerla, M.: Visualisation of wormholes in underwater sensor networks: a distributed approach. Int. J. Secur. Netw. **3**(1), 10–23 (2008)
60. Zhang, R., Zhang, Y.C.: Wormhole-resilient secure neighbor discovery in underwater acoustic networks. In: Proceedings of he IEEE INFOCOM, San Diego, CA, USA (2010)
61. Papadimitratos, P., Poturalski, M., Schaller, P., Lafourcade, P.: Secure neighborhood discovery: a fundamental element for mobile ad hoc networking. IEEE Commun. Mag. **46**(2), 132–139 (2008)
62. Basagni, S., Petrioli, C., Petroccia, R., Spaccini, D.: CARP: a channel-aware routing protocol for underwater acoustic wireless networks. Ad Hoc Netw. **34**, 92–104 (2015)
63. Rustad, H.: A lightweight protocol suite for underwater communication. In: Proceedings of the International Conference on Advanced Information Networking and Applications WS. (WAINA), Bradford, UK, pp. 1172–1178 (2009)
64. Dini, G., Duca, A.L.: SeFLOOD: a secure network discovery protocol for underwater acoustic networks. In: Proceedings of the IEEE Symposium on Computers and Communications (ISCC), Kerkyra, Greece, pp. 636–638 (2011)
65. Barker, E., Johnson, D., Smid, M.: Recommendation for pairwise key establishment schemes using discrete logarithm cryptography. Eetimes Com, 15158 (2014)
66. Blundo, C., Santis, A.D., Herzberg, A., Kutten, S., Vaccaro, U., Yung, M.: Perfectly-secure key distribution for dynamic conferences. In: Proceedings of the Annual International Cryptology Conference on Advances in Cryptology (CRYPTO), California, USA, pp. 471–486 (1993)
67. Capossele, A., Cicco, G.D., Petrioli, C.: R-CARP: a reputation based channel aware routing protocol for underwater acoustic sensor networks. In: Proceedings of the ACM International Conference Underwater Networks and Systems (WUWNet), Arlington, VA, USA (2015)
68. Dini, G., Duca, A.L.: A cryptographic suite for underwater cooperative applications. In: Proceedings of the IEEE Symposium Computers and Communications (ISCC), Kerkyra, Greece, pp. 870–875 (2011)
69. Toso, G., Munaretto, D., Conti, M., Zorzi, M.: Attack resilient underwater networks through software defined networking. In: Proceedings of the ACM International Conference on Underwater Networks and Systems (WUWNet), Rome, Italy (2014)
70. Caiti, A., Calabro, V., Dini, G., Duca, A.L., Munafo, A.: Secure cooperation of autonomous mobile sensors using an underwater acoustic network. Sensors **12**(2), 1967–1989 (2012)
71. Schneier, B.: Applied Cryptography: Protocols, Algorithms, and Source Code in C, 2nd edn. Wiley, New York (1995)
72. Karlof, C., Sastry, N., Wagner, D.: TinySec: a link layer security architecture for wireless sensor networks. In: Proceedings of the International Conference on Embedded Networked Sensor Systems, Baltimore, MD, USA, pp. 162–175 (2004)
73. Dini, G., Savino, I.M.: S2RP: a secure and scalable rekeying protocol for wireless sensor networks. In: Proceedings of the IEEE International Conference Mobile Ad-Hoc and Sensor Systems Vancouver, BC, Canada, pp. 457–466 (2006)

Appendix A
Formulas for Some Queueing Systems

A general queueing system can be modelled as Fig. A.1a, which consists of an arrival process, a queue and a service facility probably with multiple servers. This system can be expressed by the state transition diagram, which can be one-dimensional, two-dimensional or even more as illustrated in Fig. A.1b, c, respectively. The major steps to resolve such a problem include (i) determining the transient state space (K), (ii) determining state transition rates (i.e., λ_i and μ_i), and (iii) formulating P_k in equilibrium states.

Results for some simple queueing systems are summarized below. The queuing system are specified in arrival and service processes as well as the number of servers in the system with a notation $X/Y/z$, where X and Y indicate the arrival and service processes, respectively, while z denotes the number of server. By default, G, M and D represent a general process, a Poisson process and a deterministic process, respectively. The following notation is adopted. Please refer to [1] for more discussion on queueing systems.

- $\bar{N}$: average number of customers in the queueing system.
- $\bar{N}_s$: average number of customers in the service facility.
- $\bar{N}_q$: average number of customers in the queue.
- λ: average arrival rate of customers to the system.
- W: average waiting time for a customer to stay in the queue.
- $\bar{x}$: average time for a customer to stay in the server.
- T: average time for a customer to stay in the queueing system.
- p_0: probability for a server to be idle.
- ρ: fraction of time that a server is busy.
- μ: average service rate, i.e., $\mu = \frac{1}{\bar{x}}$.
- σ: coefficient of variation for service time.
- $\bar{q}$: average number of customers observed by a departure.

© Springer Nature Singapore Pte Ltd. 2018   369
S. Jiang, *Wireless Networking Principles: From Terrestrial to Underwater Acoustic*,
https://doi.org/10.1007/978-981-10-7775-3

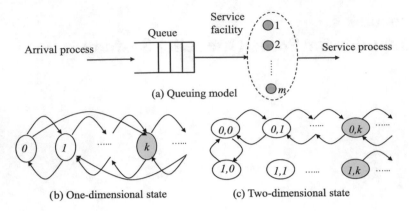

Fig. A.1 Queuing system and state-transition diagram

A.1 G/G

A.1.1 Little Formulas

The following results are applicable for any lossless queueing systems [1], in which no customer loss happens:

$$\bar{N} = \lambda \times T, \tag{A.1}$$
$$\bar{N}_q = \lambda \times W, \tag{A.2}$$
$$\bar{N}_s = \lambda \times \bar{x}, \tag{A.3}$$
$$\bar{N} = N_q + N_s, \tag{A.4}$$
$$T = \bar{x} + W. \tag{A.5}$$

A.1.2 G/G/1

$$p_0 = 1 - \rho. \tag{A.6}$$

A.2 M/G/1

$$\bar{q} = \rho + \rho^2 \frac{1 + \sigma^2}{2(1 - \rho)}, \tag{A.7}$$

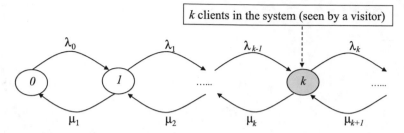

Fig. A.2 State transition diagram for M/M

$$\frac{T}{\bar{x}} = 1 + \rho \frac{1 + \sigma^2}{2(1 - \rho)},$$ (A.8)

$$\frac{W}{\bar{x}} = 1 + \rho \frac{1 + \sigma^2}{2(1 - \rho)}.$$ (A.9)

Particularly for $M/D/1$, $\sigma = 0$.

A.3 M/M

Let P_k denote the probability that there k customers seen by an arrival. As shown in Fig. A.2, we can have P_k in the equilibrium state as follows [1]:

$$P_k = P_0 \frac{\lambda_0 \lambda_1 ... \lambda_{k-1}}{\mu_1 \mu_2 ... \mu_k}$$

$$= P_0 \prod_{i=0}^{k-1} \frac{\lambda_i}{\mu_{i+1}},$$ (A.10)

where

$$P_0 = \left[1 + \sum_{k=1}^{K} \prod_{i=0}^{k-1} \frac{\lambda_i}{\mu_{i+1}} \right]^{-1},$$ (A.11)

by letting $\sum_{k=0}^{K} P_k = 1$.

We can have the following demonstration for (A.10). Following the flow conservation, we can have $P_0 \lambda_0 = P_1 \mu_1$ and $P_{k-1} \lambda_{k-1} + P_{k+1} \mu_{k+1} = P_k (\lambda_k + \mu_k)$ for $k > 0$ as illustrated in Fig. A.2. Then, we can have $P_1 = \frac{\lambda_0}{\mu_1} P_0$. Since $P_0 \lambda_0 + P_2 \mu_2 = P_1 (\lambda_1 + \mu_1)$, we can also get $P_2 = \frac{\lambda_1}{\mu_2} P_1$ by replacing P_1 therein with the derived one. Continuing such manifestation, we can have $P_{k+1} = \frac{\lambda_k}{\mu_{k+1}} P_k$.

A.3.1 M/M/1

$$p_k = (1 - \rho)\rho^k, \tag{A.12}$$

$$\bar{N} = \frac{\rho}{1 - \rho}, \tag{A.13}$$

$$W = \frac{\rho}{\mu(1 - \rho)}, \tag{A.14}$$

$$T = \frac{1}{\mu(1 - \rho)}, \tag{A.15}$$

and

$$P\{\text{the number of customers} \geq k \text{ in the quequing system}\} = \rho^k. \tag{A.16}$$

A.3.2 M/M/m

$$p_k = \begin{cases} p_0 \frac{(m\rho)^k}{k!}, & k \leq m, \\ p_0 \frac{\rho^k m^m}{m!}, & k > m, \end{cases} \tag{A.17}$$

where

$$p_0 = \left[\sum_{k=0}^{m-1} \frac{(m\rho)^k}{k!} + \frac{(m\rho)^m}{m!(1 - \rho)} \right]^{-1}. \tag{A.18}$$

The probability for an arrival to be queued is given by Erlang's C formula as follows:

$$P\{\text{an arrival to be queued}\} = \sum_{k=m}^{\infty} p_k = \frac{(m\rho)^m p_0}{m!(1 - \rho)}. \tag{A.19}$$

A.3.3 M/M/m/m

$$p_k = \begin{cases} p_0 \frac{\rho^k}{k!}, & k \leq m, \\ 0, & k > m, \end{cases} \tag{A.20}$$

where

$$p_0 = \left[\sum_{k=0}^{m} \frac{\rho^k}{k!} \right]^{-1}. \tag{A.21}$$

The Erlang's loss formula (also Erlang's B formula) is

$$p_m = \frac{\frac{\rho^m}{m!}}{\sum_{k=0}^{m} \frac{\rho^k}{k!}}. \tag{A.22}$$

Reference

1. Kleinrock, L.: Queueing Systems, volume I: Theory. Wiley, New York (1975)

Appendix B
RSA Cryptosystem

The public key cryptography is very important for information security. RSA is one of most popular such systems published by Ron **R**ivest, Adi **S**hamir and Leonard **A**dleman (RSA). The following description is based on the relevant contents available on [1] and [2].

B.1 An Example

The following example simply shows how encryption and decryption keys are generated, and how they are used in encryption and decryption.

- Let's first select two primes as follows: $p = 11$ and $q = 3$. Then let $n = p \times q = 33$ and an even $\phi = (p - 1)(q - 1) = 20$.
- Choose $e = 3 < \phi$ such that e has no common divisor with ϕ by checking the Greatest Common Divisor (gcd) as follows: $gcd(e, p - 1) = 1$, $gcd(e, q - 1) = 1$ so that $gcd(e, \phi) = 1$.
- Compute d such that $(e \times d)$ mod $\phi = 1$, i.e., $d = (r \times \phi + 1)/e$, where r is a quotient, which may be equal to 1,2,..... Simple testing with $r = 1$ gives $d = 7$.
- Then we have a pair of keys as follows: a public key $= (n, e) = (33, 3)$ while a private key $= (n, d) = (33, 7)$.

Now we look at how to use the key generated above to encrypt and decrypt a message ($m < n$).

- Using the public key for encryption while the private key for decryption: the cipher text of m is given by $c = m^e$ mod n, and the decryption is given by $m = c^d$ mod n. Given $m = 7$ with $n = 33$, $e = 3$, $d = 7$ generated above, the encryption of m with e yields $c = 7^3$ mod $33 = 13$, and the decryption of c with d gives $m = 13^7$ mod $33 = 7$.
- Using the private key for encryption while the public key for decryption: $c = m^d$ mod n and $m = c^e$ mod n. With the same setting as mentioned above, the

© Springer Nature Singapore Pte Ltd. 2018

S. Jiang, *Wireless Networking Principles: From Terrestrial to Underwater Acoustic*,

https://doi.org/10.1007/978-981-10-7775-3

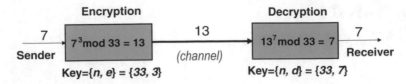

Fig. B.1 RSA for secure communication [2]

encryption with d gives $c = 7^7 \bmod 33 = 28$, while the decryption with e gives $m = 28^3 \bmod 33 = 7$.

As illustrated in Fig. B.1, both the sender and the receiver need to know n, while the sender knows the public key e and the receiver knows the private key d.

B.2 RSA Key Generation

In principle, RSA key generation follows the following mathematical relationship: $P = C^d \bmod n = (P^e)^d \bmod n = (P^d)^e \bmod n = C^e \bmod n$. General steps to generate the public and private keys are summarized below.

- Find a pair of primes p and q, and let $n = p \times q$ and $\phi = (p-1)(q-1)$. For a block size of k bits, n should also follow $2^k < n < 2^{k+1}$.
- n is a public module that is available to both the sender and the receiver. Therefore, n should be very large such that it becomes computationally infeasible to infer p and q given n.
- Select $e < \phi$ such that there are no common factors with $(p-1)$ and $(q-1)$ in order to have a non-zero remainder after a modulation of ϕ.
- Select d such that $e \times d \equiv 1 \bmod \phi$.
- Then the public key is (n, e) while the private key is (n, d).

The security of the above public key generation and corresponding encryption and decryption is based on the difficulty in factoring a large number (i.e., odd number n) into two primes p and q. For a small n, finding its small prime factors (e.g., 2, 3, 5, 7, 11, 13, 17,...) is easy, and it is not difficult either to check whether a small odd number is a prime. However, factoring a large n into two integers and checking whether they are primes take long time.

References

1. The free encyclopedia: Wikipedia. http://en.wikipedia.org/wiki/
2. RSA Algorithm: https://www.di-mgt.com.au/rsa_alg.html

Index

© Springer Nature Singapore Pte Ltd. 2018
S. Jiang, *Wireless Networking Principles: From Terrestrial to Underwater Acoustic*,
https://doi.org/10.1007/978-981-10-7775-3

Printed in the United States
By Bookmasters